Methods in Molecular Biology

Series Editor
John M. Walker
School of Life and Medical Sciences
University of Hertfordshire
Hatfield, Hertfordshire, UK

For further volumes:
http://www.springer.com/series/7651

For over 35 years, biological scientists have come to rely on the research protocols and methodologies in the critically acclaimed *Methods in Molecular Biology* series. The series was the first to introduce the step-by-step protocols approach that has become the standard in all biomedical protocol publishing. Each protocol is provided in readily-reproducible step-by-step fashion, opening with an introductory overview, a list of the materials and reagents needed to complete the experiment, and followed by a detailed procedure that is supported with a helpful notes section offering tips and tricks of the trade as well as troubleshooting advice. These hallmark features were introduced by series editor Dr. John Walker and constitute the key ingredient in each and every volume of the *Methods in Molecular Biology* series. Tested and trusted, comprehensive and reliable, all protocols from the series are indexed in PubMed.

Computational Methods for Predicting Post-Translational Modification Sites

Edited by

Dukka B. KC

Michigan Technological University, Houghton, MI, USA

Humana Press

Editor
Dukka B. KC
Michigan Technological University
Houghton, MI, USA

ISSN 1064-3745 ISSN 1940-6029 (electronic)
Methods in Molecular Biology
ISBN 978-1-0716-2316-9 ISBN 978-1-0716-2317-6 (eBook)
https://doi.org/10.1007/978-1-0716-2317-6

This Humana imprint is published by the registered company Springer Science+Business Media, LLC, part of Springer Nature.
The registered company address is: 1 New York Plaza, New York, NY 10004, U.S.A.

Dedication

To my parents Ram Prasad KC and Tika Devi KC, wife Bidya KC, and daughters Diva KC and Dika KC

Dedication

Preface

Proteins are synthesized on ribosomes generating a nascent polypeptide chain and subsequently many proteins undergo modifications. Since these modifications occur after the synthesis of the polypeptide chain, they are called "post-translational modification" (PTM). PTMs mostly come into two flavors: covalent modification to polypeptide chain and proteolytic cleavage. The living cell extends the chemical repertoire and information content of the protein by PTM. Generally, PTMs occur on amino acid side chains or at a protein's C or N termini and modify existing functional groups or introduce new ones. Proteolytic cleavage is usually an irreversible modification, while covalent modifications may be reversible, e.g., in the case of protein phosphorylation. There are more than 450 unique protein post-translational modifications including the most common and extensively studied phosphorylation. Over the past decade, the number and variety of protein post-translational modifications that have been detected and characterized have rapidly increased. PTM can result in activation, inactivation, completely altered protein function, and regulate a vast array of biological processes.

Thanks to the advances in experimental approaches, mainly mass spectrometry-based approaches, proteome-wide profiling of post-translational modifications has been made possible. However, experimental approaches are not enough to understand and characterize more than 450 different types of PTMs. With the accumulation of a large number of high-throughput datasets that can be used to develop various computational approaches, complementary computational approaches are becoming popular and increasingly critical in unraveling post-translational modifications. Needless to say, understanding and characterization of PTM is a fundamental step toward understanding the functions of a proteome.

In this book, well-established computational tools for various types of post-translational modifications are introduced. In addition, we have tried our best to cover various aspects of computational post-translational modification prediction: prediction of PTM sites, prediction/analysis of functional PTM sites, visualization of PTM, development of databases for PTMs, prediction/analysis of PTM cross-talk, tools for proteolytic cleavage among others. The book is organized taking into consideration the steps for developing a computation tool, especially machine learning tools. The book discusses these routine tasks in any Machine Learning-based approach: (1) creation of the benchmark dataset (to train and test), (2) extraction of features or encoding of the input protein sequence, and (3) development of the appropriate machine learning architecture.

The book begins (Chapter 1) by setting up the premise for the development of computational tools, specifically, introducing advances in mass spectrometry-based proteomics technologies to maximize PTM coverage so that robust datasets can be generated for computational algorithms.

In Chapter 2, an approach called "phosphopeptide library dephosphorylation followed by mass spectrometry (PLDMS)" is described that allows for the exact control of phosphorylation site incorporation to study the preferred context of protein phosphatases.

In Chapter 3, various protein descriptors that are required to convert protein sequence and structure to numerical values are described so that protein sequences can be represented well in the subsequent machine learning approaches. We then describe a new webserver and stand-alone program called FEPS (Feature Extraction from Protein Sequences) for generating features from protein sequences.

Large-scale pre-trained models have recently shown great success in the field of AI by capturing knowledge from massive labeled and unlabeled data. In Chapter 4, we present a pre-trained model learned from over 24 million unlabeled sequence fragments to predict kinase-specific site prediction of kinases.

Using multiple modes of information (e.g., protein sequence, predicted structures) is another trend in the field. In Chapter 5, we present a support vector-based tool using sequence and structure-based features for protein glycation site prediction.

Glycosylation involves the attachment of carbohydrate sugar chains, or glycans, onto an amino acid residue of a protein. In Chapter 6, we present glycan-related databases and some bioinformatics approaches that have been developed to predict the glycosylation pathway and glycosylation sites on proteins.

In addition, various tools for analysis of PTMs have also been developed. In Chapter 7, we describe FAT-PTM, the Functional Analysis Tools for Post-translational Modifications in Arabidopsis proteins.

Recently it has been observed that more and more cellular proteins undergoing redox regulation through Cys post-translational modifications and peroxiredoxins (Prxs) are actively involved in these redox regulatory processes; in that regard, in Chapter 8, we describe a random forest-based predictor and classifier for peroxiredoxins which are a protein superfamily and play a critical role in intercellular signaling processes involving redox-regulated proteins and pathways.

In Chapter 9, we describe SPRINT-GLy, a Deep Learning-based approach for the prediction of N- and O-linked glycosylation sites incorporating sequence-based, predicted structural, and physicochemical information.

In Chapter 10, we describe iPTMNET, a resource that contains rich information about PTM from curated databases as well as text mining tools along with detailed examples of the various functions defined in the RESTful API.

In Chapter 11, we describe MUscADEL, a deep learning-based approach for predicting multiple (eight types of) lysine-based PTM sites. This is one of the newer trends in the field to share parameters and knowledge between related tasks.

In Chapter 12, based on the fact that all PTMs are not functionally equivalent, we describe SAPH-ire: a multi-feature heuristic neural network model to predict functional PTM sites.

In Chapter 13, as data from various experiments poses big challenges in identifying relevant features of biological and clinical importance, we present a workflow to visualize different PTMs that can provide a new form of visualization to condensate PTM changes and PTM cross-talk.

In Chapter 14, based on the fact that different PTMs both within a single protein and across different proteins can work together or regulate reciprocally, we describe PTM-X, a suite of machine learning models to predict both intra- and inter-protein PTM cross-talk.

In Chapter 15, we summarize the recent advancements in the field of both covalent attachment-based PTM site prediction and proteolytic cleavage prediction based on deep learning. Additionally, we also briefly discuss some existing challenges and future outlooks for the field.

Consequently, this book is an important resource for anyone studying computational post-translational modification as it describes the primary methods to characterize various types of post-translational modifications. The collection of chapters in the book spans from

PTM site prediction to functional PTM site prediction to analysis and prediction of PTM cross-talk.

I hope readers receive this book as a practical guide for using bioinformatics tools related to the computational prediction of post-translational modification. In addition, I am hopeful that this book exhibits the state of the art of the current research field in the arena and also provides future trends in the field. This book would not have been possible without the generous contributions from all the authors of chapters in this book. I sincerely thank them for their contributions. I am also thankful to the series editor, Dr. John M. Walker, for patiently working with me throughout the process. In addition, I would like to thank all the members of my lab and my collaborators. Most importantly, I would like to sincerely acknowledge the inspiration, encouragement, and guidance I received from my teachers Mr. Basanta Giri, Mr. Hemanta Rana, Dr. Tatsuya Akutsu, and Dr. Dennis Livesay.

Houghton, MI, USA ***Dukka B. KC***

Acknowledgments

This work was supported in part by NSF grants DBI-1901793 and IIS-1901086 to D.B.KC.

Contents

Dedication ... *v*
Preface ... *vii*
Acknowledgments ... *xi*
Contributors ... *xv*

1 Maximizing Depth of PTM Coverage: Generating Robust MS Datasets for Computational Prediction Modeling ... 1
Anthony A. Iannetta and Leslie M. Hicks

2 PLDMS: Phosphopeptide Library Dephosphorylation Followed by Mass Spectrometry Analysis to Determine the Specificity of Phosphatases for Dephosphorylation Site Sequences ... 43
Thomas Kokot, Bernhard Hoermann, Dominic Helm, Jeremy E. Chojnacki, Mikhail M. Savitski, and Maja Köhn

3 FEPS: A Tool for Feature Extraction from Protein Sequence ... 65
Hamid Ismail, Clarence White, Hussam AL-Barakati, Robert H. Newman, and Dukka B. KC

4 A Pretrained ELECTRA Model for Kinase-Specific Phosphorylation Site Prediction ... 105
Lei Jiang, Duolin Wang, and Dong Xu

5 iProtGly-SS: A Tool to Accurately Predict Protein Glycation Site Using Structural-Based Features ... 125
Iman Dehzangi, Alok Sharma, and Swakkhar Shatabda

6 Functions of Glycosylation and Related Web Resources for Its Prediction ... 135
Kiyoko F. Aoki-Kinoshita

7 Analysis of Posttranslational Modifications in Arabidopsis Proteins and Metabolic Pathways Using the FAT-PTM Database ... 145
Madison N. Blea and Ian S. Wallace

8 Bioinformatic Analyses of Peroxiredoxins and RF-Prx: A Random Forest-Based Predictor and Classifier for Prxs ... 155
Hussam AL-Barakati, Robert H. Newman, Dukka B. KC, and Leslie B. Poole

9 Computational Prediction of N- and O-Linked Glycosylation Sites for Human and Mouse Proteins ... 177
Ghazaleh Taherzadeh, Matthew Campbell, and Yaoqi Zhou

10 iPTMnet RESTful API for Post-translational Modification Network Analysis ... 187
Sachin Gavali, Karen E. Ross, Julie Cowart, Chuming Chen, and Cathy H. Wu

11 Systematic Characterization of Lysine Post-translational Modification Sites Using MUscADEL 205
Zhen Chen, Xuhan Liu, Fuyi Li, Chen Li, Tatiana Marquez-Lago André Leier, Geoffrey I. Webb, Dakang Xu, Tatsuya Akutsu, and Jiangning Song

12 Enhancing the Discovery of Functional Post-Translational Modification Sites with Machine Learning Models – Development, Validation, and Interpretation 221
Nolan English and Matthew Torres

13 Exploration of Protein Posttranslational Modification Landscape and Cross Talk with CrossTalkMapper 261
Arthur Grimaud, Frederik Haugaard Holck, Louise Marie Buur, Rebecca Kirsch, and Veit Schwämmle

14 PTM-X: Prediction of Post-Translational Modification Crosstalk Within and Across Proteins 275
Yuxuan Li, Yuanhua Huang, and Tingting Li

15 Deep Learning–Based Advances In Protein Posttranslational Modification Site and Protein Cleavage Prediction 285
Subash C. Pakhrin, Suresh Pokharel, Hiroto Saigo, and Dukka B. KC

Index *323*

Contributors

TATSUYA AKUTSU • *Bioinformatics Center, Institute for Chemical Research, Kyoto University, Kyoto, Japan*
HUSSAM AL-BARAKATI • *Department of Computer Science, Jamoum University College, Umm Al-Qura University, Jamoum, Saudi Arabia*
KIYOKO F. AOKI-KINOSHITA • *Glycan and Life Systems Integration Center (GaLSIC), Soka University, Tokyo, Japan*
MADISON N. BLEA • *Department of Biochemistry and Molecular Biology, University of Nevada, Reno, Reno, NV, USA*
LOUISE MARIE BUUR • *Department of Biochemistry and Molecular Biology, University of Southern Denmark, Odense M, Denmark*
MATTHEW CAMPBELL • *Institute for Glycomics, Griffith University, Southport, QLD, Australia*
CHUMING CHEN • *Center for Bioinformatics and Computational Biology, Department of Computer and Information Sciences, University of Delaware, Newark, DE, USA*
ZHEN CHEN • *Key Laboratory of Rice Biology in Henan Province, Henan Agricultural University, Zhengzhou, China*
JEREMY E. CHOJNACKI • *Faculty of Biology, Institute of Biology III, University of Freiburg, Freiburg, Germany; Signalling Research Centres BIOSS and CIBSS, University of Freiburg, Freiburg, Germany*
JULIE COWART • *Center for Bioinformatics and Computational Biology, Department of Computer and Information Sciences, University of Delaware, Newark, DE, USA*
IMAN DEHZANGI • *Department of Computer Science, Rutgers University, Camden, NJ, USA; Center for Computational and Integrative Biology, Rutgers University, Camden, NJ, USA*
NOLAN ENGLISH • *School of Biological Sciences, Georgia Institute of Technology, Atlanta, GA, USA*
SACHIN GAVALI • *Center for Bioinformatics and Computational Biology, Department of Computer and Information Sciences, University of Delaware, Newark, DE, USA*
ARTHUR GRIMAUD • *Department of Biochemistry and Molecular Biology, University of Southern Denmark, Odense M, Denmark*
FREDERIK HAUGAARD HOLCK • *Department of Biochemistry and Molecular Biology, University of Southern Denmark, Odense M, Denmark*
DOMINIC HELM • *Protein Analysis Unit, German Cancer Research Center (DKFZ), Heidelberg, Germany*
LESLIE M. HICKS • *Department of Chemistry, University of North Carolina at Chapel Hill, Chapel Hill, NC, USA*
BERNHARD HOERMANN • *Faculty of Biology, Institute of Biology III, University of Freiburg, Freiburg, Germany; Signalling Research Centres BIOSS and CIBSS, University of Freiburg, Freiburg, Germany*
YUANHUA HUANG • *Department of Biomedical Sciences, The University of Hong Kong, Hong Kong, China*
ANTHONY A. IANNETTA • *Department of Chemistry, University of North Carolina at Chapel Hill, Chapel Hill, NC, USA*

HAMID ISMAIL • *Department of Animal Science, North Carolina A&T State University, Greensboro, NC, USA*
LEI JIANG • *Department of Electrical Engineering and Computer Science and Christopher S. Bond Life Sciences Center, University of Missouri, Columbia, MO, USA*
DUKKA B. KC • *Department of Computer Science, College of Computing, Michigan Technological University, Houghton, MI, USA*
REBECCA KIRSCH • *Novo Nordisk Foundation Center for Protein Research, University of Copenhagen, Copenhagen N, Denmark*
MAJA KÖHN • *Faculty of Biology, Institute of Biology III, University of Freiburg, Freiburg, Germany; Signalling Research Centres BIOSS and CIBSS, University of Freiburg, Freiburg, Germany*
THOMAS KOKOT • *Faculty of Biology, Institute of Biology III, University of Freiburg, Freiburg, Germany; Signalling Research Centres BIOSS and CIBSS, University of Freiburg, Freiburg, Germany*
ANDRÉ LEIER • *Department of Genetics, School of Medicine, University of Alabama at Birmingham, Birmingham, AL, USA; Department of Cell, Developmental and Integrative Biology, School of Medicine, University of Alabama at Birmingham, Birmingham, AL, USA*
CHEN LI • *Biomedicine Discovery Institute, Department of Biochemistry and Molecular Biology, Monash University, Melbourne, VIC, Australia*
FUYI LI • *Department of Microbiology and Immunology, The Peter Doherty Institute for Infection and Immunity, The University of Melbourne, Melbourne, VIC, Australia*
TINGTING LI • *Department of Biomedical Informatics, School of Basic Medical Sciences, Peking University Health Science Center, Beijing, China*
YUXUAN LI • *Department of Biomedical Informatics, School of Basic Medical Sciences, Peking University Health Science Center, Beijing, China*
XUHAN LIU • *Drug Discovery and Safety, Leiden Academic Centre for Drug Research, Leiden, The Netherlands*
TATIANA MARQUEZ-LAGO • *Department of Genetics, School of Medicine, University of Alabama at Birmingham, Birmingham, AL, USA; Department of Cell, Developmental and Integrative Biology, School of Medicine, University of Alabama at Birmingham, Birmingham, AL, USA*
ROBERT H. NEWMAN • *Department of Biology, North Carolina Agricultural and Technical State University, Greensboro, NC, USA*
SUBASH C. PAKHRIN • *School of Computing, College of Engineering, Wichita State University, Wichita, KS, USA*
SURESH POKHAREL • *Department of Computer Science, College of Computing, Michigan Technological University, Houghton, MI, USA*
LESLIE B. POOLE • *Department of Biochemistry, Wake Forest School of Medicine, Winston-Salem, NC, USA*
KAREN E. ROSS • *Department of Biochemistry and Molecular and Cellular Biology, Georgetown University Medical Center, Washington, DC, USA*
HIROTO SAIGO • *Faculty of Information Science and Electrical Engineering, Kyushu University, Fukuoka, Japan*
MIKHAIL M. SAVITSKI • *European Molecular Biology Laboratory, Genome Biology Unit, Heidelberg, Germany; European Molecular Biology Laboratory, Proteomics Core Facility, Heidelberg, Germany*

VEIT SCHWÄMMLE • *Department of Biochemistry and Molecular Biology, University of Southern Denmark, Odense M, Denmark*
ALOK SHARMA • *Institute for Integrated and Intelligent Systems, Griffith University, Brisbane, QLD, Australia; Department of Medical Science Mathematics, Tokyo Medical and Dental University (TMDU), Tokyo, Japan; Laboratory for Medical Science Mathematics, RIKEN Center for Integrative Medical Sciences, Yokohama, Kanagawa, Japan*
SWAKKHAR SHATABDA • *Department of Computer Science and Engineering, United International University, Dhaka, Bangladesh*
JIANGNING SONG • *Biomedicine Discovery Institute, Department of Biochemistry and Molecular Biology, Monash University, Melbourne, VIC, Australia, Monash Centre for Data Science, Faculty of Information Technology, Monash University, Melbourne, VIC, Australia*
GHAZALEH TAHERZADEH • *Department of Mathematics and Computer Science, Wilkes University, Wilkes-Barre, PA, USA*
MATTHEW TORRES • *School of Biological Sciences, Georgia Institute of Technology, Atlanta, GA, USA*
IAN S. WALLACE • *Department of Biochemistry and Molecular Biology, University of Nevada, Reno, Reno, NV, USA; Department of Chemistry, University of Nevada, Reno, Reno, NV, USA*
DUOLIN WANG • *Department of Electrical Engineering and Computer Science and Christopher S. Bond Life Sciences Center, University of Missouri, Columbia, MO, USA*
GEOFFREY I. WEBB • *Monash Centre for Data Science, Faculty of Information Technology, Monash University, Melbourne, VIC, Australia*
CLARENCE WHITE • *Computational Science and Engineering Department, North Carolina A&T State University, Greensboro, NC, USA*
CATHY H. WU • *Center for Bioinformatics and Computational Biology, Department of Computer and Information Sciences, University of Delaware, Newark, DE, USA*
DAKANG XU • *Faculty of Medical Laboratory Science, Ruijin Hospital, School of Medicine, Shanghai Jiao Tong University, Shanghai, China; Department of Molecular and Translational Science, Faculty of Medicine, Hudson Institute of Medical Research, Monash University, Melbourne, VIC, Australia*
DONG XU • *Department of Electrical Engineering and Computer Science and Christopher S. Bond Life Sciences Center, University of Missouri, Columbia, MO, USA*
YAOQI ZHOU • *Institute for Systems and Physical Biology, Shenzhen Bay Laboratory, Shenzhen, China*

Chapter 1

Maximizing Depth of PTM Coverage: Generating Robust MS Datasets for Computational Prediction Modeling

Anthony A. Iannetta and Leslie M. Hicks

Abstract

Post-translational modifications (PTMs) regulate complex biological processes through the modulation of protein activity, stability, and localization. Insights into the specific modification type and localization within a protein sequence can help ascertain functional significance. Computational models are increasingly demonstrated to offer a low-cost, high-throughput method for comprehensive PTM predictions. Algorithms are optimized using existing experimental PTM data, thus accurate prediction performance relies on the creation of robust datasets. Herein, advancements in mass spectrometry–based proteomics technologies to maximize PTM coverage are reviewed. Further, requisite experimental validation approaches for PTM predictions are explored to ensure that follow-up mechanistic studies are focused on accurate modification sites.

Key words Post-translational modifications, Liquid chromatography–tandem mass spectrometry, Bottom-up proteomics, PTM enrichment, Database searching, Bioinformatics

1 Introduction

Protein activity is regulated through post-translational modifications (PTMs), either by covalent chemical modification or hydrolytic peptide bond cleavage. These modifications facilitate rapid changes to protein structure and function, which can alter protein stability [1, 2], signaling [3, 4], and/or localization [5, 6]. More than 400 types of PTMs have been experimentally discovered [7]; however, only a small subset are prevalent and have been extensively examined [8, 9].

Liquid chromatography–tandem mass spectrometry (LC-MS/MS) has become the predominant analytical tool for investigating PTMs due to high sensitivity and the ability to perform large-scale PTM analyses across an entire proteome [10, 11]. However, PTMs are transient and present at substoichiometric levels [12], often requiring specific enrichment during sample preparation to obtain

Dukka B. KC (ed.), *Computational Methods for Predicting Post-Translational Modification Sites*,
Methods in Molecular Biology, vol. 2499, https://doi.org/10.1007/978-1-0716-2317-6_1,

modified peptides within the limit of detection/dynamic range of the mass spectrometer for thorough depth of PTM coverage. LC-MS/MS analysis of PTMs typically employs bottom-up proteomics [13, 14], where extracted proteins are enzymatically digested into smaller, more easily analyzed peptides. Enriched peptides are separated by reversed-phase liquid chromatography (RPLC) before electrospray ionization and subsequent mass analysis, often via a data dependent acquisition (DDA) mode [15]. Improvements in instrument technology and methods [16, 17] permit rapid acquisition of accurate, data-rich MS/MS spectra, leading to the detection and identification of diverse PTM-containing peptides. Peptide identification is accomplished through database searching with algorithms such as Mascot [18], SEQUEST [19], or Andromeda [20]. Here, peptide fragment spectra are compared to theoretical spectra generated from an in silico prediction based on a specified protein database, and PTM site localization can be scored based on the resulting mass shifts detected in the MS/MS spectra. Bioinformatics is not only important for peptide identification and site localization, but also for limiting false positives through the control of false discovery and localization rates [21, 22].

Mass spectrometry serves as the gold standard for large-scale, discovery based PTM profiling [23–25], yet its experiments are time-consuming, expensive, and require a high degree of technical expertise [26]. Thus, high-throughput computational tools for PTM site prediction are being developed to complement LC-MS/MS approaches. These tools leverage experimental PTM datasets to build their computational prediction models (e.g., support vector machine, random forest, and neural networks) [26–28]. Experimental datasets are split into a training set and a testing set. For algorithm training, amino acids containing a specific PTM are labeled as positive sites while those same amino acids lacking that PTM are labeled as negative sites [29]. These sites become the center of the peptide sequence and the number of surrounding amino acids included is determined by a set window size. During training, the identity and/or physicochemical properties of the specific amino acids surrounding the modification site can be considered [30]. The prediction model is then applied to the test dataset to evaluate its performance. To develop a robust and predictive model, it is necessary that experimental datasets are accurate and reliable [31, 32]. Low depth of proteome/PTM coverage or insufficient fragmentation can result in missed identification or incorrect assignment of a modified site. This can cause biologically present PTMs to be falsely included as negative sites, which can impair model accuracy. Therefore, false negatives can be present in training data if extensive PTM coverage is not obtained. Thus, optimization of sample preparation, LC-MS/MS methodology, and bioinformatics implemented to maximize PTM coverage and minimize false assignments will provide the ideal datasets for PTM

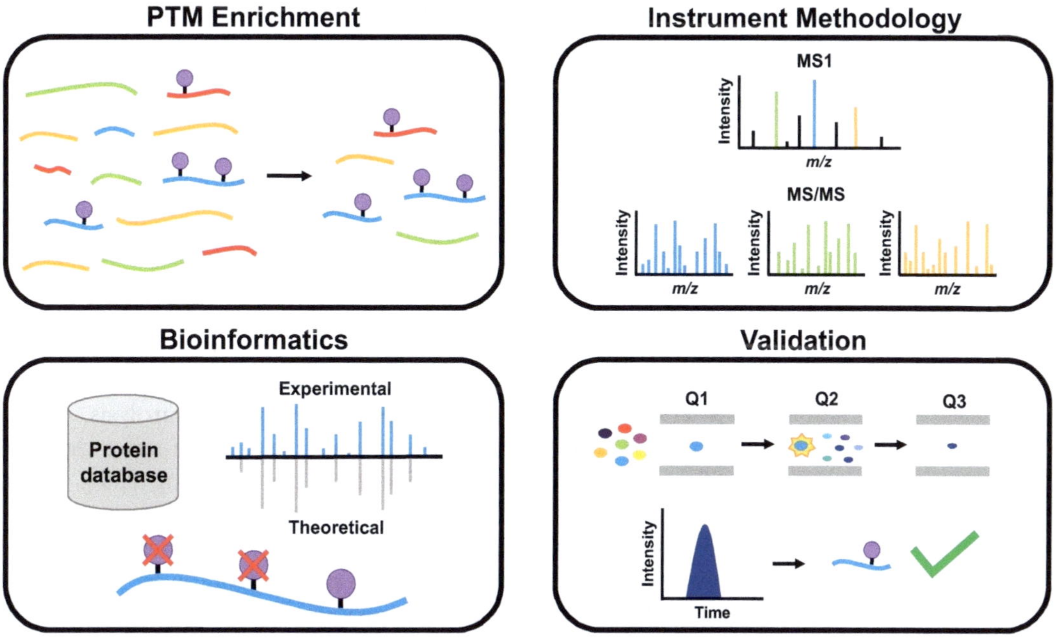

Fig. 1 Optimization of bottom-up proteomics sample preparation, LC-MS/MS methodology, and bioinformatics maximizes PTM coverage and minimizes false assignments. This provides ideal datasets for computational PTM prediction models. Validation methods can be employed to verify predicted PTMs from these tools

prediction tools. Herein, approaches to generate robust PTM datasets for computational prediction modeling using LC-MS/MS are examined (Fig. 1). Further, validation methods that should be employed to verify predicted PTMs from computational prediction tools are explored.

2 Sample Preparation

Bottom-up proteomics is the most widely used approach for LC-MS/MS–based PTM analyses [13, 14]. Proteolytic peptides are easily enriched, properly separated by RPLC, efficiently ionized, detectable with most MS hardware, and result in easily resolved characteristic fragmentation patterns [33, 34]. Enzymatic digestion of a proteome produces a complex peptide mixture with varying abundances, sizes, and modifications. To ensure robust PTM mapping, sample preparation must be tailored to ensure detection of modified peptides, often via targeted enrichment at the peptide level. Sample preparation strategies to produce ideal samples for in-depth PTM coverage via MS are elaborated in this section, including proteolysis, enrichment for selected common PTMs, and orthogonal fractionation to maximize depth of coverage. Limited in vivo proteolysis is an irreversible PTM, but techniques to identify proteolytic processed peptides and protease cleavage motifs do not follow the conventional bottom-up methods described herein, therefore those methods are not included in this review and described elsewhere [35, 36].

2.1 Proteolysis

Efficient digestion is paramount for proper PTM analysis in bottom-up proteomics. Trypsin is the preferred protease for bottom-up proteomics because of its well-defined cleavage specificity at the C-terminus of arginine and lysine residues [37]. This is an advantage because (1) these amino acids are positively charged in a low pH environment facilitating peptide ionization, and (2) lysine and arginine are common within proteins, ensuring that most resulting peptides are within an easily resolvable mass range (typically ~1–3 kDa) [38]. Native trypsin is sensitive to autolysis, which generates products that do not maintain the same strict cleavage specificity necessary for effective database searching [39]. Thus, mass spectrometry-grade trypsin modified to be extremely resistant to autolysis without compromising its enzymatic efficiency is most widely used in proteomics [40]. Trypsin proteolysis is typically performed for 12–16 h using a protease–protein ratio of 1:20 to 1:100 (w/w) and can be reversibly inactivated by adjusting the pH to <4. Faster digestions (<30 min) can be achieved with immobilized trypsin [41], resulting in similar proteome/PTM coverage compared to solution-phase trypsin digestion [42].

In cases where trypsin digestion results in non–ideal sized peptides (very small or large) [43], or when efficient cleavage is hindered due to presence of a PTM at or near the cleavage site [44], the combined or parallel use of multiple proteases has helped improve protein sequence coverage [45, 46]. Dual LysC/trypsin digestion of HeLa cells yielded a 30% increase in the number of identified proteins when performing a global proteome analysis and a twofold increase in the number of identified phosphorylation sites after TiO_2 enrichment vs. a single step trypsin digestion [47]. A twofold increase was also observed in the number of identified human phosphorylation sites with tandem Glu-C/trypsin digestion compared to trypsin alone [48]. While the application of multiple proteases can increase protein/PTM sequence coverage, bottom-up PTM studies continue to predominantly solely employ trypsin. This is likely due to increased experiment cost, decreased throughput, and more complex bioinformatics due to the inclusion of multiple protease specificities.

2.2 PTM Enrichment

Most PTMs are transient and substoichiometric, resulting in low abundance of modified peptides within a digested proteome [12]. Enrichment prior to LC-MS/MS analysis is commonly used to decrease sample complexity and improve PTM detection, leading to greater depth of coverage [49, 50]. Herein, enrichment techniques developed for protein phosphorylation, oxidation, glycosylation, ubiquitination, acetylation, and methylation are discussed. These are among the top ten frequent PTMs across popularly studied organisms [8, 9] and heavily associated with human disease [51].

2.2.1 Phosphorylation

Protein phosphorylation is a prevalent PTM responsible for regulating a variety of cellular processes from cell growth to programmed cell death [52, 53]. The addition or removal of a phosphate moiety by a kinase or phosphatase is responsible for conformational changes on direct substrates which affect downstream signaling networks. The significant role phosphorylation has in processes such as protein synthesis, metabolism, cell division, and signal transduction has increased interest in understanding the biological significance of this modification [54–56]. In eukaryotes, the distribution of experimentally identified phosphosites is 67% pSer, 24% pThr, and 9% pTyr [57]. Most developed canonical phosphopeptide enrichment techniques are efficient for phosphoserine/threonine-containing peptides and are elaborated in detail herein. However, orthogonal, specific enrichment is needed for phosphotyrosine-containing peptides, typically via antibodies or superbinder Src homology 2 domains [58, 59]. Likewise, noncanonical phosphorylation (e.g., His, Lys, Arg, and Cys phosphorylation) is involved in important signaling mechanisms [60, 61], as shown in bacterial two-component signaling systems [62]. Because phosphoramidate and phosphothiolate bonds are acid labile and cannot survive acid-based enrichment methods used for canonical phosphorylation, methods utilizing strong anion exchange chromatography at near-neutral pH have been designed to enrich for both canonical and noncanonical phosphorylated peptides [63].

Immobilized metal affinity chromatography (IMAC) and metal oxide affinity chromatography (MOAC) are the most common enrichments for bottom-up analysis of phosphorylated Ser/Thr residues (Fig. 2a) [64–67]. In IMAC, phosphopeptides are enriched due to the affinity between positively charged immobilized metal ions and the negatively charged phosphate group on phosphorylated peptides [68]. Multiply phosphorylated peptides have stronger interactions with the metal cation and are preferentially enriched in IMAC compared to peptides containing a single phosphorylation [69, 70]. The metal ion selected for the enrichment influences the selectivity for basic or acidic phosphopeptides (Fe^{3+}-IMAC enriches more acidic phosphopeptides than Ti^{4+}-IMAC) [71, 72]. Polymer-based metal ion affinity capture (PolyMAC) is a metal ion-functionalized soluble nanopolymer (Ti^{4+} is the most extensively used ion) designed to rapidly and reproducibly enrich phosphopeptides, boasting 95% selectivity and 90% recovery [73, 74].

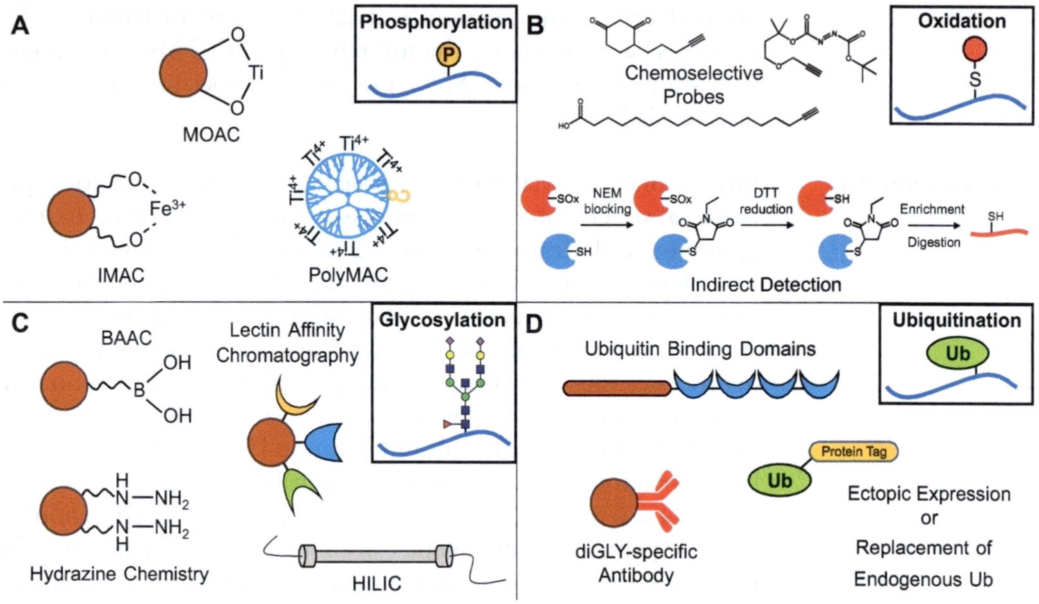

Fig. 2 Developed enrichment techniques for protein (**a**) phosphorylation, (**b**) oxidation, (**c**) glycosylation, and (**d**) ubiquitination

For MOAC, the interaction of phosphate-containing peptides and the metal oxide occur by reversible Lewis acid-base chemistry [75]. TiO_2 was the first metal oxide utilized for phosphopeptide enrichment and is extensively used [76, 77], but ZrO_2, Al_2O_3, Fe_3O_4, and other metal oxides have been developed for this purpose as well [78–80]. TiO_2 predominately enriches singly phosphorylated peptides, while IMAC and PolyMAC tend to enrich more multiply phosphorylated peptides [81, 82]. Therefore, methods utilizing multiple phosphopeptide enrichment strategies have been developed to take advantage of this orthogonality and create more complete PTM datasets [83, 84], although additional processing steps can contribute to increased identification variability due to sample loss. Furthermore, an extensive comparison between IMAC and MOAC with strong anion exchange fractionation suggests that the orthogonality between techniques is overstated and results more from unoptimized enrichment and instrument methods [85].

Specificity: Both IMAC and MOAC suffer from nonspecific binding of acidic peptides, but this can be alleviated with acidic loading and washing buffer to protonate peptidyl carboxylic acids [86]. Additionally, organic acid additives such as phthalic acid or lactic acid can be included in MOAC buffers to outcompete nonphosphorylated peptides for adsorption to the metal oxide, increasing enrichment specificity [87]. IMAC and MOAC can reach up to >90% selectivity for phosphopeptides when measures are included to decrease nonspecific binding of unphosphorylated peptides [85].

Enrichment of intact phosphoproteins has been attempted using IMAC and MOAC, but these methods suffer from low specificity and efficiency [88]. Other affinity-based methods have been developed including nanoparticles functionalized with Zn (II)-dipicolylamine or aluminum glycinate for the selective capture and enrichment of phosphoproteins [89, 90]. However, these approaches are not yet common for large-scale studies of the phosphoproteome and require additional investigation to determine their utility.

2.2.2 Oxidation

Cysteine residues in proteins are readily oxidized by reactive oxygen/nitrogen species (ROS/RNS), which are naturally present but increase under abiotic and biotic stress [91]. This can result in reversible Cys oxidation, serving as a regulator of protein signaling and function [92]. Further interactions with ROS/RNS can lead to irreversible Cys oxidation, indicating loss of protein function and degradation [93]. Further understanding of ROS/RNS-induced cysteine oxidation can reveal the connection between oxidative stress and associated signaling responses.

Measurements of the in vivo redox status of proteins is challenging because the associated modifications are labile and prone to artifactual induction during sample preparation. Enrichment of redox PTMs can be categorized into two classes: (1) indirect detection via reduction of reversible redox modifications and enrichment of previously oxidized sites, and (2) direct detection via tagging and enrichment of redox modifications using chemoselective probes (Fig. 2b).

Indirect differential alkylation-based methods using the biotin-switch technique (BST) or thiol-affinity chromatography for enrichment at the protein or peptide level are prevalent in redox proteomics [94–97]. Both techniques involve blocking reduced thiols with an alkylating agent (e.g., *N*-ethylmaleimide or iodoacetamide), reducing reversibly oxidized cysteines using a nonspecific reducing agent (e.g., dithiothreitol or tris(2-carboxyethyl) phosphine), and enriching previously oxidized thiols. In BST, nascent thiols are typically labeled with biotin-HPDP (*N*-[6-(biotinamido)hexyl]-3′-(2′-pyridyldithio)propionamide) [98]. This effectively converts these free cysteines to biotinylated cysteines and labeled residues can be enriched using biotin–avidin affinity purification. Oxidized cysteine resin-assisted capture (OxRAC), a method using thiol-affinity chromatography, implements thiopropyl Sepharose 6B for enrichment via reversible thiol-disulfide exchange reactions [99–101]. Selective reduction can be achieved with both techniques by replacing DTT with a chemical agent that selectively reduces specific thiol modifications (e.g., reduction of S-nitrosylation with ascorbate) [102].

Direct labeling has been achieved via numerous probes to covalently tag and enrich specific oxidative modifications at the protein level. Sulfenic acids are most commonly labelled and enriched using dimedone-based tags [103]. A clickable, electrophilic diazene probe has been developed to directly capture sulfinic acids [104]. Direct analysis of *S*-nitrosylated proteins has been achieved using triaryl phosphine probes [105], while protein *S*-acylation has been studied using terminal alkyne-tagged fatty acid probes to label residues, specifically protein palmitoylation [106]. Although chemoselective probes are useful for the detection of specific redox modifications, many suffer from both slow reactivity and efficiency and have limited utility in providing deep coverage on Cys residues with a variety of oxidative modifications [107]. Therefore, selection of the ideal approach for redox proteome mapping must be considered within the context of the computational prediction algorithm to be developed and biological question(s) addressed.

2.2.3 Glycosylation

Protein glycosylation mainly refers to covalent attachment of oligosaccharides (glycans) to protein Asn (*N*-glycosylation) or Ser/Thr (*O*-glycosylation) residues. Glycosylation is ubiquitous on membrane-bound proteins and involved in diverse functions including protein folding, immune modulation, and cell-to-cell recognition [108, 109]. Glycans are highly complex structures assembled from ten possible monosaccharides; competitions between various glycosyltransferases produces branched structures with different anomeric linkages [110]. Glycosylation can occur cotranslationally to prevent premature degradation, and this type of modification is indistinguishable from post-translational glycan modifications when performing LC-MS/MS analysis [111]. Tandem metabolic labeling strategies have been developed to specifically enrich for cotranslationally *O*-GlcNAc modified nascent proteins [111]. Other more well-established glycopeptide enrichment techniques will enrich for both co- and post-translationally modified glycopeptides, and should be considered when developing computational PTM algorithms.

Common strategies for the enrichment of glycopeptides include (1) hydrazide chemistry, (2) lectin affinity chromatography, (3) boronic acid affinity chromatography (BAAC), and (4) hydrophilic interaction liquid chromatography (HILIC) (Fig. 2c). These methods are based on glycan recognition or take advantage of the physiochemical differences between glycopeptides and unmodified peptides. They also differ in their specificity; some methods enrich specific glycans while others allow all glycopeptides to be enriched.

Hydrazide chemistry utilizes sodium periodate to oxidize glycopeptides/glycoproteins, converting their carbohydrate *cis*-diols to aldehydes and allowing them to bind to hydrazide resin through hydrazone bonds [112]. After trypsin digestion of immobilized proteins to remove nonglycosylated peptides, the captured glycopeptides are released from the support. Peptides bound by *N*-linked glycans can be enzymatically released by PNGase F [112], but there are no enzymes that can efficiently release *O*-linked glycopeptides. Therefore, peptides bound by *O*-linked glycans must be liberated through chemical methods, such as β-elimination after the addition of hydroxylamine [113]. Due to the covalent capture of glycopeptides, the selectivity of hydrazide chemistry enrichment is typically the highest among the listed methods (typically around ~90%) [114]. However, drawbacks include high cost, lengthy procedures, and the loss of information regarding the glycans at each modification site [12, 115]. This method allows for the identification of modified sites, but the release of the formerly *N*-linked or *O*-linked glycosylated peptide leaves the glycan coupled to the hydrazide support. Glycans with terminal sialic acids can be liberated from the support via acid hydrolysis of the sialic acids, preserving the modification on glycopeptides, but this prevents the enrichment of glycans without those terminal monosaccharides [116].

Lectin affinity chromatography uses lectin proteins that contain one or more specific carbohydrate recognition domains for glycopeptide/glycoprotein enrichment [117]. Lectins recognize specific glycans through the reversible binding of the anomeric hydroxyl groups on carbohydrates to the hydrophilic groups from lectin amino acid residues [118]. This specificity allows lectin affinity chromatography to be tuned for the selective capture of specific glycoforms based on their lectins [119]. This increased specificity manifests as a disadvantage for global glycosylation profiling, although studies have used lectin combinations for broader glycan enrichment [120, 121]. Another disadvantage of lectin affinity approaches is that lectins have weak and variable affinities for glycans (mM to μM), which can result in sample loss during washing steps [122].

In BAAC, glycopeptides are enriched through covalent bonds between boronic acid and carbohydrate *cis*-diols generating stable cyclic esters at a high pH [123]. These boronate esters are reversible under acidic conditions allowing for the release of the enriched glycopeptides after washing and removing any noncovalent molecules. BAAC contains simple capture and elution processes, does not require desalting before MS analysis, allows for a broader range of *N*- and *O*-linked glycopeptides to be captured compared to other methods, and preserves the glycans in the released glycopeptides allowing for the analysis of the carbohydrate chains at each specific modification site [12]. However, BAAC limitations include its low selectivity (~40% of identified peptides are glycopeptides) and sensitivity (nmol vs. fmol for hydrazide chemistry) in complex mixtures due to associated nonspecific binding and low binding capacities [24, 124].

HILIC enrichment of glycopeptides relies on the polarity differences between unmodified peptides and the more hydrophilic glycopeptides due to the presence of the attached carbohydrate [125]. This method is compatible with MS (uses mobile phases similar to low pH RPLC), high throughput, reproducible, has low cost, and enriches glycopeptides in an unbiased manner preserving the attached glycan [126, 127]. This allows for the characterization of these glycan structures at their modification sites. The major drawback for HILIC is its poor specificity for glycopeptides due to the large steric hindrances of the modified materials used for the stationary phase [115].

The diversity of protein glycosylation has led to the creation of various computational models that can predict sites modified by a specific glycan [128, 129], modification sites based on a certain glycan type (*N*-glycans vs. *O*-glycans) [130, 131], or all glycosites [132, 133]. The enrichment strategy implemented to generate a robust protein glycosylation dataset should be tailored to the type of prediction model that is being created. The selectivity of lectin affinity chromatography makes it the ideal enrichment technique for models predicting sites modified by a specific glycan. Hydrazide chemistry is the preferred enrichment strategy for models predicting sites based on a certain glycan type since it can be modified to liberate peptides previously containing an *N*- or *O*-linked glycan based on the elution chosen. Both BAAC and HILIC have the ability to simultaneously enrich *N*- and *O*-linked glycopeptides, so these enrichment techniques would benefit computational models that have complete glycosite prediction capabilities.

2.2.4 Ubiquitination

The enzymatic attachment of ubiquitin to lysine residues is best characterized as a trigger for protein degradation, but ubiquitination has been implicated in other processes such as DNA repair and gene silencing [134, 135]. Several strategies have been developed for the enrichment and investigation of ubiquitinated proteins (Fig. 2d).

One common enrichment approach uses ubiquitin-binding domains (UBDs) that have binding specificities to particular ubiquitin linkages [136], and methods utilizing multiple UBD-containing proteins have been developed to fully profile the ubiquitinome [137]. These tandem ubiquitin-binding entities protect against proteasome and deubiquitylating enzyme activities, allowing for the recovery of short-lived ubiquitin-conjugated proteins [137, 138], but have weak affinities toward ubiquitin (2–500 μM) and also enrich proteins containing ubiquitin-like modifiers [139].

Alternatively, epitope-tagged ubiquitin is incorporated either through ectopic expression or the replacement of endogenous ubiquitin [140, 141]. In ectopic expression, the genome is edited to incorporate an epitope tag into expressed ubiquitin. The ubiquitin replacement method proceeds via knockdown of endogenous ubiquitin and introduction of epitope-tagged ubiquitin into the cell. The selection of the epitope tag for these purification methods affects the composition of nonspecific background binding proteins. 6xHis-Ub and avidin tagged-Ub can be purified under harsh denaturation conditions, but these enrichments suffer from background proteins that contain histidine or are biotinylated in vivo [142]. Enrichments employing other epitope tags, such as FLAG, do not have these disadvantages, but require purification under more mild conditions and therefore will contain a general background of unmodified proteins [143].

Ubiquitin is cleaved with trypsin leaving a diglycine (diGLY) modification on the resulting tryptic peptide modified lysine residue. An antibody specific toward this specific diGLY motif was generated for enrichment [144, 145], providing the simplest method to analyze the ubiquitinome. However, these antibodies cannot recognize ubiquitin-modified N-termini, are biased toward motifs of acidic amino acids surrounding the diglycine-modified lysine, and also purify peptides from proteins modified by ubiquitin-like modifiers [144, 145]. Even with these disadvantages, enrichments using this antibody are common for examining ubiquitination [144, 146, 147].

2.2.5 Acetylation and Methylation

Protein acetylation occurs through the acetyltransferase-catalyzed or nonenzymatic transfer of an acetyl group from acetyl-coenzyme A to protein N-termini or lysine residues. Since acetyl-CoA is consumed during protein acetylation, this process interferes with metabolic processes and energy homeostasis [148]. Acetylation also affects protein function by regulating stability, enzymatic activity, and subcellular localization [149–152]. Likewise, protein methylation is catalyzed by methyltransferases and predominantly occurs on arginine and lysine residues. Histone lysine methylation regulates gene expression [153], while methylation of other proteins is involved in processes such as DNA repair, signal transduction, protein synthesis, and RNA metabolism [154–157]. Immunoprecipitation techniques currently dominate enrichment approaches for in-depth analyses of proteome-wide lysine acetylation and arginine/lysine methylation [158, 159].

2.3 Orthogonal Fractionation

Even with continuous MS hardware advances, the dynamic range of peptide abundances in a one-dimensional separation of a PTM-enriched proteome sample precludes identification of low abundance peptides [160]. Orthogonal separations (offline and online) integrated into proteomics workflows can decrease sample complexity and allow for improved coverage. Strong cation exchange (SCX) chromatography is common as it provides high orthogonality to low pH RPLC [161]. High pH RPLC has become the preferred method for offline fractionation due to its superior chromatographic resolution, peak capacity, and retention time reproducibility over other methods, although it is not completely orthogonal to low pH RPLC [162, 163]. Many fractions can be collected and subsequently concatenated, ensuring that combined fractions have little overlap in the retention time of eluting peptides [164, 165], thus improving high pH RPLC orthogonality and throughput.

To ameliorate sample losses, online two-dimensional LC (2D-LC) separations can be employed. Multidimensional protein identification technology (MudPIT), in which peptides are separated first by ion exchange and then by RP resins in the same capillary, is widely used in bottom-up proteomics for 2D-LC and has been extended to PTM analyses [166, 167]. High pH RP and HILIC can also be employed in the first-dimension separation. RP-RP HPLC results in more peptide identifications than SCX-RP HPLC, possibly due to its higher separation efficiency [168]. HILIC provides higher resolution than ion exchange chromatography and enrichment for polar PTMs such as glycosylation and phosphorylation [169, 170]. While orthogonal fractionation decreases sample complexity and increases PTM identifications, there are trade-offs with decreased throughput that derive from the analysis of separate fractions. For example, SCX fractionation of a TiO_2-enriched phosphoproteome resulted in a ~ 50% increase in phosphosite identifications, with a 2–2.5 fold increase in needed acquisition time [171]. Even so, orthogonal fractionation is a principal factor in obtaining the greatest depth of PTM coverage.

3 Instrument Methodology

Robust PTM detection is dominantly achieved via LC-MS/MS. Enriched peptides are separated by RPLC using a gradient elution with increasing nonpolar solvents, where hydrophobic peptides are retained longer and elute later than hydrophilic peptides. Upon elution, peptides are ionized by electrospray ionization and subjected to one of the various MS acquisition modes, such as data dependent acquisition (DDA) [15]. Here, an MS1 survey scan identifies a set number of peptide ions that reach certain intensity and charge thresholds. These ions are sequentially isolated and fragmented to examine each peptide separately. Once the MS/MS fragment spectra for those set of peptides are collected, the cycle repeats with another survey scan. The efficient collection of intense, data-rich fragment spectra dictates the maximum number of peptide identifications. Therefore, advances in mass spectrometer hardware (including acquisition speed/duty cycle) and employment of robust fragmentation methods are required for the production of in depth PTM datasets for computational prediction models.

3.1 Improvements in Modern Technology

Instrument sensitivity and speed are paramount for maximum peptide detection and identification. The current limit of detection for peptides using modern instrumentation is at the attomole level, and these advances in MS sensitivity allowed the identification of ~4000 human phosphopeptides after phosphopeptide enrichment of 25 μg protein [172]. Nearly all PTM analyses utilize hybrid mass spectrometers that combine ion filters and high resolving power (RP) mass analyzers, enabling peptide ion selection and MS/MS fragmentation in a quadrupole or ion trap while high resolution mass spectra are acquired in a time-of-flight (TOF) or Orbitrap mass analyzer.

Quadrupoles are excellent mass filters due to their near-instantaneous mass selection [173]. Enhanced quadrupole transition rates improve ion transmission, which increases precursor ion abundance during accumulation and fragment spectrum intensity. This leads to better peptide identification and modification localization.

Faster duty cycles in high RP mass analyzers have allowed for an increased collection of MS/MS spectra and peptide identifications. Mass spectrometers from the Sciex TripleTOF series are commonly used QTOF instruments, while mass spectrometers from the Thermo Scientific Q Exactive series are commonly used quadrupole-Orbitrap instruments. With mass spectrometers containing a TOF mass analyzer, ions are accumulated for a fixed amount of time and then scanned at a defined speed (e.g., 100 Hz). In a 2 s cycle, the accumulation time for the survey scan is typically ~250 ms, while the accumulation time for each MS/MS scan is ~85 ms if 20 MS/MS scans are collected per survey scan. For mass spectrometers containing an Orbitrap, ion accumulation occurs in the C-trap until a sufficient ion population is obtained or the maximum injection time is reached. Since resolution is correlated with scan speed in Orbitrap mass analyzers, MS1 survey scans are typically collected at a higher resolving power (e.g., 120,000 with 2.5 Hz scan speed), while MS/MS scans are collected at a lower resolving power (e.g., 30,000 with 15 Hz scan speed). The average cycle time when peptides are eluting is ~2 s if 20 MS/MS scans are collected per survey scan. With instruments containing either mass analyzer, >40,000 MS/MS spectra can be collected over a typical 1–2 h acquisition time resulting in a data file size of >2 GB. When comparing the mass analyzers, TOF mass analyzers provide the greatest mass range (>1,000,000 *m/z* vs. 50,000 *m/z* for Orbitrap, although mass range of hybrid instruments is limited by the coupled quadrupole) and fast scan speeds (100 Hz vs. maximum of 40 Hz for Orbitrap) for increased acquisition of MS/MS spectra [174], but are inferior to the Orbitrap in resolving power (35,000 vs. 500,000 for Orbitrap) and mass accuracy (<3 ppm vs. <1 ppm for Orbitrap) [175]. Hybrid Orbitrap instruments also have the capabilities for parallel acquisition, where a fragmented peptide can be analyzed while fragment ions from the subsequent precursor can accumulate and prepare for injection into the Orbitrap [176]. This allows for more MS/MS spectra to be collected, boosting the number of identified peptides. Advances in Orbitrap geometry have allowed for faster transients, allowing for high resolution acquisition in a shorter amount of time [177].

For PTM-enriched and fractionated samples, scan speed is not a limiting factor due to the low number of coeluting peptides [178]. Therefore, MS/MS spectra quality can be improved by increasing Orbitrap analyzer RP and strengthening ion intensity with longer ion accumulation times to enhance identification rates, even though this lengthens instrument cycle times [178]. Also, narrower precursor isolation windows (0.8 vs. 1.5 *m/z*) can be utilized to prevent the coisolation/fragmentation of multiple peptide ions, although this can be disadvantageous as fragment spectra will be less intense, leading to lower quality spectra.

Other MS improvements include advances in instrument optics that have decreased ion movement and transfer time [179]. Additionally, developed instrument algorithms improve charge state assignment, increasing the number of available peptide precursors for selection in DDA experiments [179].

3.2 Fragmentation Methods

Collision induced dissociation (CID) [and the more recent higher-energy collisional dissociation (HCD)] is the most common fragmentation technique used for the collection of MS/MS spectra in PTM analyses (Fig. 3a). In these methods, peptide ions are subjected to collisions with neutral gases (e.g., nitrogen, helium, or argon) leading to an increase in their internal energy and subsequent peptide fragmentation (producing *b/y* ions). However, these fragmentation techniques can preferentially cleave labile PTMs (e.g., phosphorylation and glycosylation) over the peptide backbone [180]. For example, CID/HCD MS/MS spectra of phosphopeptides contain many instances of neutral losses associated with the loss of the phosphate group [181]. The presence of these neutral losses in MS/MS spectra is based on the modification. For example, neutral losses appear in 97% and 100% of MS/MS spectra for peptides containing pSer/pThr and glycans, respectively [182, 183]. Peptide identification can proceed by including probable neutral losses in the search based on the modification (e.g., −98 Da for a loss of H_3PO_4 in peptides containing pSer/pThr), but neutral losses hinder PTM localization, as unmodified fragments and fragments resulting from a neutral loss of the PTM have the same mass and are indistinguishable.

Electron transfer dissociation (ETD) involves an ion/ion reaction between a peptide ion and an anion reagent with low electron affinity (Fig. 3b) [184]. An electron is transferred from the anion reagent to the peptide ion causing fragmentation along the peptide

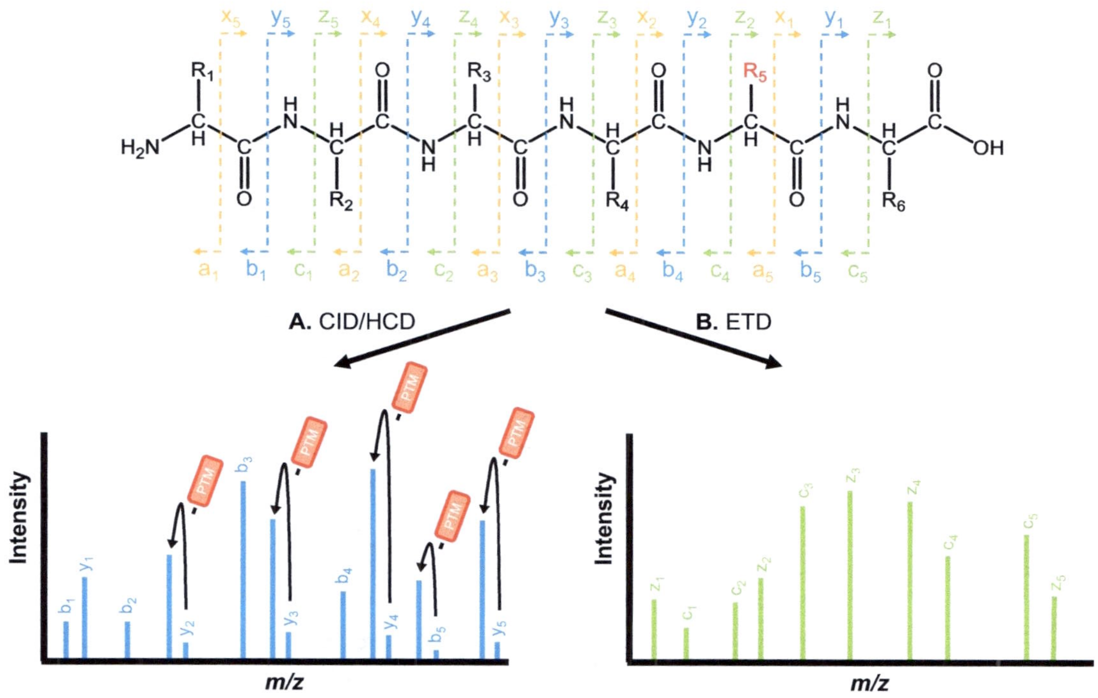

Fig. 3 Methods for MS/MS fragmentation. The red amino acid side chain represents a modified residue in the displayed peptide. (**a**) In collision induced dissociation (CID) and higher-energy collisional dissociation (HCD) peptide ions are subjected to collisions with neutral gases, producing *b/y* ions (blue). Neutral losses resulting from the cleavage of labile PTMs can occur in these methods. (**b**) In electron transfer dissociation (ETD), reactions between peptide ions and an anion reagent result in fragmentation along the peptide backbone producing *c/z* ions (green). ETD fragmentation efficiency is dependent on peptide charge state and ETD reaction times are longer than CID/HCD activation

backbone (producing *c*/*z* ions). An advantage of ETD over CID/HCD is that PTMs are preserved on resulting fragment ions, aiding modification localization [185]. However, fragmentation efficiency is highly dependent on peptide charge state (poor fragmentation with $z < 3$), and ETD reaction times are longer than what is required for CID/HCD activation, decreasing the number of peptides that can be analyzed [186, 187]. Dual fragmentation, where ETD is first applied to peptide ions before unreacted ions and resulting fragments are subjected to HCD, deemed EThcD, combines the advantages of both techniques and produces data-rich MS/MS spectra containing both *b*/*y* and *c*/*z* ion series, aiding in peptide identification and PTM localization [188].

4 Bioinformatics

Data obtained from analyzing PTM-containing peptides often include >40,000 MS/MS spectra, necessitating post-acquisition bioinformatic tools to identify peptides and localize PTMs. Typically, automated peptide identification is accomplished through database searching, where experimental fragment spectra are compared against a sequenced protein database, and modifications that should be considered are specified in the search parameters [18–20, 189–194]. If the peptide sequence contains only one possible modification site, then PTM localization is trivial. PTM localization becomes more difficult when multiple possible sites exist, so additional algorithms facilitate determination of site localization confidence [195–199]. Both peptide identification and PTM site localization tools rely on probabilistic scoring of candidate modified peptides based on the fragment ions detected, therefore algorithms to minimize false positives are required [197, 200].

4.1 Peptide Identification

Bottom-up experiments acquired through DDA result in thousands of MS/MS spectra, each nominally representing the fragmentation of a single peptide. The most common approach to identify these peptides relies on automated database searching using publicly available databases from genome/transcriptome sequencing (Table 1).

Database searching algorithms are numerous (Table 2) [18–20, 189–194], but they generally operate in a similar fashion. The selected database is processed in silico (based on defined parameters such as protease specificity, possible modifications, fragmentation method, parent/fragment mass tolerances, etc.) into all possible peptides that may be found in the analyzed sample. Theoretical fragment spectra of these candidate peptides are generated and then compared and scored against each experimental spectrum. The output contains the collected MS/MS spectra, each with a list of ranked peptides, where the highest scoring peptide is assigned resulting in a set of peptide-spectrum matches (PSMs).

Table 1
General and organism-specific websites housing publicly available genomic/proteomic databases. Protein databases are used in database searching to identify peptides from acquired MS/MS spectra

Organism	Organism specific website	Website address
Multiple	UniProt	https://www.uniprot.org/
Multiple	NCBI	https://www.ncbi.nlm.nih.gov/
Arabidopsis thaliana	The Arabidopsis Information Resource	https://www.arabidopsis.org/
Caenorhabditis elegans	WormBase	https://wormbase.org/
Chlamydomonas reinhardtii	Chlamydomonas Resource Center	https://www.chlamycollection.org/
Danio rerio	Zebrafish Information Network	http://zfin.org/
Drosophila melanogaster	FlyBase	http://flybase.org/
Escherichia coli	EcoCyc	https://ecocyc.org/
Homo sapiens	The Human Protein Atlas	https://www.proteinatlas.org/
Mus musculus	Mouse Genome Informatics	http://www.informatics.jax.org/
Saccharomyces cerevisiae	*Saccharomyces* Genome Database	https://www.yeastgenome.org/

Assessing confidence. The number of spectra that can be collected by modern instruments and the size of protein databases prevents the ability to quickly and objectively perform manual validation of each assignment. Thus, target-decoy searching is employed to estimate a false-discovery rate (FDR) [201]. Here, each MS/MS spectrum is also searched against a decoy database that is generated by reversing or randomizing the protein sequences in the specified database [200]. The expectation is that there should be no PSMs in the decoy database search, so these matches provide an excellent estimation of the number of false positives that are present within the target search. Based on the FDR obtained from the decoy search, the peptide score threshold can be adjusted in the target search to reach a certain FDR (commonly 1%) [202, 203].

Bioinformatics in proteomics still has the potential to improve. One significant issue that persists is the number of unassigned MS/MS spectra in a given experiment (60–70% is possible) [204]. Even with continuous advances in sample preparation and MS technology/methodology, PTM coverage will be limited if a majority of spectra cannot be identified. Since only specific PTMs are considered during database searching, many MS/MS spectra can be left unassigned if they are derived from peptides containing unexpected modifications/mass shifts. Consideration of multiple PTMs during database searching greatly increases the peptide search space,

Table 2
Programs and algorithms commonly used for database searching. Each has its own compatibility with various analysis software and PTM localization algorithms

Database Searching Algorithm	Freeware or Proprietary?	Website	Analysis Software Compatibility	Localization Algorithm Compatibility	Tailored for Open Searches?
Andromeda	Freeware	https://www.maxquant.org/	MaxQuant PeptideShaker Scaffold	AScore LuciPHOr2 PtmRS PTM-Score	No
Comet	Freeware	http://comet-ms.sourceforge.net/	PeptideShaker ProHits Trans-Proteomic Pipeline	AScore LuciPHOr2 PtmRS SLoMo	No
Mascot	Proprietary	https://www.matrixscience.com/	PeptideShaker ProHits Proteome Discoverer Scaffold Trans-Proteomic Pipeline	AScore LuciPHOr2 MD Score PtmRS SLoMo	No
MetaMorpheus	Freeware	https://github.com/smith-chem-wisc/MetaMorpheus	PeptideShaker	AScore LuciPHOr2 PtmRS	Yes
MS Amanda	Freeware	https://ms.imp.ac.at/?goto=msamanda	PeptideShaker Proteome Discoverer Scaffold	AScore LuciPHOr2 MD Score PtmRS	No
MSFragger	Freeware	http://msfragger.nesvilab.org/	FragPipe ProteomeDiscoverer	LuciPHOr2 MD Score PtmRS	Yes
MS-GF+	Freeware	https://omics.pnl.gov/software/ms-gf	PeptideShaker ProHits Scaffold Trans-Proteomic Pipeline	AScore LuciPHOr2 PtmRS SLoMo	No
SEQUEST	Proprietary	http://fields.scripps.edu/yates/wp/?page_id=17	Proteome Discoverer Scaffold Trans-Proteomic Pipeline	AScore LuciPHOr2 MD Score PtmRS SLoMo	No
X!Tandem	Freeware	https://www.thegpm.org/tandem/	PeptideShaker ProHits Scaffold Trans-Proteomic Pipeline	AScore LuciPHOr2 PtmRS SLoMo	No

causing an increase in search time and false positives. Instead, searches with wide precursor mass tolerances can reveal additional modifications when unmodified peptides are partially matched to fragment spectra and precursor mass shifts are examined [205], although these methods are computationally demanding and require extensive search times. Faster search algorithms have made these open searches more feasible and can improve peptide identification when unknown modifications are present [189, 193]. Another issue that contributes to unassigned MS/MS spectra is the production of uncharacteristic peptide fragment ions during fragmentation, as database searching only considers characteristic backbone fragment ions when assigning PSMs. Deep learning approaches for the prediction of complete peptide fragment spectra (both backbone and uncharacteristic ions) are starting to be explored [206], but have not yet been developed for modified peptides. Additional contributions from the computational community could help tailor these models for PTM-containing peptides and increase the depth of PTM coverage.

4.2 Site Localization

While database searching is a powerful technique for accurately identifying modified peptides, additional bioinformatics tools are needed to assess PTM localization confidence. Application of said algorithms can provide additional selection criteria/confidence in final datasets used for computational PTM prediction models. Numerous scoring algorithms (such as AScore [195], PTM-Score [20], SLoMo [196], LuciPHOr2 [197], Mascot Delta Score [198], and PtmRS [199]) have been implemented in this regard, and all of them are free to utilize except for Mascot Delta Score. While certain platforms were initially created to localize phosphorylation, all localization algorithms mentioned here can handle any PTM that can be included as a variable modification during database searching.

AScore, SLoMo, PTM-Score, and PtmRS have similar localization procedures with minor differences in how they select experimental and theoretical fragment ions for comparison. In all algorithms, theoretical fragment ion masses for each considered modification site permutation are generated and compared to experimental masses from the collected MS/MS spectrum. The analyzed experimental masses are limited to a set number of peaks (determined by the algorithm) in each 100 *m/z* window. This helps to eliminate noise from less intense peaks that can prevent proper modification localization. Peptide scores for all considered permutations are calculated using a binomial distribution probability, the one with the highest score is matched with the MS/MS spectrum, and the localization probability is calculated from the difference in permutation peptide scores. AScore and SLoMo use the same localization scoring output, where scores ≥ 19 indicate that there is a 99% or better possibility that the modification was correctly localized. Scores between 15 and 19 indicate >90% certainty, while scores between 3 and 15 have a success rate around 80% [195]. PTM-Score groups identified peptides with modifications into four categories labeled class I–IV [207]. Class I peptide are those most confidently localized with a localization probability ≥0.75. PtmRS outputs localization site probabilities between 0 and 1, where 0.99 is typically used as the cutoff for confident PTM localization, corresponding to a false localization rate (FLR) of 1% [199].

The algorithms for Mascot Delta (MD) Score and LuciPHOr2 calculate the localization probability through separate comparisons of different modification site permutations to the experimental fragment spectrum. The MD Score is derived through the Mascot Peptide Score difference of the top two alternative PTM site assignments from the same matched peptide sequence [198]. The typical threshold for confident PTM site localization is a MD Score ≥ 10, which corresponds to a localization probability of 91%. The LuciPHOr2 algorithm reports a LuciPHOr delta score, which is the score difference of the top two possible modification sites in the peptide [197]. Unlike the other algorithms, which set arbitrary localization probability thresholds to determine confident localization, LuciPHOr2 estimates the FLRs for every delta score through a target-decoy procedure, which dictates the delta score threshold required for a specific FLR (commonly 1%) [197].

Comparing different PTM site localization algorithms can be challenging as they are specific to certain database searching platforms, therefore performances of full bioinformatic pipelines are typically evaluated. Fifteen bioinformatic pipelines were compared using a phosphopeptide enriched sample from HeLa cells and synthetic phosphopeptides [208]. When analyzing the complex phosphoproteomics data, the SEQUEST/PtmRS and X!Tandem/D-Score combinations resulted in the most unique localized phosphopeptides. For the synthetic phosphopeptides analysis, the Comet/PtmRS combination resulted in the most PSMs with zero phosphorylation localization misassignments.

5 Computational Prediction Models: Considerations

Computational prediction models leverage experimental LC-MS/MS-based PTM datasets for algorithm training and testing. Publicly available PTM-specific databases can be found in Table 3. In these training sets, amino acids containing a specific PTM are labeled as positive sites while those same amino acids lacking that PTM are labeled as negative sites [29]. These sites become the center of the peptide sequence and the number of surrounding amino acids included is determined by a set window size. In the training set, typically the number of negative inputs greatly exceeds

Table 3
PTM specific databases that can be leveraged for computational prediction models

PTM	PTM Database	Database Website Address
Multiple	Uniprot	https://www.uniprot.org/
	PhosphoSitePlus	https://www.phosphosite.org/
	dbPTM	http://dbptm.mbc.nctu.edu.tw/
	PHOSIDA	http://141.61.102.18/phosida/index.aspx
	PLMD	http://plmd.biocuckoo.org/
	BioGRID	https://thebiogrid.org/
	qPTM	http://qptm.omicsbio.info/
Phosphorylation	PhosPhAt	http://phosphat.uni-hohenheim.de/
	P3DB	http://www.p3db.org/
	Phospho.ELM	http://phospho.elm.eu.org/
	dbPAF	http://dbpaf.biocuckoo.org/
	EPSD	http://epsd.biocuckoo.cn/
	dbPSP	http://dbpsp.biocuckoo.cn/
	FPD	http://bis.zju.edu.cn/FPD/
Oxidation	iCysMod	http://icysmod.omicsbio.info/
	dbSNO	http://140.138.144.145/~dbSNO/index.php
	GPS-SNO	http://sno.biocuckoo.org/
	GPS-Palm	http://gpspalm.biocuckoo.cn/
Glycosylation	*N*-GlycositeAtlas[a]	http://nglycositeatlas.biomarkercenter.org/
	O-GlycBase	http://www.cbs.dtu.dk/databases/OGLYCBASE/
	Unipep	http://www.unipep.org/
	GlyConnect	https://glyconnect.expasy.org/
Ubiquitination	SCUD	http://scud.kaist.ac.kr/
	UUCD	http://uucd.biocuckoo.org/
	mUbiSiDa	http://reprod.njmu.edu.cn/cgi-bin/mubisida/mUbiSiDa.php
Methylation/acetylation	WERAM	http://weram.biocuckoo.org/

[a] Requires free registration

the positive inputs so methods are utilized to balance these groups [209]. After balancing the two groups, peptide sequences are transformed into vectors that can be input for algorithm training. During training, the identity of the specific amino acids surrounding the modification site can be the only factor used or various amino acid physiochemical properties (e.g., hydrophobicity and size) can be considered as well [30]. The prediction model is then assessed against the testing dataset to evaluate its predicative performance.

False negatives in training datasets: The accuracy of prediction algorithms relies on extensive positive and negative datasets with few false positives/negatives [31, 32]. For positive datasets this is accomplished through peptide identification and modification localization validation (decreasing FDR and FLR), but it can be more difficult to create a sufficient negative dataset for algorithm testing. A biologically present PTM might be falsely included in the negative dataset if the peptide containing that modification is outside the mass spectrometer limit of detection or exhibits poor fragmentation, leading to lack of identification or modification site misassignment. Therefore, false negatives can be present in the training data if extensive PTM coverage is not obtained via proteomics.

PTM cross talk: While PTMs have been largely investigated independently, proteins can be modified by multiple PTMs that can interact or cross talk, where one PTM regulates the presence of another PTM. Serial PTM enrichment methods have been developed to identify multiply modified proteins and investigate PTM cross talk [99, 210]. However, peptides resulting from bottom-up strategies are unlikely to reveal multiply modified proteoforms directly. Top-down or middle-down proteomics strategies, where proteins are analyzed intact or with minimal cleavage have been employed to increase the chance of finding multiple PTMs on the same peptide/protein [211, 212]. However, these approaches cannot obtain the same depth of PTM coverage as bottom-up techniques, which limits global PTM characterization. Computational PTM cross talk predictions based on these generated datasets can be challenging because the observation of multiple PTMs in close proximity does not guarantee that cross talk is occurring. Therefore kinetic-based MS assays should be used to validate cross talk between two PTMs [213].

Organism specific training data: Availability of species-specific PTM datasets is ideal to create robust prediction algorithms. While cross-species training and testing has been accomplished, it shows decreased success when compared to species-specific algorithms [214, 215]. This occurs because modified amino acid residues may not be evolutionarily conserved across species. The development of exceptional species-specific PTM prediction models relies on robust LC-MS/MS datasets. Table 4 contains the results of a thorough literature investigation for datasets containing the most unique modification sites across all model organisms, which can be used to train and test species-specific PTM prediction algorithms.

Table 4
Datasets containing the most unique modification sites across all model organisms. These can be used to train and test species-specific PTM prediction algorithms

Organism	PTM	Identified Sites	Reference
Arabidopsis thaliana	Phosphorylation	15,445[a]	[216]
	Reversible oxidation	2113	[217]
	S-sulfenylation	1537	[218]
	S-nitrosylation	1195	[219]
	Glycosylation	2186	[220]
	Ubiquitination	3009	[221]
	Acetylation	2152	[222]
	Methylation	617	[223]
Caenorhabditis elegans	Phosphorylation	18,187	[224]
	Glycosylation	1794	[220]
Chlamydomonas reinhardtii	Phosphorylation	15,862	[225]
	Reversible oxidation	9295	[101]
	S-nitrosylation	392	[226]
	Glycosylation	181	[227]
Danio rerio	Phosphorylation	11,474	[228]
	Glycosylation	2254	[220]
	Acetylation	377	[229]
Drosophila melanogaster	Phosphorylation	13,720	[230]
	Reversible oxidation	1191	[231]
	Glycosylation	2229	[220]
	Ubiquitination	14,018	[232]
	Acetylation	1981	[233]
Escherichia coli	Phosphorylation	380	[234]
	Reversible oxidation	1229	[235]
	Acetylation	8284	[236]
	Methylation	37	[237]
Homo sapiens	Phosphorylation	38,229	[238]
	Reversible oxidation	10,440	[239]
	S-nitrosylation	8304	[240]
	Glycosylation	4783	[241]
	Ubiquitination	63,455	[242]
	Acetylation	8115	[243]
	Methylation	8030	[244]
Mus musculus	Phosphorylation	51,821	[245]
	Reversible oxidation	2162	[246]
	S-nitrosylation	620	[247]
	Glycosylation	8386	[248]
	Ubiquitination	20,085	[145]
	Acetylation	2187	[249]
	Methylation	818	[158]
Saccharomyces cerevisiae	Phosphorylation	13,763	[250]
	Glycosylation	2600	[251]
	Ubiquitination	2299	[252]
	Acetylation	2878	[253]
	Methylation	167	[254]

[a] Only number of identified modified peptides were reported, not modified sites

Data availability: Mass spectrometrists need to ensure that their published MS-based PTM datasets along with the protein database used for database searching are made publicly available in repositories (e.g., PRIDE [255]) to provide computational scientists resources for algorithm development. This is the current standard for publication in most proteomics journals. Importantly, all spreadsheets submitted should have descriptive metadata so that bioinformaticians incorporating experimental data into algorithms can easily understand the dataset. The PTM site residue and position as well as the protein accession should be clearly annotated for each entry. Peptide identification and modification site localization scores should also be included so that computational researchers can recognize the most confidently identified PTM sites.

6 Validation of Computational Predictions

Computational models efficiently reveal putative protein modifications. Although these predictions are valuable resources that can guide mechanistic studies, proposed PTMs need to be validated to correctly focus on a specific set of modified sites and gain proper biological insights. This is also true for hypothesis-generating proteomics experiments. PTM validation has historically relied on Western blotting to confirm identification of modified sites [256, 257], but requires the availability of costly modification-specific antibodies that lack cross-reactivity to other nonspecific modified sites [258, 259]. Therefore, validation approaches utilizing mass spectrometry have become increasingly popular due to its sensitivity and ability to accurately detect site-specific modifications [260–262].

Targeted MS: Targeted LC-MS methods provide superior reproducibility, sensitivity, and dynamic range compared to untargeted, discovery based methods [261], making it the preferred approach for PTM validation from prediction algorithms. Selected reaction monitoring (SRM) utilizing triple quadrupole mass spectrometers is a popular method for targeted proteomics [263, 264]. In SRM, the first quadrupole selects for a predetermined peptide ion, fragmentation of this ion occurs in the second quadrupole, and the third quadrupole filters for one or more specific product ions yielded from this fragmentation (Fig. 4a). SRM assay development is typically accomplished using synthetic peptides to determine the optimal transition (precursor and fragment ion pair) and the ideal collision energy for the most sensitive method. It is necessary to choose fragment ions that can localize the modification site when analyzing peptides containing multiple sites that can be modified. Since only preselected transitions are analyzed, the intensities of

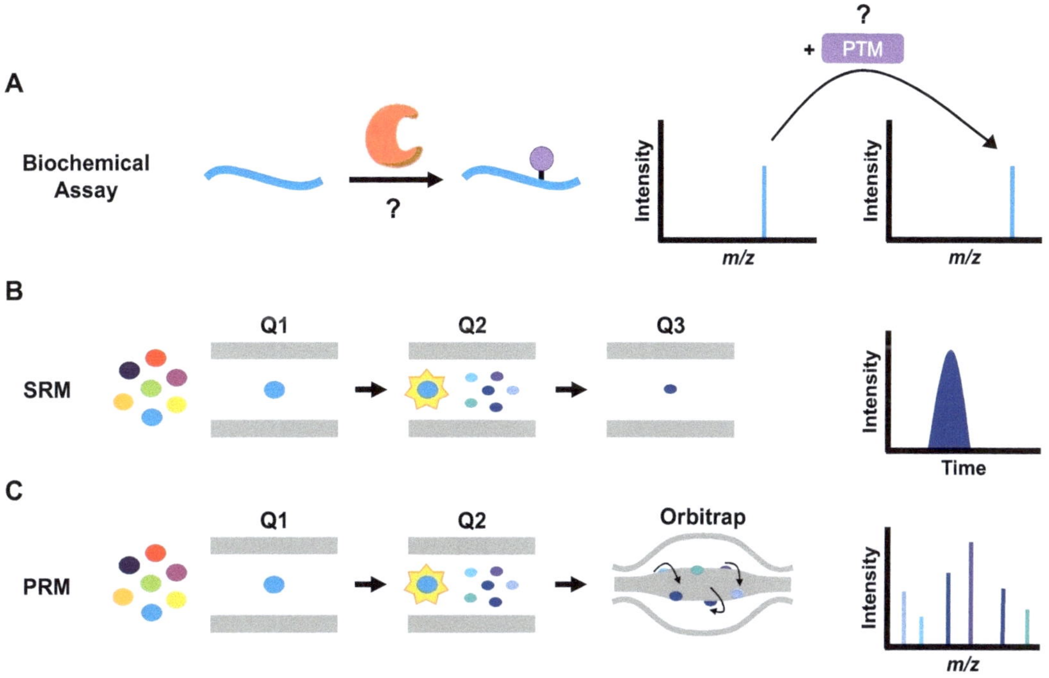

Fig. 4 Validation methods for PTM site predictions. (**a**) Selected reaction monitoring (SRM) is a targeted MS method on triple quadrupole mass spectrometers where the first quadrupole selects for a predetermined peptide ion, fragmentation of this ion occurs in the second quadrupole, and the third quadrupole filters for one or more specific product ions yielded from this fragmentation. (**b**) Parallel reaction monitoring (PRM) is another targeted MS method on high resolution mass spectrometers where the precursor ion is selected for fragmentation and a full-scan fragment spectrum is collected in an Orbitrap or time-of-flight mass analyzer. (**c**) If a modification is enzymatically driven, biochemical assays involving a peptide library, purified enzyme, and necessary cofactors can be employed to validate PTMs. The treated and untreated peptides are assessed by MS to determine if a mass shift to the intact peptide occurred, corresponding to the addition of a PTM

background noise ions are significantly reduced and the limit of detection for peptides are decreased by up to 100-fold compared to conventional full scan MS analysis [265]. SRM can be expanded to multiple reaction monitoring (MRM) where the MS method is configured to cycle through several transitions, allowing for the analysis of multiple peptides. This is possible since the required detection time for each transition is on the millisecond time scale. Furthermore, the method can be constructed to monitor certain transitions only at their peptide retention times, permitting the study of more peptides in a single LC-MS/MS run [266].

While SRM is considered the most sensitive MS-based method, the low resolution of quadrupole mass analyzers affects precursor and fragment filtering, leading to the selection of interferents and incorrect peptide assignment [267]. An alternative to SRM is parallel reaction monitoring (PRM), which is possible with the advent of high resolution and high accuracy mass analyzers. Here, the precursor ion is selected for fragmentation and a full-scan fragment spectrum is collected (Fig. 4b). This mirrors precursor selection/fragmentation and MS/MS spectrum collection in untargeted DDA methods, where peptide identification using the full fragment spectrum increases the confidence in accurate peptide identification and PTM site assignment [268, 269]. PRM on high resolution instruments can achieve tighter *m/z* selection windows, lower limits of detection, and obtain data over a wider dynamic range than SRM on triple quadrupole instruments [270], but both targeted MS approaches continue to be important tools for site-specific PTM validation.

In vitro biochemical assays: If the predicted PTM is enzymatically added, biochemical assays can be performed to validate the modification (Fig. 4c). These assays involve incubation of a purified or heterologously expressed enzyme, extracted or synthetic peptide library containing the proposed modification sites, and necessary cofactors [74, 271]. If a peptide library is prepared from a cell lysate then endogenous modifications and the complexity of the resulting sample need to be considered to ensure proper enzymatic accessibility and detection of proposed modification sites. Therefore, enzymatic removal of endogenous modifications and PTM enrichment following the in vitro biochemical assay are typically performed. The enzyme-treated and untreated peptides are analyzed by MS to determine if the modifications are present. This is determined through the detection of a corresponding mass shift of the intact peptide and fragment ions that can unambiguously localize the modification to a specific site. These assays are more useful for prediction models that are specific to one or a class of enzymes, since knowledge of the enzyme that catalyzes the specified modification is required.

7 Conclusions

Computational PTM prediction models leverage mass spectrometry datasets to provide high-throughput, cost-effective means for identifying protein modifications. These models rely on robust datasets for accurate predictions, so it is essential that methods are tailored to maximize PTM coverage.

Extracted proteomes are proteolyzed and PTM enrichment and orthogonal fractionation techniques are incorporated to target PTM-containing peptides for LC-MS/MS analysis. Advances in MS technologies have improved instrument sensitivity and speed, allowing for increased sampling of modified peptides. Extensive identification of these peptides depends on the collection of intense, data-rich MS/MS spectra so fragmentation methods have been developed to improve peptide fragmentation efficiency and produce modification-containing fragment ions.

Bioinformatics algorithms using protein database searching facilitate peptide identification and site localization in the highly complex datasets acquired. These algorithms utilize probabilities to select the highest scoring peptide or modification site match, therefore methods to limit false positives in the final PTM dataset are included. PTM prediction algorithms leverage these data for training and testing to develop robust computational models, but bioinformaticians should verify that datasets contain confidently localized sites before incorporating those datasets into their algorithms to ensure maximum model accuracy. Furthermore, obtaining reliable negative datasets is difficult as biologically present PTMs can evade detection or be wrongly assigned to another site due to low depth of PTM coverage or insufficient fragmentation.

One process that has been used to alleviate these negative dataset issues made use of available structural data to identify negative sites that are buried within the protein structure [272]. These sites are sterically unavailable and therefore most likely unmodified. Continuous use of complementary information can assist the creation of robust negative datasets.

With PTM prediction models, the computational community should strive to: (1) provide transparency in model operation, (2) produce/maintain publicly available tools, (3) ensure easy function of models, and (4) explain the significance of site prediction probability cutoffs.

Inputs for these computational models typically describe the presence or absence of modifications on different protein residues. However, it is the dynamic modulation of PTMs that controls protein function and regulates important biological processes across different cellular states. Quantitative LC-MS/MS-based proteomics with isobaric labeling techniques continues to improve PTM coverage and reproducibility [273, 274]. With an abundance of these publicly available quantitative datasets, new PTM prediction models should be developed to exploit this quantitative information. This quantitative data can produce specific modifications sites that are increased or decreased in abundance for a given condition based on statistical significance. These curated lists can be leveraged as positive sites to provide a framework for computational modeling that predicts likely PTMs across different drug treatments or disease states.

Acknowledgments

This work was supported by a National Science Foundation CAREER award (MCB-1552522) awarded to L.M.H. We thank Prof. Dukka B. KC, Amanda Smythers, and Patric Sadecki for critical reading of the manuscript.

References

1. Qi HH, Ongusaha PP, Myllyharju J et al (2008) Prolyl 4-hydroxylation regulates argonaute 2 stability. Nature 455(7211): 421–424. https://doi.org/10.1038/nature07186
2. Sahar S, Zocchi L, Kinoshita C et al (2010) Regulation of BMAL1 protein stability and circadian function by GSK3β-mediated phosphorylation. PLoS One 5(1):e8561. https://doi.org/10.1371/journal.pone.0008561
3. Deribe YL, Pawson T, Dikic I (2010) Post-translational modifications in signal integration. Nat Struct Mol Biol 17(6):666–672. https://doi.org/10.1038/nsmb.1842
4. Liu J, Qian C, Cao X (2016) Post-translational modification control of innate immunity. Immunity 45(1):15–30. https://doi.org/10.1016/j.immuni.2016.06.020
5. Ahearn IM, Haigis K, Bar-Sagi D et al (2012) Regulating the regulator: post-translational modification of RAS. Nat Rev Mol Cell Biol 13(1):39–51. https://doi.org/10.1038/nrm3255
6. Xie Y, Kang R, Sun X et al (2015) Posttranslational modification of autophagy-related proteins in macroautophagy. Autophagy 11(1):28–45. https://doi.org/10.4161/15548627.2014.984267
7. Smith LM, Kelleher NL (2018) Proteoforms as the next proteomics currency: identifying precise molecular forms of proteins can improve our understanding of function. Science 359(6380):1106–1107. https://doi.org/10.1126/science.aat1884
8. Khoury GA, Baliban RC, Floudas CA (2011) Proteome-wide post-translational modification statistics: frequency analysis and curation of the swiss-prot database. Sci Rep 1(1):1–5. https://doi.org/10.1038/srep00090
9. Audagnotto M, Dal Peraro M (2017) Protein post-translational modifications: in silico prediction tools and molecular modeling. Comput Struct Biotechnol J 15:307–319. https://doi.org/10.1016/j.csbj.2017.03.004
10. Choudhary C, Mann M (2010) Decoding signalling networks by mass spectrometry-based proteomics. Nat Rev Mol Cell Biol 11(6):427–439. https://doi.org/10.1038/nrm2900
11. Olsen JV, Mann M (2013) Status of large-scale analysis of posttranslational modifications by mass spectrometry. Mol Cell Proteomics 12(12):3444–3452. https://doi.org/10.1074/mcp.O113.034181
12. Zhang Y, Zhang C, Jiang H et al (2015) Fishing the PTM proteome with chemical approaches using functional solid phases. Chem Soc Rev 44(22):8260–8287. https://doi.org/10.1039/c4cs00529e
13. Zhang Y, Fonslow BR, Shan B et al (2013) Protein analysis by shotgun/bottom-up proteomics. Chem Rev 113(4):2343–2394. https://doi.org/10.1021/cr3003533
14. Gillet LC, Leitner A, Aebersold R (2016) Mass spectrometry applied to bottom-up proteomics: entering the high-throughput era for hypothesis testing. Annu Rev Anal Chem 9(1):449–472. https://doi.org/10.1146/annurev-anchem-071015-041535
15. Stahl DC, Swiderek KM, Davis MT et al (1996) Data-controlled automation of liquid chromatography/tandem mass spectrometry analysis of peptide mixtures. J Am Soc Mass Spectrom 7(6):532–540. https://doi.org/10.1016/1044-0305(96)00057-8
16. Singh C, Zampronio CG, Creese AJ et al (2012) Higher energy collision dissociation (HCD) product ion-triggered electron transfer dissociation (ETD) mass spectrometry for the analysis of N-linked glycoproteins. J Proteome Res 11(9):4517–4525. https://doi.org/10.1021/pr300257c
17. Frese CK, Altelaar AFM, Hennrich ML et al (2011) Improved peptide identification by targeted fragmentation using CID, HCD and ETD on an LTQ-orbitrap velos. J Proteome Res 10(5):2377–2388. https://doi.org/10.1021/pr1011729
18. Perkins DN, Pappin DJC, Creasy DM et al (1999) Probability-based protein identification by searching sequence databases using mass spectrometry data. Electrophoresis 20(18):3551–3567. 10.1002/(SICI)1522-2683(19991201)20:18<3551::AID-ELPS3551>3.0.CO;2-2
19. Tabb DL, Eng JK, Yates JR (2001) Protein identification by SEQUEST. Springer, Berlin, pp 125–142
20. Cox J, Neuhauser N, Michalski A et al (2011) Andromeda: a peptide search engine integrated into the MaxQuant environment. J Proteome Res 10(4):1794–1805. https://doi.org/10.1021/pr101065j
21. Nesvizhskii AI (2010) A survey of computational methods and error rate estimation procedures for peptide and protein identification in shotgun proteomics. J Proteomics 73(11):

2092–2123. https://doi.org/10.1016/j.jprot.2010.08.009

22. Chalkley RJ, Clauser KR (2012) Modification site localization scoring: strategies and performance. Mol Cell Proteomics 11(5):3–14. https://doi.org/10.1074/mcp.R111.015305
23. Needham EJ, Parker BL, Burykin T et al (2019) Illuminating the dark phosphoproteome. Sci Signal 12(565):8645. https://doi.org/10.1126/scisignal.aau8645
24. Xiao H, Chen W, Smeekens JM et al (2018) An enrichment method based on synergistic and reversible covalent interactions for large-scale analysis of glycoproteins. Nat Commun 9(1):1–12. https://doi.org/10.1038/s41467-018-04081-3
25. Dang L, Jia L, Zhi Y et al (2019) Mapping human N-linked glycoproteins and glycosylation sites using mass spectrometry. Trends Anal Chem 114:143–150. https://doi.org/10.1016/j.trac.2019.02.009
26. Al-Barakati HJ, EW MC, Hicks LM et al (2018) SVM-SulfoSite: a support vector machine based predictor for sulfenylation sites. Sci Rep 8(1):1–9. https://doi.org/10.1038/s41598-018-29,126-x
27. Gnad F, Ren S, Cox J et al (2007) PHOSIDA (phosphorylation site database): management, structural and evolutionary investigation, and prediction of phosphosites. Genome Biol 8(11):R250. https://doi.org/10.1186/gb-2007-8-11-r250
28. Wang D, Liang Y, Xu D (2019) Capsule network for protein post-translational modification site prediction. Bioinformatics 35(14):2386–2394. https://doi.org/10.1093/bioinformatics/bty977
29. Hasan MM (2017) Prediction of protein post-translational modification sites: an overview. Ann Proteomics Bioinforma 2(1):49–57. https://doi.org/10.29328/journal.apb.1001005
30. Zhang N, Li BQ, Gao S et al (2012) Computational prediction and analysis of protein γ-carboxylation sites based on a random forest method. Mol Biosyst 8(11):2946–2955. https://doi.org/10.1039/c2mb25185j
31. Wang J-R, Huang W-L, Tsai M-J et al (2017) ESA-UbiSite: accurate prediction of human ubiquitination sites by identifying a set of effective negatives. Bioinformatics 33(5):btw701. https://doi.org/10.1093/bioinformatics/btw701
32. Chen QY, Tang J, Du PF (2017) Predicting protein lysine phosphoglycerylation sites by hybridizing many sequence based features. Mol Biosyst 13(5):874–882. https://doi.org/10.1039/c6mb00875e
33. Wysocki VH, Resing KA, Zhang Q et al (2005) Mass spectrometry of peptides and proteins. Methods 35(3):211–222. https://doi.org/10.1016/j.ymeth.2004.08.013
34. Issaq HJ, Conrads TP, Janini GM et al (2002) Methods for fractionation, separation and profiling of proteins and peptides. Electrophoresis 23(17):3048–3061. https://doi.org/10.1002/1522-2683(200209)23:17 < 3048::AID-ELPS3048 > 3.0.CO;2-L
35. Canbay V, auf dem Keller U (2021) New strategies to identify protease substrates. Curr Opin Chem Biol 60:89–96. https://doi.org/10.1016/j.cbpa.2020.09.009
36. Iannetta AA, Rogers HT, Al-Mohanna T et al (2021) Profiling thimet oligopeptidase-mediated proteolysis in *Arabidopsis thaliana*. Plant J 106(2):336–350. https://doi.org/10.1111/tpj.15165
37. Olsen JV, Ong SE, Mann M (2004) Trypsin cleaves exclusively C-terminal to arginine and lysine residues. Mol Cell Proteomics 3(6):608–614. https://doi.org/10.1074/mcp.T400003-MCP200
38. Lodish H, Berk A, Zipursky SL et al (2000) Hierarchical structure of proteins. In: Molecular cell biology. W. H. Freeman, New York
39. Keil-Dlouhá V, Zylber N, Imhoff JM et al (1971) Proteolytic activity of pseudotrypsin. FEBS Lett 16(4):291–295. https://doi.org/10.1016/0014-5793(71)80373-3
40. Rice RH, Means GE, Brown WD (1977) Stabilization of bovine trypsin by reductive methylation. BBA Protein Struct 492(2):316–321. https://doi.org/10.1016/0005-2795(77)90082-4
41. Ma J, Liang Z, Qiao X et al (2008) Organic-inorganic hybrid silica monolith based immobilized trypsin reactor with high enzymatic activity. Anal Chem 80(8):2949–2956. https://doi.org/10.1021/ac702343a
42. Sun L, Zhu G, Yan X et al (2014) Uncovering immobilized trypsin digestion features from large-scale proteome data generated by high-resolution mass spectrometry. J Chromatogr A 1337:40–47. https://doi.org/10.1016/j.chroma.2014.02.014
43. Tran BQ, Hernandez C, Waridel P et al (2011) Addressing trypsin bias in large scale (pPhospho)proteome analysis by size exclusion chromatography and secondary digestion of large post-trypsin peptides. J Proteome Res 10(2):800–811. https://doi.org/10.1021/pr100951t

44. Imre T, Schlosser G, Pocsfalvi G et al (2005) Glycosylation site analysis of human alpha-1-acid glycoprotein (AGP) by capillary liquid chromatography—electrospray mass spectrometry. J Mass Spectrom 40(11): 1472–1483. https://doi.org/10.1002/jms.938
45. Guo X, Trudgian DC, Lemoff A et al (2014) Confetti: a multiprotease map of the HeLa proteome for comprehensive proteomics. Mol Cell Proteomics 13(6):1573–1584. https://doi.org/10.1074/mcp.M113.035170
46. Swaney DL, Wenger CD, Coon JJ (2010) Value of using multiple proteases for large-scale mass spectrometry-based proteomics. J Proteome Res 9(3):1323–1329. https://doi.org/10.1021/pr900863u
47. Wiśniewski JR, Mann M (2012) Consecutive proteolytic digestion in an enzyme reactor increases depth of proteomic and phosphoproteomic analysis. Anal Chem 84(6): 2631–2637. https://doi.org/10.1021/ac300006b
48. Bian Y, Ye M, Song C et al (2012) Improve the coverage for the analysis of phosphoproteome of HeLa cells by a tandem digestion approach. J Proteome Res 11(5): 2828–2837. https://doi.org/10.1021/pr300242w
49. Zhao Y, Jensen ON (2009) Modification-specific proteomics: strategies for characterization of post-translational modifications using enrichment techniques. Proteomics 9(20):4632–4641. https://doi.org/10.1002/pmic.200900398
50. Huang J, Wang F, Ye M et al (2014) Enrichment and separation techniques for large-scale proteomics analysis of the protein post-translational modifications. J Chromatogr A 1372:1–17. https://doi.org/10.1016/j.chroma.2014.10.107
51. Xu H, Wang Y, Lin S et al (2018) PTMD: a database of human disease-associated post-translational modifications. Genomics Proteomics Bioinformatics 16(4):244–251. https://doi.org/10.1016/j.gpb.2018.06.004
52. Ward PS, Thompson CB (2012) Signaling in control of cell growth and metabolism. Cold Spring Harb Perspect Biol 4(7):1–15. https://doi.org/10.1101/cshperspect.a006783
53. Dhanasekaran DN, Premkumar Reddy E (2017) JNK-signaling: a multiplexing hub in programmed cell death. Genes Cancer 8(9–10):682–694. https://doi.org/10.18632/genesandcancer.155
54. Humphrey SJ, James DE, Mann M (2015) Protein phosphorylation: a major switch mechanism for metabolic regulation. Trends Endocrinol Metab 26(12):676–687. https://doi.org/10.1016/j.tem.2015.09.013
55. Dennis MD, Jefferson LS, Kimball SR (2012) Role of p70S6K1-mediated phosphorylation of eIF4B and PDCD4 proteins in the regulation of protein synthesis. J Biol Chem 287(51):42890–42899. https://doi.org/10.1074/jbc.M112.404822
56. Beilharz K, Nováková L, Fadda D et al (2012) Control of cell division in *Streptococcus pneumoniae* by the conserved Ser/Thr protein kinase StkP. Proc Natl Acad Sci U S A 109(15):E905–E913. https://doi.org/10.1073/pnas.1119172109
57. Lin S, Wang C, Zhou J et al (2021) EPSD: a well-annotated data resource of protein phosphorylation sites in eukaryotes. Brief Bioinform 22(1):298–307. https://doi.org/10.1093/bib/bbz169
58. Rush J, Moritz A, Lee KA et al (2005) Immunoaffinity profiling of tyrosine phosphorylation in cancer cells. Nat Biotechnol 23(1): 94–101. https://doi.org/10.1038/nbt1046
59. Kaneko T, Huang H, Cao X et al (2012) Superbinder SH2 domains act as antagonists of cell signaling. Sci Signal 5(243):ra68. https://doi.org/10.1126/scisignal.2003021
60. Fuhs SR, Hunter T (2017) pHisphorylation: the emergence of histidine phosphorylation as a reversible regulatory modification. Curr Opin Cell Biol 45:8–16. https://doi.org/10.1016/j.ceb.2016.12.010
61. Cieśla J, Fraczyk T, Rode W (2011) Phosphorylation of basic amino acid residues in proteins: important but easily missed. Acta Biochim Pol 58(2):137–148. https://doi.org/10.18388/abp.2011_2258
62. Falke JJ, Bass RB, Butler SL et al (1997) The two-component signaling pathway of bacterial chemotaxis: a molecular view of signal transduction by receptors, kinases, and adaptation enzymes. Annu Rev Cell Dev Biol 13: 457–512. https://doi.org/10.1146/annurev.cellbio.13.1.457
63. Hardman G, Perkins S, Brownridge PJ et al (2019) Strong anion exchange-mediated phosphoproteomics reveals extensive human non-canonical phosphorylation. EMBO J 38(21):e100847. https://doi.org/10.15252/embj.2018100847
64. Humphrey SJ, Azimifar SB, Mann M (2015) High-throughput phosphoproteomics reveals in vivo insulin signaling dynamics. Nat

Biotechnol 33(9):990–995. https://doi.org/10.1038/nbt.3327

65. Liu JJ, Sharma K, Zangrandi L et al (2018) In vivo brain GPCR signaling elucidated by phosphoproteomics. Science 360(6395): eaao4927. https://doi.org/10.1126/science.aao4927

66. Tan H, Yang K, Li Y et al (2017) Integrative proteomics and phosphoproteomics profiling reveals dynamic signaling networks and bioenergetics pathways underlying T cell activation. Immunity 46(3):488–503. https://doi.org/10.1016/j.immuni.2017.02.010

67. Aasebø E, Mjaavatten O, Vaudel M et al (2016) Freezing effects on the acute myeloid leukemia cell proteome and phosphoproteome revealed using optimal quantitative workflows. J Proteomics 145:214–225. https://doi.org/10.1016/j.jprot.2016.03.049

68. Andersson L, Porath J (1986) Isolation of phosphoproteins by immobilized metal (Fe3+) affinity chromatography. Anal Biochem 154(1):250–254. https://doi.org/10.1016/0003-2697(86)90523-3

69. Jensen SS, Larsen MR (2007) Evaluation of the impact of some experimental procedures on different phosphopeptide enrichment techniques. Rapid Commun Mass Spectrom 21(22):3635–3645. https://doi.org/10.1002/rcm.3254

70. Ficarro SB, McCleland ML, Stukenberg PT et al (2002) Phosphoproteome analysis by mass spectrometry and its application to Saccharomyces cerevisiae. Nat Biotechnol 20(3): 301–305. https://doi.org/10.1038/nbt0302-301

71. Lai AC-Y, Tsai C-F, Hsu C-C et al (2012) Complementary Fe^{3+} and Ti^{4+} immobilized metal ion affinity chromatography for purification of acidic and basic phosphopeptides. Rapid Commun Mass Spectrom 26(18): 2186–2194. https://doi.org/10.1002/rcm.6327

72. Tsai CF, Hsu CC, Hung JN et al (2014) Sequential phosphoproteomic enrichment through complementary metal-directed immobilized metal ion affinity chromatography. Anal Chem 86(1):685–693. https://doi.org/10.1021/ac4031175

73. Iliuk AB, Martin VA, Alicie BM et al (2010) In-depth analyses of kinase-dependent tyrosine phosphoproteomes based on metal ion-functionalized soluble nanopolymers. Mol Cell Proteomics 9(10):2162–2172. https://doi.org/10.1074/mcp.M110.000091

74. Xue L, Wang WH, Iliuk A et al (2012) Sensitive kinase assay linked with phosphoproteomics for identifying direct kinase substrates. Proc Natl Acad Sci U S A 109(15): 5615–5620. https://doi.org/10.1073/pnas.1119418109

75. Leitner A (2010) Phosphopeptide enrichment using metal oxide affinity chromatography. Trends Anal Chem 29(2):177–185. https://doi.org/10.1016/j.trac.2009.08.007

76. Pinkse MWH, Uitto PM, Hilhorst MJ et al (2004) Selective isolation at the femtomole level of phosphopeptides from proteolytic digests using 2D-NanoLC-ESI-MS/MS and titanium oxide precolumns. Anal Chem 76(14):3935–3943. https://doi.org/10.1021/ac0498617

77. Werth EG, McConnell EW, Couso Lianez I et al (2019) Investigating the effect of target of rapamycin kinase inhibition on the *Chlamydomonas reinhardtii* phosphoproteome: from known homologs to new targets. New Phytol 221(1):247–260. https://doi.org/10.1111/nph.15339

78. Ma WF, Zhang C, Zhang YT et al (2014) Magnetic MSP@ZrO2 microspheres with yolk-shell structure: designed synthesis and application in highly selective enrichment of phosphopeptides. Langmuir 30(22): 6602–6611. https://doi.org/10.1021/la501381v

79. Huang SY, Chen YC (2013) Magnetic nanoparticle-based platform for characterization of histidine-rich proteins and peptides. Anal Chem 85(6):3347–3354. https://doi.org/10.1021/ac4000128

80. Li Y, Leng T, Lin H et al (2007) Preparation of Fe3O4@ZrO2 core—shell microspheres as affinity probes for selective enrichment and direct determination of phosphopeptides using matrix-assisted laser desorption ionization mass spectrometry. J Proteome Res 6(11):4498–4510. https://doi.org/10.1021/pr070167s

81. Mann M, Ong SE, Grønborg M et al (2002) Analysis of protein phosphorylation using mass spectrometry: deciphering the phosphoproteome. Trends Biotechnol 20(6): 261–268. https://doi.org/10.1016/S0167-7799(02)01944-3

82. Bodenmiller B, Mueller LN, Mueller M et al (2007) Reproducible isolation of distinct, overlapping segments of the phosphoproteome. Nat Methods 4(3):231–237. https://doi.org/10.1038/nmeth1005

83. Thingholm TE, Jensen ON, Robinson PJ et al (2008) SIMAC (sequential elution from IMAC), a phosphoproteomics strategy for the rapid separation of monophosphorylated from multiply phosphorylated peptides. Mol Cell Proteomics 7(4):661–671. https://doi.org/10.1074/mcp.M700362-MCP200
84. Peng J, Zhang H, Li X et al (2016) Dual-metal centered zirconium-organic framework: a metal-affinity probe for highly specific interaction with phosphopeptides. ACS Appl Mater Interfaces 8(51):35012–35020. https://doi.org/10.1021/acsami.6b12630
85. Ruprecht B, Koch H, Medard G et al (2015) Comprehensive and reproducible phosphopeptide enrichment using iron immobilized metal ion affinity chromatography (Fe-IMAC) columns. Mol Cell Proteomics 14(1):205–215. https://doi.org/10.1074/mcp.M114.043109
86. Beltran L, Cutillas PR (2012) Advances in phosphopeptide enrichment techniques for phosphoproteomics. Amino Acids 43(3):1009–1024. https://doi.org/10.1007/s00726-012-1288-9
87. Sugiyama N, Masuda T, Shinoda K et al (2007) Phosphopeptide enrichment by aliphatic hydroxy acid-modified metal oxide chromatography for nano-LC-MS/MS in proteomics applications. Mol Cell Proteomics 6(6):1103–1109. https://doi.org/10.1074/mcp.T600060-MCP200
88. Dubrovska A, Souchelnytskyi S (2005) Efficient enrichment of intact phosphorylated proteins by modified immobilized metal-affinity chromatography. Proteomics 5(18):4678–4683. https://doi.org/10.1002/pmic.200500002
89. Hwang L, Ayaz-Guner S, Gregorich ZR et al (2015) Specific enrichment of phosphoproteins using functionalized multivalent nanoparticles. J Am Chem Soc 137(7):2432–2435. https://doi.org/10.1021/ja511833y
90. Liu W, Zheng J, Li S et al (2015) Aluminium glycinate functionalized silica nanoparticles for highly specific separation of phosphoproteins. J Mater Chem B 3(31):6528–6535. https://doi.org/10.1039/c5tb01055a
91. Schieber M, Chandel NS (2014) ROS function in redox signaling and oxidative stress. Curr Biol 24(10):R453–R462. https://doi.org/10.1016/j.cub.2014.03.034
92. Poole LB, Nelson KJ (2008) Discovering mechanisms of signaling-mediated cysteine oxidation. Curr Opin Chem Biol 12(1):18–24. https://doi.org/10.1016/j.cbpa.2008.01.021
93. Couturier J, Chibani K, Jacquot JP et al (2013) Cysteine-based redox regulation and signaling in plants. Front Plant Sci 4:105. https://doi.org/10.3389/fpls.2013.00105
94. Paulech J, Solis N, Edwards AVG et al (2013) Large-scale capture of peptides containing reversibly oxidized cysteines by thiol-disulfide exchange applied to the myocardial redox proteome. Anal Chem 85(7):3774–3780. https://doi.org/10.1021/ac400166e
95. Guo J, Gaffrey MJ, Su D et al (2014) Resin-assisted enrichment of thiols as a general strategy for proteomic profiling of cysteine-based reversible modifications. Nat Protoc 9(1):64–75. https://doi.org/10.1038/nprot.2013.161
96. Leonard SE, Carroll KS (2011) Chemical "omics" approaches for understanding protein cysteine oxidation in biology. Curr Opin Chem Biol 15(1):88–102. https://doi.org/10.1016/j.cbpa.2010.11.012
97. Murray CI, Van Eyk JE (2012) Chasing cysteine oxidative modifications: proteomic tools for characterizing cysteine redox status. Circ Cardiovasc Genet 5(5):591. https://doi.org/10.1161/CIRCGENETICS.111.961425
98. Jaffrey SR, Erdjument-Bromage H, Ferris CD et al (2001) Protein S-nitrosylation: a physiological signal for neuronal nitric oxide. Nat Cell Biol 3(2):193–197. https://doi.org/10.1038/35055104
99. McConnell EW, Werth EG, Hicks LM (2018) The phosphorylated redox proteome of Chlamydomonas reinhardtii: revealing novel means for regulation of protein structure and function. Redox Biol 17:35–46. https://doi.org/10.1016/j.redox.2018.04.003
100. Ford MM, Smythers AL, McConnell EW et al (2019) Inhibition of TOR in Chlamydomonas reinhardtii leads to rapid cysteine oxidation reflecting sustained physiological changes. Cells 8(10):1171. https://doi.org/10.3390/cells8101171
101. Smythers AL, McConnell EW, Lewis HC et al (2020) Photosynthetic metabolism and nitrogen reshuffling are regulated by reversible cysteine thiol oxidation following nitrogen deprivation in Chlamydomonas. Plants 9(6):784. https://doi.org/10.3390/plants9060784
102. Murray CI, Uhrigshardt H, O'Meally RN et al (2012) Identification and quantification of S-nitrosylation by cysteine reactive tandem mass tag switch assay. Mol Cell Proteomics 11(2):1–12. https://doi.org/10.1074/mcp.M111.013441

103. Poole LB, Zeng BB, Knaggs SA et al (2005) Synthesis of chemical probes to map sulfenic acid modifications on proteins. Bioconjug Chem 16(6):1624–1628. https://doi.org/10.1021/bc050257s
104. Akter S, Fu L, Jung Y et al (2018) Chemical proteomics reveals new targets of cysteine sulfinic acid reductase. Nat Chem Biol 14(11): 995–1004. https://doi.org/10.1038/s41589-018-0116-2
105. Seneviratne U, Nott A, Bhat VB et al (2016) S-nitrosation of proteins relevant to Alzheimer's disease during early stages of neurodegeneration. Proc Natl Acad Sci U S A 113(15):4152–4157. https://doi.org/10.1073/pnas.1521318113
106. Martin BR, Cravatt BF (2009) Large-scale profiling of protein palmitoylation in mammalian cells. Nat Methods 6(2):135–138. https://doi.org/10.1038/nmeth.1293
107. Held JM (2020) Redox systems biology: harnessing the sentinels of the cysteine redoxome. Antioxid Redox Signal 32(10): 659–676. https://doi.org/10.1089/ars.2019.7725
108. Shental-Bechor D, Levy Y (2008) Effect of glycosylation on protein folding: a close look at thermodynamic stabilization. Proc Natl Acad Sci U S A 105(24):8256–8261. https://doi.org/10.1073/pnas.0801340105
109. Rudd PM, Elliott T, Cresswell P et al (2001) Glycosylation and the immune system. Science 291(5512):2370–2376. https://doi.org/10.1126/science.291.5512.2370
110. Moremen KW, Tiemeyer M, Nairn AV (2012) Vertebrate protein glycosylation: diversity, synthesis and function. Nat Rev Mol Cell Biol 13(7):448–462. https://doi.org/10.1038/nrm3383
111. Zhu Y, Willems LI, Salas D et al (2020) Tandem bioorthogonal labeling uncovers endogenous cotranslationally O-GlcNAc modified nascent proteins. J Am Chem Soc 142(37): 15729–15739. https://doi.org/10.1021/jacs.0c04121
112. Zhang H, Li X-J, Martin DB et al (2003) Identification and quantification of N-linked glycoproteins using hydrazide chemistry, stable isotope labeling and mass spectrometry. Nat Biotechnol 21(6):660–666. https://doi.org/10.1038/nbt827
113. Kameyama A, Thet Tin WW, Toyoda M et al (2019) A practical method of liberating O-linked glycans from glycoproteins using hydroxylamine and an organic superbase. Biochem Biophys Res Commun 513(1): 186–192. https://doi.org/10.1016/j.bbrc.2019.03.144
114. Cao J, Shen C, Wang H et al (2009) Identification of N-glycosylation sites on secreted proteins of human hepatocellular carcinoma cells with a complementary proteomics approach. J Proteome Res 8(2):662–672. https://doi.org/10.1021/pr800826u
115. Xia C, Jiao F, Gao F et al (2018) Two-dimensional MoS2-based zwitterionic hydrophilic interaction liquid chromatography material for the specific enrichment of glycopeptides. Anal Chem 90(11): 6651–6659. https://doi.org/10.1021/acs.analchem.8b00461
116. Nilsson J, Rüetschi U, Halim A et al (2009) Enrichment of glycopeptides for glycan structure and attachment site identification. Nat Methods 6(11):809–811. https://doi.org/10.1038/nmeth.1392
117. Calvano CD, Zambonin CG, Jensen ON (2008) Assessment of lectin and HILIC based enrichment protocols for characterization of serum glycoproteins by mass spectrometry. J Proteomics 71(3):304–317. https://doi.org/10.1016/j.jprot.2008.06.013
118. Zhang B, Yu RZ, Yu YH et al (2018) Lectin inspired polymers based on the dipeptide Ser-Asp for glycopeptide enrichment. Analyst 143(21):5090–5093. https://doi.org/10.1039/c8an01258j
119. Waniwan JT, Chen YJ, Capangpangan R et al (2018) Glycoproteomic alterations in drug-resistant nonsmall cell lung cancer cells revealed by lectin magnetic nanoprobe-based mass spectrometry. J Proteome Res 17(11): 3761–3773. https://doi.org/10.1021/acs.jproteome.8b00433
120. Zeng Z, Hincapie M, Pitteri SJ et al (2011) A proteomics platform combining depletion, multi-lectin affinity chromatography (M-LAC), and isoelectric focusing to study the breast cancer proteome. Anal Chem 83(12):4845–4854. https://doi.org/10.1021/ac2002802
121. Plavina T, Wakshull E, Hancock WS et al (2007) Combination of abundant protein depletion and multi-lectin affinity chromatography (M-LAC) for plasma protein biomarker discovery. J Proteome Res 6(2): 662–671. https://doi.org/10.1021/pr060413k
122. Imberty A, Mitchell EP, Wimmerová M (2005) Structural basis of high-affinity glycan

recognition by bacterial and fungal lectins. Curr Opin Struct Biol 15(5):525–534. https://doi.org/10.1016/j.sbi.2005.08.003

123. Sparbier K, Koch S, Kessler I et al (2005) Selective isolation of glycoproteins and glycopeptides for MALDI-TOF MS detection supported by magnetic particles. J Biomol Tech 16(4):407–411

124. Qu Y, Liu J, Yang K et al (2012) Boronic acid functionalized core-shell polymer nanoparticles prepared by distillation precipitation polymerization for glycopeptide enrichment. Chem A Eur J 18(29):9056–9062. https://doi.org/10.1002/chem.201103514

125. Wohlgemuth J, Karas M, Jiang W et al (2010) Enhanced glyco-profiling by specific glycopeptide enrichment and complementary monolithic nano-LC (ZIC-HILIC/RP18e)/ESI-MS analysis. J Sep Sci 33(6–7):880–890. https://doi.org/10.1002/jssc.200900771

126. Gaunitz S, Nagy G, Pohl NLB et al (2017) Recent advances in the analysis of complex glycoproteins. Anal Chem 89(1):389–413. https://doi.org/10.1021/acs.analchem.6b04343

127. Ahn YH, Kim JY, Yoo JS (2015) Quantitative mass spectrometric analysis of glycoproteins combined with enrichment methods. Mass Spectrom Rev 34(2):148–165. https://doi.org/10.1002/mas.21428

128. Ma J, Hart GW (2014) O-GlcNAc profiling: from proteins to proteomes. Clin Proteomics 11(1):8. https://doi.org/10.1186/1559-0275-11-8

129. Steentoft C, Vakhrushev SY, Joshi HJ et al (2013) Precision mapping of the human O-GalNAc glycoproteome through SimpleCell technology. EMBO J 32(10):1478–1488. https://doi.org/10.1038/emboj.2013.79

130. Akmal MA, Rasool N, Khan YD (2017) Prediction of N-linked glycosylation sites using position relative features and statistical moments. PLoS One 12(8):e0181966. https://doi.org/10.1371/journal.pone.0181966

131. Hassan H, Badr A, Abdelhalim MB (2015) Prediction of O-glycosylation sites using random forest and GA-tuned PSO technique. Bioinform Biol Insights 9:103–109. https://doi.org/10.4137/BBI.S26864

132. Chauhan JS, Rao A, Raghava GPS (2013) In silico platform for prediction of N-, O- and C-glycosites in eukaryotic protein sequences. PLoS One 8(6):e67008. https://doi.org/10.1371/journal.pone.0067008

133. Hamby SE, Hirst JD (2008) Prediction of glycosylation sites using random forests. BMC Bioinformatics 9(1):500. https://doi.org/10.1186/1471-2105-9-500

134. Johnson ES (2002) Ubiquitin branches out. Nat Cell Biol 4(12):295. https://doi.org/10.1038/ncb1202-e295

135. Sun L, Chen ZJ (2004) The novel functions of ubiquitination in signaling. Curr Opin Cell Biol 16(2):119–126. https://doi.org/10.1016/j.ceb.2004.02.005

136. Akimov V, Rigbolt KTG, Nielsen MM et al (2011) Characterization of ubiquitination dependent dynamics in growth factor receptor signaling by quantitative proteomics. Mol Biosyst 7(12):3223–3233. https://doi.org/10.1039/c1mb05185g

137. Hjerpe R, Aillet F, Lopitz-Otsoa F et al (2009) Efficient protection and isolation of ubiquitylated proteins using tandem ubiquitin-binding entities. EMBO Rep 10(11):1250–1258. https://doi.org/10.1038/embor.2009.192

138. Scott D, Oldham NJ, Strachan J et al (2015) Ubiquitin-binding domains: mechanisms of ubiquitin recognition and use as tools to investigate ubiquitin-modified proteomes. Proteomics 15(5–6):844–861. https://doi.org/10.1002/pmic.201400341

139. Hicke L, Schubert HL, Hill CP (2005) Ubiquitin-binding domains. Nat Rev Mol Cell Biol 6(8):610–621. https://doi.org/10.1038/nrm1701

140. Danielsen JMR, Sylvestersen KB, Bekker-Jensen S et al (2011) Mass spectrometric analysis of lysine ubiquitylation reveals promiscuity at site level. Mol Cell Proteomics 10(3):1–12. https://doi.org/10.1074/mcp.M110.003590

141. Peng J, Schwartz D, Elias JE et al (2003) A proteomics approach to understanding protein ubiquitination. Nat Biotechnol 21(8):921–926. https://doi.org/10.1038/nbt849

142. Pirone L, Xolalpa W, Sigursson JO et al (2017) A comprehensive platform for the analysis of ubiquitin-like protein modifications using in vivo biotinylation. Sci Rep 7:40756. https://doi.org/10.1038/srep40756

143. Mattern M, Sutherland J, Kadimisetty K et al (2019) Using ubiquitin binders to decipher the ubiquitin code. Trends Biochem Sci 44(7):599–615. https://doi.org/10.1016/j.tibs.2019.01.011

144. Kim W, Bennett EJ, Huttlin EL et al (2011) Systematic and quantitative assessment of the

ubiquitin-modified proteome. Mol Cell 44(2):325–340. https://doi.org/10.1016/j.molcel.2011.08.025

145. Wagner SA, Beli P, Weinert BT et al (2012) Proteomic analyses reveal divergent ubiquitylation site patterns in murine tissues. Mol Cell Proteomics 11(12):1578–1585. https://doi.org/10.1074/mcp.M112.017905

146. Wagner SA, Beli P, Weinert BT et al (2011) A proteome-wide, quantitative survey of in vivo ubiquitylation sites reveals widespread regulatory roles. Mol Cell Proteomics 10(10): M111.013284. https://doi.org/10.1074/mcp.m111.013284

147. Udeshi ND, Svinkina T, Mertins P et al (2013) Refined preparation and use of anti-diglycine remnant (k-ε-gg) antibody enables routine quantification of 10,000s of ubiquitination sites in single proteomics experiments. Mol Cell Proteomics 12(3):825–831. https://doi.org/10.1074/mcp.O112.027094

148. Drazic A, Myklebust LM, Ree R et al (2016) The world of protein acetylation. Biochim Biophys Acta Proteins Proteomics 1864(10): 1372–1401. https://doi.org/10.1016/j.bbapap.2016.06.007

149. Inuzuka H, Gao D, Finley LWS et al (2012) Acetylation-dependent regulation of Skp2 function. Cell 150(1):179–193. https://doi.org/10.1016/j.cell.2012.05.038

150. Li T, Diner BA, Chen J et al (2012) Acetylation modulates cellular distribution and DNA sensing ability of interferon-inducible protein IFI16. Proc Natl Acad Sci U S A 109(26): 10558–10563. https://doi.org/10.1073/pnas.1203447109

151. Cao W, Bao C, Padalko E et al (2008) Acetylation of mitogen-activated protein kinase phosphatase-1 inhibits Toll-like receptor signaling. J Exp Med 205(6):1491–1503. https://doi.org/10.1084/jem.20071728

152. Narita T, Weinert BT, Choudhary C (2019) Functions and mechanisms of non-histone protein acetylation. Nat Rev Mol Cell Biol 20(3):156–174. https://doi.org/10.1038/s41580-018-0081-3

153. Dillon SC, Zhang X, Trievel RC et al (2005) The SET-domain protein superfamily: protein lysine methyltransferases. Genome Biol 6(8): 1–10. https://doi.org/10.1186/gb-2005-6-8-227

154. Dhayalan A, Kudithipudi S, Rathert P et al (2011) Specificity analysis-based identification of new methylation targets of the SET7/9 protein lysine methyltransferase. Chem Biol 18(1):111–120. https://doi.org/10.1016/j.chembiol.2010.11.014

155. Mazur PK, Reynoird N, Khatri P et al (2014) SMYD3 links lysine methylation of MAP3K2 to Ras-driven cancer. Nature 510(7504): 283–287. https://doi.org/10.1038/nature13320

156. Bedford MT, Clarke SG (2009) Protein arginine methylation in mammals: who, what, and why. Mol Cell 33(1):1–13. https://doi.org/10.1016/j.molcel.2008.12.013

157. Kwak YT, Guo J, Prajapati S et al (2003) Methylation of SPT5 regulates its interaction with RNA polymerase II and transcriptional elongation properties. Mol Cell 11(4): 1055–1066. https://doi.org/10.1016/S1097-2765(03)00101-1

158. Guo A, Gu H, Zhou J et al (2014) Immunoaffinity enrichment and mass spectrometry analysis of protein methylation. Mol Cell Proteomics 13(1):372–387. https://doi.org/10.1074/mcp.O113.027870

159. Cao XJ, Arnaudo AM, Garcia BA (2013) Large-scale global identification of protein lysine methylation in vivo. Epigenetics 8(5): 477–485. https://doi.org/10.4161/epi.24547

160. Nielsen ML, Savitski MM, Zubarev RA (2006) Extent of modifications in human proteome samples and their effect on dynamic range of analysis in shotgun proteomics. Mol Cell Proteomics 5(12):2384–2391. https://doi.org/10.1074/mcp.M600248-MCP200

161. Liu T, Qian WJ, Gritsenko MA et al (2005) Human plasma N-glycoproteome analysis by immunoaffinity subtraction, hydrazide chemistry, and mass spectrometry. J Proteome Res 4(6):2070–2080. https://doi.org/10.1021/pr0502065

162. Gilar M, Olivova P, Daly AE et al (2005) Orthogonality of separation in two-dimensional liquid chromatography. Anal Chem 77(19):6426–6434. https://doi.org/10.1021/ac050923i

163. Song C, Ye M, Han G et al (2010) Reversed-phase-reversed-phase liquid chromatography approach with high orthogonality for multidimensional separation of phosphopeptides. Anal Chem 82(1):53–56. https://doi.org/10.1021/ac9023044

164. Yang F, Shen Y, Camp DG et al (2012) High-pH reversed-phase chromatography with fraction concatenation for 2D proteomic analysis. Expert Rev Proteomics 9(2):129–134. https://doi.org/10.1586/epr.12.15

165. Wang Y, Yang F, Gritsenko MA et al (2011) Reversed-phase chromatography with multiple fraction concatenation strategy for proteome profiling of human MCF10A cells. Proteomics 11(10):2019–2026. https://doi.org/10.1002/pmic.201000722
166. Washburn MP, Wolters D, Yates JR (2001) Large-scale analysis of the yeast proteome by multidimensional protein identification technology. Nat Biotechnol 19(3):242–247. https://doi.org/10.1038/85686
167. Pankow S, Bamberger C, Yates JR (2019) A posttranslational modification code for CFTR maturation is altered in cystic fibrosis. Sci Signal 12(562):eaan7984. https://doi.org/10.1126/scisignal.aan7984
168. Delmotte N, Lasaosa M, Tholey A et al (2007) Two-dimensional reversed-phase x ion-pair reversed-phase HPLC: an alternative approach to high-resolution peptide separation for shotgun proteome analysis. J Proteome Res 6(11):4363–4373. https://doi.org/10.1021/pr070424t
169. Di Palma S, Boersema PJ, Heck AJR et al (2011) Zwitterionic hydrophilic interaction liquid chromatography (ZIC-HILIC and ZIC-cHILIC) provide high resolution separation and increase sensitivity in proteome analysis. Anal Chem 83(9):3440–3447. https://doi.org/10.1021/ac103312e
170. Boersema PJ, Divecha N, Heck AJR et al (2007) Evaluation and optimization of ZIC-HILIC-RP as an alternative MudPIT strategy. J Proteome Res 6(3):937–946. https://doi.org/10.1021/pr060589m
171. Dehghani A, Gödderz M, Winter D (2018) Tip-based fractionation of batch-enriched phosphopeptides facilitates easy and robust phosphoproteome analysis. J Proteome Res 17(1):46–54. https://doi.org/10.1021/acs.jproteome.7b00256
172. Humphrey SJ, Karayel O, James DE et al (2018) High-throughput and high-sensitivity phosphoproteomics with the EasyPhos platform. Nat Protoc 13(9):1897–1916. https://doi.org/10.1038/s41596-018-0014-9
173. Scheltema RA, Hauschild JP, Lange O et al (2014) The Q exactive HF, a benchtop mass spectrometer with a pre-filter, high-performance quadrupole and an ultra-high-field orbitrap analyzer. Mol Cell Proteomics 13(12):3698–3708. https://doi.org/10.1074/mcp.M114.043489
174. Andrews GL, Simons BL, Young JB et al (2011) Performance characteristics of a new hybrid quadrupole time-of-flight tandem mass spectrometer (TripleTOF 5600). Anal Chem 83(13):5442–5446. https://doi.org/10.1021/ac200812d
175. Scigelova M, Makarov A (2006) Orbitrap mass analyzer—overview and applications in proteomics. Proteomics 1(1–2):16–21. https://doi.org/10.1002/pmic.200600528
176. Michalski A, Damoc E, Hauschild JP et al (2011) Mass spectrometry-based proteomics using Q exactive, a high-performance benchtop quadrupole orbitrap mass spectrometer. Mol Cell Proteomics 10(9):M111.011015. https://doi.org/10.1074/mcp.M111.011015
177. Michalski A, Damoc E, Lange O et al (2012) Ultra high resolution linear ion trap orbitrap mass spectrometer (orbitrap elite) facilitates top down LC MS/MS and versatile peptide fragmentation modes. Mol Cell Proteomics 11(3):1–11. https://doi.org/10.1074/mcp.O111.013698
178. Batth TS, Francavilla C, Olsen JV (2014) Off-line high-pH reversed-phase fractionation for in-depth phosphoproteomics. J Proteome Res 13(12):6176–6186. https://doi.org/10.1021/pr500893m
179. Kelstrup CD, Bekker-Jensen DB, Arrey TN et al (2018) Performance evaluation of the Q exactive HF-X for shotgun proteomics. J Proteome Res 17(1):727–738. https://doi.org/10.1021/acs.jproteome.7b00602
180. Boersema PJ, Mohammed S, Heck AJR (2009) Phosphopeptide fragmentation and analysis by mass spectrometry. J Mass Spectrom 44(6):861–878. https://doi.org/10.1002/jms.1599
181. Engholm-Keller K, Larsen MR (2013) Technologies and challenges in large-scale phosphoproteomics. Proteomics 13(6):910–931. https://doi.org/10.1002/pmic.201200484
182. Villén J, Beausoleil SA, Gygi SP (2008) Evaluation of the utility of neutral-loss-dependent MS3 strategies in large-scale phosphorylation analysis. Proteomics 8(21):4444–4452. https://doi.org/10.1002/pmic.200800283
183. Riley NM, Malaker SA, Driessen MD et al (2020) Optimal dissociation methods differ for N- and O-glycopeptides. J Proteome Res 19(8):3286–3301. https://doi.org/10.1021/acs.jproteome.0c00218
184. Syka JEP, Coon JJ, Schroeder MJ et al (2004) Peptide and protein sequence analysis by electron transfer dissociation mass spectrometry. Proc Natl Acad Sci U S A 101(26):9528–9533. https://doi.org/10.1073/pnas.0402700101
185. Molina H, Horn DM, Tang N et al (2007) Global proteomic profiling of

phosphopeptides using electron transfer dissociation tandem mass spectrometry. Proc Natl Acad Sci U S A 104(7):2199–2204. https://doi.org/10.1073/pnas.0611217104
186. Good DM, Wirtala M, McAlister GC et al (2007) Performance characteristics of electron transfer dissociation mass spectrometry. Mol Cell Proteomics 6(11):1942–1951. https://doi.org/10.1074/mcp.M700073-MCP200
187. Rose CM, Rush MJP, Riley NM et al (2015) A calibration routine for efficient ETD in large-scale proteomics. J Am Soc Mass Spectrom 26(11):1848–1857. https://doi.org/10.1007/s13361-015-1183-1
188. Frese CK, Altelaar AFM, Van Den Toorn H et al (2012) Toward full peptide sequence coverage by dual fragmentation combining electron-transfer and higher-energy collision dissociation tandem mass spectrometry. Anal Chem 84(22):9668–9673. https://doi.org/10.1021/ac3025366
189. Kong AT, Leprevost FV, Avtonomov DM et al (2017) MSFragger: ultrafast and comprehensive peptide identification in mass spectrometry-based proteomics. Nat Methods 14(5):513–520. https://doi.org/10.1038/nmeth.4256
190. Dorfer V, Pichler P, Stranzl T et al (2014) MS Amanda, a universal identification algorithm optimized for high accuracy tandem mass spectra. J Proteome Res 13(8):3679–3684. https://doi.org/10.1021/pr500202e
191. Craig R, Beavis RC (2004) TANDEM: matching proteins with tandem mass spectra. Bioinformatics 20(9):1466–1467. https://doi.org/10.1093/bioinformatics/bth092
192. Kim S, Pevzner PA (2014) MS-GF+ makes progress towards a universal database search tool for proteomics. Nat Commun 5:5277. https://doi.org/10.1038/ncomms6277
193. Solntsev SK, Shortreed MR, Frey BL et al (2018) Enhanced global post-translational modification discovery with MetaMorpheus. J Proteome Res 17(5):1844–1851. https://doi.org/10.1021/acs.jproteome.7b00873
194. Eng JK, Jahan TA, Hoopmann MR (2013) Comet: an open-source MS/MS sequence database search tool. Proteomics 13(1):22–24. https://doi.org/10.1002/pmic.201200439
195. Beausoleil SA, Villén J, Gerber SA et al (2006) A probability-based approach for high-throughput protein phosphorylation analysis and site localization. Nat Biotechnol 24(10):1285–1292. https://doi.org/10.1038/nbt1240
196. Bailey CM, Sweet SMM, Cunningham DL et al (2009) SLoMo: automated site localization of modifications from ETD/ECD mass spectra. J Proteome Res 8(4):1965–1971. https://doi.org/10.1021/pr800917p
197. Fermin D, Avtonomov D, Choi H et al (2015) LuciPHOr2: site localization of generic post-translational modifications from tandem mass spectrometry data. Bioinformatics 31(7):1141–1143. https://doi.org/10.1093/bioinformatics/btu788
198. Savitski MM, Lemeer S, Boesche M et al (2011) Confident phosphorylation site localization using the mascot delta score. Mol Cell Proteomics 10(2):S1–S12. https://doi.org/10.1074/mcp.M110.003830
199. Taus T, Köcher T, Pichler P et al (2011) Universal and confident phosphorylation site localization using phosphoRS. J Proteome Res 10(12):5354–5362. https://doi.org/10.1021/pr200611n
200. Elias JE, Gygi SP (2007) Target-decoy search strategy for increased confidence in large-scale protein identifications by mass spectrometry. Nat Methods 4(3):207–214. https://doi.org/10.1038/nmeth1019
201. Elias JE, Gygi SP (2010) Target-decoy search strategy for mass spectrometry-based proteomics. Methods Mol Biol 60:455–471. https://doi.org/10.1007/978-1-60,761-444-9_5
202. Olsen JV, Vermeulen M, Santamaria A et al (2010) Quantitative phosphoproteomics reveals widespread full phosphorylation site occupancy during mitosis. Sci Signal 3(104):ra3. https://doi.org/10.1126/scisignal.2000475
203. Udeshi ND, Mani DR, Eisenhaure T et al (2012) Methods for quantification of in vivo changes in protein ubiquitination following proteasome and deubiquitinase inhibition. Mol Cell Proteomics 11(5):148–159. https://doi.org/10.1074/mcp.M111.016857
204. Dupree EJ, Jayathirtha M, Yorkey H et al (2020) A critical review of bottom-up proteomics: the good, the bad, and the future of this field. Proteomes 8(3):1–26. https://doi.org/10.3390/proteomes8030014
205. Chick JM, Kolippakkam D, Nusinow DP et al (2015) A mass-tolerant database search identifies a large proportion of unassigned spectra in shotgun proteomics as modified peptides. Nat Biotechnol 33(7):743–749. https://doi.org/10.1038/nbt.3267
206. Liu K, Li S, Wang L et al (2020) Full-spectrum prediction of peptides tandem mass spectra using deep neural network.

Anal Chem 92(6):4275–4283. https://doi.org/10.1021/acs.analchem.9b04867

207. Olsen JV, Blagoev B, Gnad F et al (2006) Global, in vivo, and site-specific phosphorylation dynamics in signaling networks. Cell 127(3):635–648. https://doi.org/10.1016/j.cell.2006.09.026

208. Locard-Paulet M, Bouyssié D, Froment C et al (2020) Comparing 22 popular phosphoproteomics pipelines for peptide identification and site localization. J Proteome Res 19(3):1338–1345. https://doi.org/10.1021/acs.jproteome.9b00679

209. Chen Z, Chen YZ, Wang XF et al (2011) Prediction of ubiquitination sites by using the composition of K-Spaced amino acid pairs. PLoS One 6(7):e22930. https://doi.org/10.1371/journal.pone.0022930

210. Swaney DL, Beltrao P, Starita L et al (2013) Global analysis of phosphorylation and ubiquitylation cross-talk in protein degradation. Nat Methods 10(7):676–682. https://doi.org/10.1038/nmeth.2519

211. Sidoli S, Garcia BA (2017) Middle-down proteomics: a still unexploited resource for chromatin biology. Expert Rev Proteomics 14(7):617–626. https://doi.org/10.1080/14789450.2017.1345632

212. Zhou M, Paša-Tolić L, Stenoien DL (2017) Profiling of histone post-translational modifications in mouse brain with high-resolution top-down mass spectrometry. J Proteome Res 16(2):599–608. https://doi.org/10.1021/acs.jproteome.6b00694

213. Leney AC, El Atmioui D, Wu W et al (2017) Elucidating crosstalk mechanisms between phosphorylation and O-GlcNAcylation. Proc Natl Acad Sci U S A 114(35):E7255–E7261. https://doi.org/10.1073/pnas.1620529114

214. Hasan MM, Khatun MS, Mollah MNH et al (2017) A systematic identification of species-specific protein succinylation sites using joint element features information. Int J Nanomedicine 126:303–6315. https://doi.org/10.2147/IJN.S140875

215. Liu Z, Dong W, Jiang W et al (2019) csDMA: an improved bioinformatics tool for identifying DNA 6 mA modifications via Chou's 5-step rule. Sci Rep 9(1):1–9. https://doi.org/10.1038/s41598-019-49,430-4

216. Roitinger E, Hofer M, Köcher T et al (2015) Quantitative phosphoproteomics of the ataxia telangiectasia-mutated (ATM) and ataxia telangiectasia-mutated and Rad3-related (ATR) dependent DNA damage response in Arabidopsis thaliana. Mol Cell Proteomics 14(3):556–571. https://doi.org/10.1074/mcp.M114.040352

217. McConnell EW, Berg P, Westlake TJ et al (2019) Proteome-wide analysis of cysteine reactivity during effector-triggered immunity. Plant Physiol 179(4):1248–1264. https://doi.org/10.1104/pp.18.01194

218. Huang J, Willems P, Wei B et al (2019) Mining for protein S-sulfenylation in Arabidopsis uncovers redox-sensitive sites. Proc Natl Acad Sci U S A 116(42):20256–20261. https://doi.org/10.1073/pnas.1906768116

219. Hu J, Huang X, Chen L et al (2015) Site-specific nitrosoproteomic identification of endogenously *S*-nitrosylated proteins in Arabidopsis. Plant Physiol 167(4):1731–1746. https://doi.org/10.1104/pp.15.00026

220. Zielinska DF, Gnad F, Schropp K et al (2012) Mapping N-glycosylation sites across seven evolutionarily distant species reveals a divergent substrate proteome despite a common core machinery. Mol Cell 46(4):542–548. https://doi.org/10.1016/j.molcel.2012.04.031

221. Walton A, Stes E, Cybulski N et al (2016) It's time for some "site"-seeing: novel tools to monitor the ubiquitin landscape in *Arabidopsis thaliana*. Plant Cell 28(1):6–16. https://doi.org/10.1105/tpc.15.00878

222. Hartl M, Füßl M, Boersema PJ et al (2017) Lysine acetylome profiling uncovers novel histone deacetylase substrate proteins in *Arabidopsis*. Mol Syst Biol 13(10):949. https://doi.org/10.15252/msb.20177819

223. Liang Q, Geng Q, Jiang L et al (2020) Protein methylome analysis in Arabidopsis reveals regulation in RNA-related processes. J Proteomics 213:103601. https://doi.org/10.1016/j.jprot.2019.103601

224. Huang J, Wu Z, Zhang X (2020) Short-term mild temperature-stress-induced alterations in the C. elegans phosphoproteome. Int J Mol Sci 21(17):6409. https://doi.org/10.3390/ijms21176409

225. Wang H, Gau B, Slade WO et al (2014) The global phosphoproteome of chlamydomonas reinhardtii reveals complex organellar phosphorylation in the flagella and thylakoid membrane. Mol Cell Proteomics 13(9):2337–2353. https://doi.org/10.1074/mcp.M114.038281

226. Morisse S, Zaffagnini M, Gao XH et al (2014) Insight into protein *S*-nitrosylation in Chlamydomonas reinhardtii. Antioxid Redox Signal 21(9):1271–1284. https://doi.org/10.1089/ars.2013.5632

227. Schulze S, Oltmanns A, Machnik N et al (2018) *N*-glycoproteomic characterization of mannosidase and xylosyltransferase mutant strains of *Chlamydomonas reinhardtii*. Plant Physiol 176(3):1952–1964. https://doi.org/10.1104/pp.17.01450
228. Yan J, Long Y, Zhou T et al (2020) Dynamic phosphoproteome profiling of zebrafish embryonic fibroblasts during cold acclimation. Proteomics 20(2):1900257. https://doi.org/10.1002/pmic.201900257
229. Kwon OK, Kim S, Lee S (2016) Global proteomic analysis of lysine acetylation in zebrafish (*Danio rerio*) embryos. Electrophoresis 37(23–24):3137–3145. https://doi.org/10.1002/elps.201600210
230. Bo Zhai JV, Beausoleil SA, Mintseris J et al (2008) Phosphoproteome analysis of drosophila melanogaster embryos. J Proteome Res 7(4):1675–1682. https://doi.org/10.1021/pr700696a
231. Menger KE, James AM, Cochemé HM et al (2015) Fasting, but not aging, dramatically alters the redox status of cysteine residues on proteins in Drosophila melanogaster. Cell Rep 11(12):1856–1865. https://doi.org/10.1016/j.celrep.2015.05.033
232. Sap KA, Bezstarosti K, Dekkers DHW et al (2017) Quantitative proteomics reveals extensive changes in the ubiquitinome after perturbation of the proteasome by targeted dsRNA-mediated subunit knockdown in drosophila. J Proteome Res 16(8):2848–2862. https://doi.org/10.1021/acs.jproteome.7b00156
233. Weinert BT, Wagner SA, Horn H et al (2011) Proteome-wide mapping of the Drosophila acetylome demonstrates a high degree of conservation of lysine acetylation. Sci Signal 4(183):ra48. https://doi.org/10.1126/scisignal.2001902
234. Semanjski M, Germain E, Bratl K et al (2018) The kinases HipA and HipA7 phosphorylate different substrate pools in Escherichia coli to promote multidrug tolerance. Sci Signal 11(547):5750. https://doi.org/10.1126/scisignal.aat5750
235. Shakir S, Vinh J, Chiappetta G (2017) Quantitative analysis of the cysteine redoxome by iodoacetyl tandem mass tags. Anal Bioanal Chem 409(15):3821–3830. https://doi.org/10.1007/s00216-017-0326-6
236. Weinert BT, Iesmantavicius V, Wagner SA et al (2013) Acetyl-phosphate is a critical determinant of lysine acetylation in E. coli. Mol Cell 51(2):265–272. https://doi.org/10.1016/j.molcel.2013.06.003
237. Zhang M, Xu J-Y, Hu H et al (2018) Systematic proteomic analysis of protein methylation in prokaryotes and eukaryotes revealed distinct substrate specificity. Proteomics 18(1):1700300. https://doi.org/10.1002/pmic.201700300
238. Sharma K, D'Souza RCJ, Tyanova S et al (2014) Ultradeep human phosphoproteome reveals a distinct regulatory nature of Tyr and Ser/Thr-based signaling. Cell Rep 8(5):1583–1594. https://doi.org/10.1016/j.celrep.2014.07.036
239. Huang H, Petersen MH, Ibañez-Vea M et al (2016) Simultaneous enrichment of cysteine containing peptides and phosphopeptides using a cysteine-specific phosphonate adaptable tag (CysPAT) in combination with titanium dioxide (TiO2) chromatography. Mol Cell Proteomics 15(10):3282–3296. https://doi.org/10.1074/mcp.M115.054551
240. Mnatsakanyan R, Markoutsa S, Walbrunn K et al (2019) Proteome-wide detection of S-nitrosylation targets and motifs using bioorthogonal cleavable-linker-based enrichment and switch technique. Nat Commun 10(1):1–12. https://doi.org/10.1038/s41467-019-10,182-4
241. Zhu J, Sun Z, Cheng K et al (2014) Comprehensive mapping of protein N-glycosylation in human liver by combining hydrophilic interaction chromatography and hydrazide chemistry. J Proteome Res 13(3):1713–1721. https://doi.org/10.1021/pr401200h
242. Akimov V, Barrio-Hernandez I, Hansen SVF et al (2018) Ubisite approach for comprehensive mapping of lysine and n-terminal ubiquitination sites. Nat Struct Mol Biol 25(7):631–640. https://doi.org/10.1038/s41594-018-0084-y
243. Schölz C, Weinert BT, Wagner SA et al (2015) Acetylation site specificities of lysine deacetylase inhibitors in human cells. Nat Biotechnol 33(4):415–425. https://doi.org/10.1038/nbt.3130
244. Larsen SC, Sylvestersen KB, Mund A et al (2016) Proteome-wide analysis of arginine monomethylation reveals widespread occurrence in human cells. Sci Signal 9(443):rs9. https://doi.org/10.1126/scisignal.aaf7329
245. Wang Z, Ma J, Miyoshi C et al (2018) Quantitative phosphoproteomic analysis of the molecular substrates of sleep need. Nature 558(7710):435–439. https://doi.org/10.1038/s41586-018-0218-8

246. Wang J, Choi H, Chung NC et al (2018) Integrated dissection of cysteine oxidative post-translational modification proteome during cardiac hypertrophy. J Proteome Res 17(12):4243–4257. https://doi.org/10.1021/acs.jproteome.8b00372
247. Zareba-Koziol M, Bartkowiak-Kaczmarek A, Figiel I et al (2019) Stress-induced changes in the S-palmitoylation and S-nitrosylation of synaptic proteins. Mol Cell Proteomics 18(10):1916–1938. https://doi.org/10.1074/mcp.RA119.001581
248. Fang P, Wang XJ, Xue Y et al (2016) In-depth mapping of the mouse brain N-glycoproteome reveals widespread N-glycosylation of diverse brain proteins. Oncotarget 7(25):38796–38809. https://doi.org/10.18632/oncotarget.9737
249. Rardin MJ, Newman JC, Held JM et al (2013) Label-free quantitative proteomics of the lysine acetylome in mitochondria identifies substrates of SIRT3 in metabolic pathways. Proc Natl Acad Sci U S A 110(16):6601–6606. https://doi.org/10.1073/pnas.1302961110
250. Li J, Paulo JA, Nusinow DP et al (2019) Investigation of proteomic and phosphoproteomic responses to signaling network perturbations reveals functional pathway organizations in yeast. Cell Rep 29(7):2092–2104.e4. https://doi.org/10.1016/j.celrep.2019.10.034
251. Neubert P, Halim A, Zauser M et al (2016) Mapping the O-mannose glycoproteome in saccharomyces cerevisiae. Mol Cell Proteomics 15(4):1323–1337. https://doi.org/10.1074/mcp.M115.057505
252. Iesmantavicius V, Weinert BT, Choudhary C (2014) Convergence of ubiquitylation and phosphorylation signaling in rapamycin-treated yeast cells. Mol Cell Proteomics 13(8):1979–1992. https://doi.org/10.1074/mcp.O113.035683
253. Henriksen P, Wagner SA, Weinert BT et al (2012) Proteome-wide analysis of lysine acetylation suggests its broad regulatory scope in saccharomyces cerevisiae. Mol Cell Proteomics 11(11):1510–1522. https://doi.org/10.1074/mcp.M112.017251
254. Zhang Y, Pan Y, Liu W et al (2016) In vivo protein allylation to capture protein methylation candidates. Chem Commun 52(40):6689–6692. https://doi.org/10.1039/c6cc02386j
255. Perez-Riverol Y, Csordas A, Bai J et al (2019) The PRIDE database and related tools and resources in 2019: improving support for quantification data. Nucleic Acids Res 47(D1):D442–D450. https://doi.org/10.1093/nar/gky1106
256. Zhang Z, Tan M, Xie Z et al (2011) Identification of lysine succinylation as a new post-translational modification. Nat Chem Biol 7(1):58–63. https://doi.org/10.1038/nchembio.495
257. Newton K, Matsumoto ML, Wertz IE et al (2008) Ubiquitin chain editing revealed by polyubiquitin linkage-specific antibodies. Cell 134(4):668–678. https://doi.org/10.1016/j.cell.2008.07.039
258. Ghosh R, Gilda JE, Gomes AV (2014) The necessity of and strategies for improving confidence in the accuracy of western blots. Expert Rev Proteomics 11(5):549–560. https://doi.org/10.1586/14789450.2014.939635
259. Pillai-Kastoori L, Schutz-Geschwender AR, Harford JA (2020) A systematic approach to quantitative Western blot analysis. Anal Biochem 593(15):113608. https://doi.org/10.1016/j.ab.2020.113608
260. Yu Y, Anjum R, Kubota K et al (2009) A site-specific, multiplexed kinase activity assay using stable-isotope dilution and high-resolution mass spectrometry. Proc Natl Acad Sci U S A 106(28):11606–11611. https://doi.org/10.1073/pnas.0905165106
261. Gillette MA, Carr SA (2013) Quantitative analysis of peptides and proteins in biomedicine by targeted mass spectrometry. Nat Methods 10(1):28–34. https://doi.org/10.1038/NMETH.2309
262. Morris M, Knudsen GM, Maeda S et al (2015) Tau post-translational modifications in wild-type and human amyloid precursor protein transgenic mice. Nat Neurosci 18(8):1183–1189. https://doi.org/10.1038/nn.4067
263. Shi T, Song E, Nie S et al (2016) Advances in targeted proteomics and applications to biomedical research. Proteomics 16(15–16):2160–2182. https://doi.org/10.1002/pmic.201500449
264. Ebhardt HA, Root A, Sander C et al (2015) Applications of targeted proteomics in systems biology and translational medicine. Proteomics 15(18):3193–3208. https://doi.org/10.1002/pmic.201500004
265. Zhao Y, Brasier AR (2013) Applications of selected reaction monitoring (SRM)-mass spectrometry (MS) for quantitative measurement of signaling pathways. Methods 61(3):313–322. https://doi.org/10.1016/j.ymeth.2013.02.001

266. Gallien S, Duriez E, Crone C et al (2012) Targeted proteomic quantification on quadrupole-orbitrap mass spectrometer. Mol Cell Proteomics 11(12):1709–1723. https://doi.org/10.1074/mcp.O112.019802

267. Ronsein GE, Pamir N, von Haller PD et al (2015) Parallel reaction monitoring (PRM) and selected reaction monitoring (SRM) exhibit comparable linearity, dynamic range and precision for targeted quantitative HDL proteomics. J Proteomics 113:388–399. https://doi.org/10.1016/j.jprot.2014.10.017

268. Rauniyar N (2015) Parallel reaction monitoring: a targeted experiment performed using high resolution and high mass accuracy mass spectrometry. Int J Mol Sci 16(12):28566–28581. https://doi.org/10.3390/ijms161226120

269. Schilling B, MacLean B, Held JM et al (2015) Multiplexed, scheduled, high-resolution parallel reaction monitoring on a full scan QqTOF instrument with integrated data-dependent and targeted mass spectrometric workflows. Anal Chem 87(20):10222–10229. https://doi.org/10.1021/acs.analchem.5b02983

270. Peterson AC, Russell JD, Bailey DJ et al (2012) Parallel reaction monitoring for high resolution and high mass accuracy quantitative, targeted proteomics. Mol Cell Proteomics 11(11):1475–1488. https://doi.org/10.1074/mcp.O112.020131

271. Xu G, Jaffrey SR (2013) Proteomic identification of protein ubiquitination events. Biotechnol Genet Eng Rev 29(1):73–109. https://doi.org/10.1080/02648725.2013.801232

272. Blom N, Sicheritz-Pontén T, Gupta R et al (2004) Prediction of post-translational glycosylation and phosphorylation of proteins from the amino acid sequence. Proteomics 4(6):1633–1649. https://doi.org/10.1002/pmic.200300771

273. Unwin RD, Griffiths JR, Whetton AD (2010) Simultaneous analysis of relative protein expression levels across multiple samples using iTRAQ isobaric tags with 2D nano LC-MS/MS. Nat Protoc 51:1574–1582. https://doi.org/10.1038/nprot.2010.123

274. Thompson A, Wölmer N, Koncarevic S et al (2019) TMTpro: design, synthesis, and initial evaluation of a proline-based isobaric 16-Plex Tandem Mass Tag Reagent Set. Anal Chem 91(24):15941–15950. https://doi.org/10.1021/acs.analchem.9b04474

Chapter 2

PLDMS: Phosphopeptide Library Dephosphorylation Followed by Mass Spectrometry Analysis to Determine the Specificity of Phosphatases for Dephosphorylation Site Sequences

Thomas Kokot, Bernhard Hoermann, Dominic Helm, Jeremy E. Chojnacki, Mikhail M. Savitski, and Maja Köhn

Abstract

A detailed understanding of the sequence preference surrounding phosphorylation sites is essential for deciphering the function of the human phosphoproteome. Whereas the mechanisms for substrate site recognition by kinases are relatively well understood, the selection mechanisms for the corresponding phosphatases pose several obstacles. However, multiple pieces of evidence point towards a role of the amino acid sequence in the direct vicinity of the phosphorylation site for recognition by phosphatase enzymes. Peptide library–based studies for enzymes attaching posttranslational modifications (PTMs) are relatively straight forward to carry out. However, studying enzymes removing PTMs pose a challenge in that libraries with a PTM attached are needed as a starting point. Here, we present our methodology using large synthetic phosphopeptide libraries to study the preferred sequence context of protein phosphatases. The approach, termed "phosphopeptide library dephosphorylation followed by mass spectrometry" (PLDMS), allows for the exact control of phosphorylation site incorporation and the synthetic route is capable of covering several thousand peptides in a single tube reaction. Furthermore, it enables the user to analyze MS data tailored to the needs of a specific library and thereby increase data quality. We therefore expect a wide applicability of this technique for a range of enzymes catalyzing the removal of PTMs.

Key words Posttranslational Modification, Protein Phosphorylation, Protein Phosphatases, Phosphoprotein Phosphatases, Solid-Phase Peptide Synthesis, Peptide Libraries, Dephosphorylation Assay, Enzyme Kinetics, Mass Spectrometry, R Analysis

1 Introduction

Posttranslational modifications (PTMs) of protein residues are found in all kingdoms of life and represent the cell's fastest option to adapt signaling pathways under changing conditions [1], may it

Thomas Kokot and Bernhard Hoermann contributed equally to this work.

Dukka B. KC (ed.), *Computational Methods for Predicting Post-Translational Modification Sites*,
Methods in Molecular Biology, vol. 2499, https://doi.org/10.1007/978-1-0716-2317-6_2,

be for example environmental input, growth, aging, or memory formation. Consequently, dysregulation of these amino acid side chain modifications poses an equally important risk for the formation and advance of diseases. Although, concomitant with technological advances in mass spectrometry, hundreds of thousands of protein residues modified by PTMs have been identified to date [2, 3], a current challenge is the discrimination between PTMs that act as tightly controlled switches in cellular signaling pathways and those that are not regulated and might exhibit less potential to act as regulating PTMs [3]. Since the precise attachment and detachment in time and space is a central necessity for PTM regulation, the enzymes catalyzing these steps are of the utmost importance.

Protein phosphorylation, that is the addition or removal of a phosphoryl group from amino acid side chains, is among the most frequently identified PTMs, and recent studies suggest the presence of more than 116,000 phosphorylation sites (p-sites) in the human body [3]. Among protein phosphorylation, phosphoserine (pSer) is by far the most often found p-site, with phosphothreonine and phosphotyrosine (pThr and pTyr) representing only 4–5% of all p-sites [3, 4], and pTyr being an evolutionary rather young "offspring" of pSer/pThr phosphorylation [5]. Importantly, several pieces of evidence also point towards a regulatory, yet more labile, noncanonical phosphorylation of amino acids such as histidine, aspartic acid, and glutamic acid [6]. Whether *N*-linked phosphorylation also plays an important role but is underrepresented in current research efforts due to the intrinsically labile nature of the modification and the resulting technical difficulties, is a question that remains to be answered.

Kinases, the enzymes attaching phosphoryl groups, have been in the focus of research for decades [7–9]. The number of these types of proteins has massively increased during evolution in parallel with the accumulating p-sites in complex organisms. During this process, the different classes of protein kinases have diverged in a way such that they can recognize a p-site sequence motif, that is the amino acids surrounding the p-site, ranging from highly basic to acidic and from hydrophilic to hydrophobic patterns in order to attach phosphoryl groups onto protein residues in a specific and controlled manner [10]. A significant contribution to these insights comes from approaches using relatively short, linear sets of peptides or highly complex mixtures of peptides, termed peptide libraries [11–13]. These approaches have allowed researchers to decipher the exact motif preference for multiple kinases and have paved the way towards a deepened understanding of kinase specificity on the protein level and ultimately represent a cornerstone in the development of therapeutic applications.

In contrast, the mechanism of substrate specificity of counteracting phosphatases is significantly less understood. This is despite the fact that their value as therapeutic targets has been highlighted

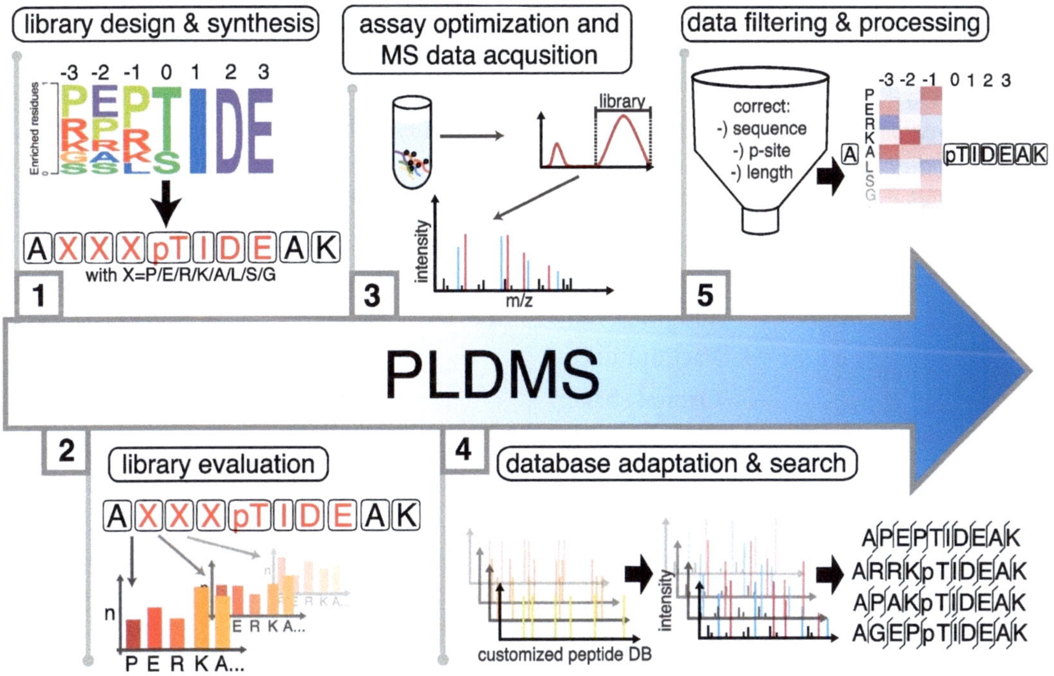

Fig. 1 Chronological overview of the complete PLDMS approach divided into its five main steps. After designing, synthesizing, and evaluating an individual library (1 and 2), the targeted dephosphorylation rate by the phosphatase has to be established (3). After successful library treatment and measurement follows the library-tailored database adaption and search for its sequences (4). In the final step, the exported search results are manually filtered in the R environment and biological analysis can be carried out (5)

in numerous cases [14–18]. The mechanism of action for phosphoprotein phosphatases (PPPs) catalyzing the selective dephosphorylation of pSer/pThr is particularly complex. For most members of this family a set of regulatory, substrate specifying subunits binds to a catalytic core protein in order to form holodimers or holotrimers [19, 20]. This results in a multilayer system for substrate specification, consisting of regulatory proteins and contributions from the catalytic subunit. We have recently furthered the applicability of synthetic phosphopeptide libraries by developing the PLDMS approach (Fig. 1) to study the influence of the catalytic subunits of PPPs protein phosphatase-1 and -2A (PP1 and PP2A) [21]. A special focus in this methodology was given to three aspects: (1) the synthesis of a tailored phosphopeptide library, derived from computational analysis of known phosphatase sites, (2) an assay design that is widely applicable to enzymes catalyzing PTM removal, and (3) a pipeline for analysis of MS raw data and library-specific data filtering that comes with specific advantages compared to proteome-based datasets. Here, we present the underlying methodology in detail, hoping to facilitate the use of PLDMS for other researchers.

2 Materials

2.1 Fluorenylmethyloxycarbonyl (Fmoc) Solid-Phase Peptide Synthesis (SPPS)

1. To guarantee reproducibility during optimization experiments and quality control, synthesis should be carried out using an automated peptide synthesis robot. We used a MultiPep RSi peptide synthesizer from Intavis.
2. Precision balance, antistatic spatula, and balance equipment.
3. 5 mL syringes with a polyethylene filter at the bottom, plunger, and cap.
4. Vacuum line.
5. Orbital shaker.
6. Rotary evaporator.
7. Precoupled H-Lys(Boc)-2-ClTrt resin can be used as a resin (*see* **Note 1**).
8. Amino acid (AA)–*N*-hydroxybenzotriazole hydrate (HOBt) mixture: 0.5 M Fmoc protected amino acid (mixture or single amino acid) with standard-side chain protecting groups such as Pbf, OtBu, Trt, tBu, and Boc, which can be cleaved by TFA, dissolved in 0.5 M HOBt should be used for library synthesis (*see* **Notes 2** and **3**).
9. *N*,*N*-Dimethylformamide (DMF).
10. 0.5 M *O*-benzotriazole-*N*,*N*,*N*′,*N*′-tetramethyl-uronium hexafluorophosphate (HBTU) in DMF (*see* **Note 3**).
11. 0.5 M *N*-methylmorpholine (NMM) in DMF (*see* **Note 3**).
12. Capping solution: 5% acetic anhydride (Ac_2O), 5% 2,6-lutidine in DMF (*see* **Note 3**).
13. Fmoc deprotection reagent: 20% piperidine in DMF.
14. Complete cleavage cocktail: 92.5% trifluoroacetic acid (TFA), 5% triisopropylsilane (TIPS), and 2.5% H_2O.
15. Toluene.

2.2 Dephosphorylation Assays

1. Synthesized library and recombinant phosphatase of interest.
2. Dimethyl sulfoxide (DMSO).
3. Transparent flat-bottom 96-well microtiter plates.
4. Standard microplate reader capable of measuring absorbance at 360 nm and 620 nm in 96-well plates.
5. Reaction buffer: 20 mM Tris–HCl (pH 7.5 at RT), 100 mM NaCl, 1 mM Dithiothreitol (DTT), 0.15 units purine nucleoside phosphorylase (PNP), 0.2 mM 2-amino-6-mercapto-7-methylpurine riboside (MESG).

6. Dephosphorylation buffer: 20 mM Tris–HCl (pH 7.5 at room temperature), 100 mM NaCl, 1 mM DTT.
7. Stop solution: 0.37% HCl.

2.3 Size-Exclusion Chromatography (SEC) Purification

1. ÄKTA or equivalent liquid chromatography system coupled to an automated fractionation system.
2. Superdex Peptide 10/300 GL column.
3. Running buffer: 0.1 M ammonium acetate (pH 5 at room temperature).
4. 96-well plates for elution fractionation.
5. Vacuum concentrator.

2.4 Mass Spectrometry Analysis

1. Aqueous solvent A: 0.1% formic acid (FA) in LC-MS grade water.
2. Organic solvent B: 0.1% FA in LC-MS grade acetonitrile (ACN).
3. Trapping/preconcentration solvent, for example, 0.05% TFA in LC-MS grade water.
4. Optional: Trapping/preconcentration column, for example, Acclaim PepMap300 C18, 5 μm, 300 Å wide pore.
5. Analytical column, for example, nanoEase HSS C18 T3, 75 μm × 25 cm, 1.8 μm, 100 Å.
6. UPLC system, for example, here Ultimate 3000 RSLC nano.
7. High-resolution mass spectrometer, for example, Q Exactive Plus.

2.5 Data Processing Software

1. IsobarQuant 1.1.0 [22].
2. Mascot search engine 2.4 (Matrix Science) (*see* **Note 4**).
3. Sequence data base in fasta-format (*see* Subheading 3.5 for details).

2.6 Data Analysis

1. Output txt-file of the data processing procedure.
2. List (txt-file) of all theoretical sequences of the library.
3. R software environment (here R version 3.6.3) (*see* **Note 5**).
4. The tidyverse collection of R packages (here version 1.3.0), and the stringi package (here version 1.5.3), both from the CRAN repository (*see* **Note 5**).
5. The Biostrings package from Bioconductor (here version 2.54.0) (*see* **Note 5**).

3 Methods

If not stated differently, all steps of the following procedures are to be performed at room temperature (*see* **Note 6**).

3.1 Library Design

In order to design phosphopeptide libraries tailored towards studying the amino acid preference of a selected phosphatase, special attention should be given to the selection of amino acids, library size, peptide length, as well as to the validation of these factors after library synthesis (Fig. 2). AAs surrounding the p-site should be empirically chosen based on the following considerations.

1. Modern, state-of-the-art mass spectrometers analyze thousands of peptides per hour. Yet, depending on the natural occurrence of amino acids surrounding p-sites, this can still lead to poor amino acid coverage for rare amino acids in the course of a several hour MS run and limit conclusions on p-sites containing these residues. When given the necessary consideration at the initial step of library design, this problem can be overcome during the creation of the synthetic libraries, by allowing counts high enough for statistical analysis to be obtained for all amino acids, independent of their natural occurrence. To this end, it is important to balance the number of different amino acids with the sequencing speed/capacity of the MS system (*see* **Note 7**).
2. One way to focus the peptide library is to consider the AAs naturally found around p-site of the phosphatase in question. Bioinformatic analysis of AA distribution around known substrate sites and insights from available literature in terms of motifs and sequence conservation can support this step, if such data is available and supported by statistical analysis.
3. Since amino acids can be grouped based on their biophysical properties, we suggest to also include at least one amino acid from each of the following groups: from hydrophobic, hydrophilic, positively charged, negatively charged, small, and bulky. If patterns for the AA representative for one of these groups are found, more targeted follow-up experiments can help to verify potential trends (*see* **Note 8**).
4. The length of the peptides is another important factor that will influence the complexity of the resulting library and thus the quality of data filtering after data processing. We suggest a minimal length of 9 AA with a *C*-terminal lysine and a minimum charge state of +2 for good compatibility with MS acquisition and database search. In general, peptide libraries can be separated into libraries of fixed length or variable length (Fig. 2). In the case of variable length, any trends consistently found over a wider range of peptide lengths are more likely to

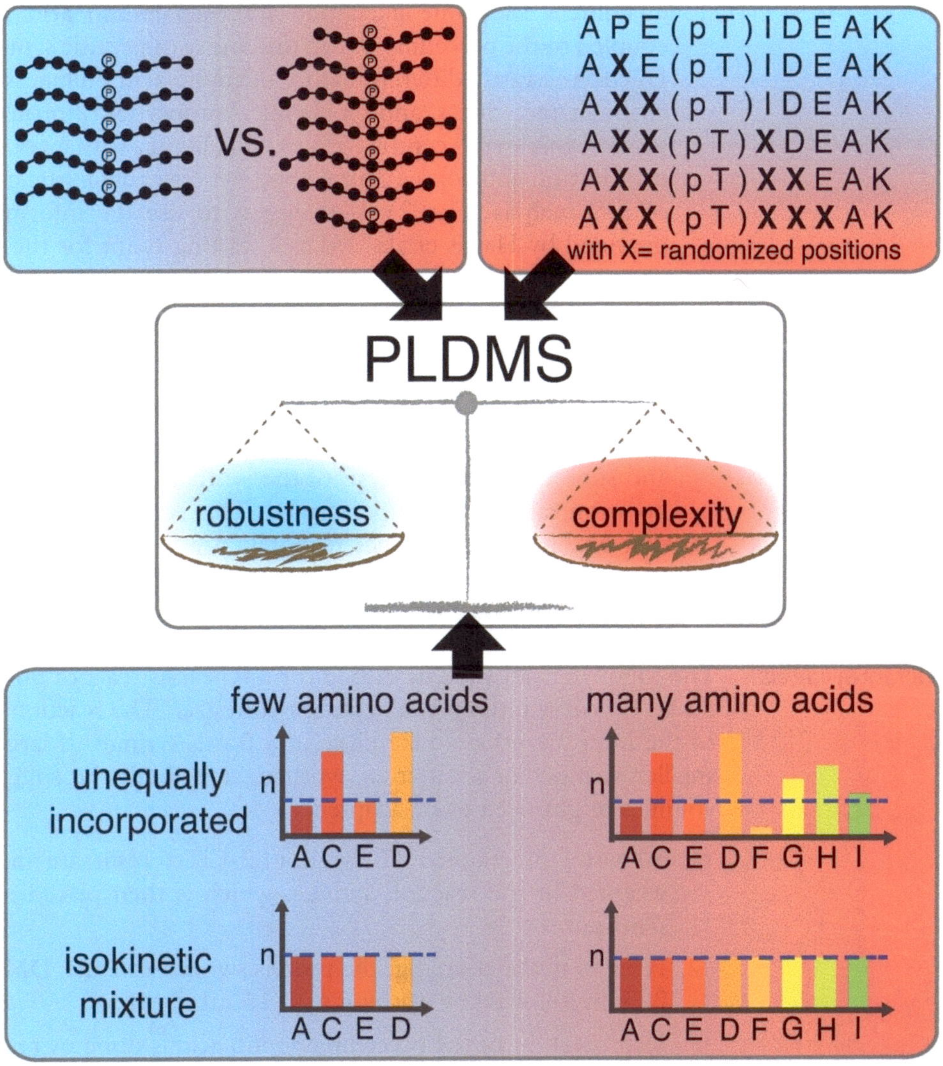

Fig. 2 The phosphopeptide library design is based on balancing robustness (blue) versus complexity (red) of three major aspects. Varying peptide lengths within a library increases the complexity due to rising efforts in the empirical filtering procedure (top left). With extending randomized positions, the complexity increases, simultaneously robustness gets reduced (top right). The incorporation rate as well as the number of the amino acids in the isokinetic mixture also determine the library complexity (bottom)

be reproducible and physiologically relevant. In the case of fixed length, a significantly better data filtering is possible and preferred substrate sites for phosphatases can be analyzed in greater detail due to better data filtering (*see* **Note 9**).

5. Depending on the number of amino acids chosen for randomized positions, this tailored set of amino acids needs to be optimized toward an isokinetic mixture, that is a mix of side chain-protected amino acids that is not equimolar, but rather

guarantees the same likelihood for each amino acid to be coupled to the preceding one in the course of peptide synthesis. Due to the vastly different nature of amino acid bulkiness, side chain characteristics, and biochemical properties of protecting groups, this step is central for a peptide library with a good AA distribution and consequently good AA coverage in the course of MS analysis (Fig. 2). We suggest to use the information provided by Harris et al. [23] as a starting point for the most commonly used protected amino acids in SPPS. Importantly, in our experience, leaving out AAs from the full set of all naturally occurring amino acids to reduce library complexity influences the incorporation efficiency of all remaining AAs in a highly intricate manner. Therefore, several rounds of optimization of SPPS with randomized permutations using an isokinetic mix and subsequent MS analysis of the synthesis to analyze incorporation rates must be performed. This is the only approach that enables the researcher to obtain a well-distributed phosphopeptide library that sufficiently covers all amino acids under investigation (*see* **Note 10**).

3.2 Library Synthesis

The following steps are described for a synthesis scale of 50 μmol for an Intavis MultiPep RSi peptide synthesizer. The reaction steps of the library synthesis take place in a 5 mL syringe. If larger or smaller syringes or a different synthesizer is used, the suggested volumes might need to be adapted.

1. 50 μmol precoupled H-Lys(Boc)-2-ClTrt resin are directly weighed in the reaction syringe, which is then placed in the synthesizer.
2. Prior to initial coupling, the resin is swollen in 1 mL DMF for 20 min followed by washing with 15 mL DMF.
3. Deprotection of the preceding amino acid is done by reacting the derivative-coupled resin twice with 800 μL of the Fmoc deprotection reagent for 3 min and then 8 min.
4. Next, the resin is washed with 20 mL DMF.
5. Before adding to the resin, 400 μL of 0.5 M HBTU, 100 μL of 0.5 M NMM, 5 μL NMP, and 420 μL of AA–HOBt mixture are mixed in a separate tube for preactivation. The coupling mixture is subsequently added into the syringe containing the resin. The reaction then takes place for 45 min (*see* **Note 11**).
6. The previous step is repeated once in order to guarantee maximal coupling efficiency.
7. Next, capping is performed by using 1 mL of the capping solution. It is added to the resin for 5 min.
8. The resin is washed with 20 mL DMF.

9. After Fmoc capping, the next AA can be coupled by repeating **steps 3–8**. After the last amino acid is coupled, the peptide can be fully deprotected and cleaved from the column as described in the following steps.
10. Deprotect the resin twice again with the Fmoc deprotection reagent for 3 and 8 min.
11. Wash with 30 mL DMF.
12. Wash with 28 mL dichloromethane (DCM).
13. Wash with 20 mL DMF.
14. Remove the syringe from the synthesizer, cap the tip, and insert a plunger into the barrel to seal the resin from the atmosphere. When ready, aspirate 1 mL of the complete cleavage cocktail into the syringe. Replace the cap and mix for 4 h on a shaking plate (*see* **Note 12**).
15. Next, dispense the resulting library-containing cleavage cocktail into a suitable flask. Save the resin syringe in case of a second round of cleavage is needed or desired. TFA is then removed by coevaporation with toluene. This is achieved by adding 15 mL of toluene to the library cleavage cocktail and rotating the flask under vacuum using a rotation evaporator at 40 °C.
16. To remove any residual solvent or moisture, the dried library should be further dried under high vacuum overnight.
17. Store the library sealed with Parafilm at −20 °C.

Importantly, it is not possible to validate the success of the synthesis of a phosphopeptide library by standard MALDI or HPLC-ESI machines, since the library is a complex mixture of thousands of different masses. Therefore, more elaborate LC-MS/MS analysis is needed to ascertain the degree of success of the synthesis (Subheadings 3.4–3.7). Before analyzing the library through an LC-MS/MS approach, a dephosphorylation assay as described in Subheading 3.3 should be performed as a robust first step in library analysis.

3.3 Library Evaluation and Optimization

To confirm the incorporation of phosphate-containing amino acids into the synthesized library peptide sequence we recommend using the chosen phosphatase in a test dephosphorylation reaction using MESG. If the amount of recombinant phosphatase available is limited, this step can also be carried out using an unspecific and commercially available phosphatase, such as alkaline phosphatase:

1. Dissolve a small amount (1–2 mg) of dried library in 25% DMSO in dH_2O for a final concentration of 10 mM. Ensure the measured library mass is as accurate as possible using a fine balance. The library's average peptide weight is used for the molarity calculation.

2. Prepare a library dilution of 500 μM for the following assays by mixing 1 μL of the library with 19 μL of dH_2O per processed sample.
3. Prepare 75 μL of the reaction buffer per needed well. Depending on the phosphatase used, additional, specific buffer components (cofactors) may be needed for full enzyme activity. Keep on ice.
4. Transfer the reaction buffer (75 μL) into a well of a transparent 96-well plate. Add 20 μL of the prediluted library (500 μM) to the 75 μL of the reaction buffer. Incubate for 5 min at 25 °C in a plate reader.
5. Start a dephosphorylation reaction by adding 5 μL phosphatase to the assay mixture for a total volume of 100 μL with a final working concentration of 25 nM phosphatase (*see* **Note 13**). The phosphatase should be added directly before starting the plate reader measurement.
6. Measure the release of inorganic phosphate by observing the conversion of MESG to 2-amino-6-mercapto-7-methylpurine riboside and the corresponding change in absorbance at 360 nm at 25 °C in the plate reader.
7. The measured absorbance corresponds to the relative library dephosphorylation. When no further change in the absorbance at 360 nm can be observed, the maximal amount of library has been dephosphorylated (corresponding to 100%). The difference between the initial measured value (corresponding to 0%; *see* **Note 14**) and the plateau of the maximum absorbance represents the range of library dephosphorylation (Fig. 3). The timepoint at which 30%–40% of the library has been dephosphorylated is a suitable timepoint for terminating the reaction for further MS analysis. At this timepoint, enough of the library has been dephosphorylated to identify differences in phosphorylation rates and potential trends.

3.4 Library Measurement

In general, LC-MS/MS parameters are subject to change and require adaptations with respect to the physicochemical properties and the complexity of the designed library. Thus, they may differ strongly from standard proteomic settings.

1. If successful release of phosphate is observed in the previous confirmation assay, the library can be analyzed by LC-MS/MS. To this end, again a small aliquot of dried library is dissolved in aqueous solvent A prior to the MS measurement (*see* **Note 15**).
2. Prior to analytical separation, a trapping cartridge/preconcentration step of about 5 min is advisable.
3. Analytical separation is then carried out using a gradient including (solvent A and solvent B) of a sufficient length of

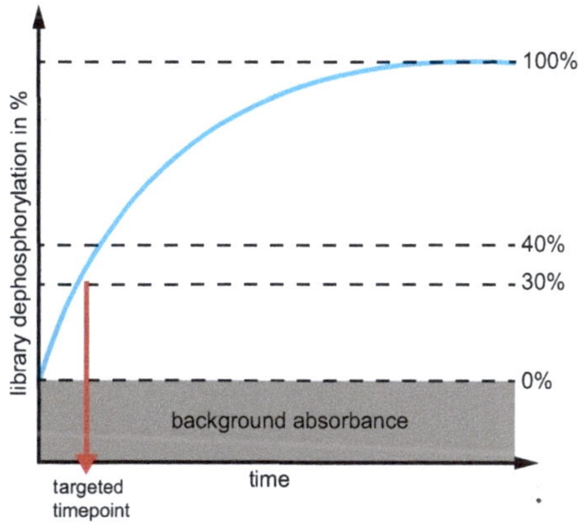

Fig. 3 Schematic library dephosphorylation (blue line) reaction monitored at 360 nm as described in Subheading 3.3. The first measured datapoint represents the completely phosphorylated library, whereas the saturation at the plateau depicts the completely dephosphorylated library. These two levels determine the range of library dephosphorylation. Using these min and max values, the timepoint of the desired library dephosphorylation rate (here 30–40%) can be calculated and consequently the moment for reaction termination can be determined

time in respect to the library complexity. For libraries as shown in Fig. 4, a total analysis time of 120 min was used. During this time the concentration of solvent B was ramped in multiple steps as follows: 4–20% for 88 min, 20–40% for 20 min and finally 40–80% for 0.5 min with a constant flow rate of 0.3 μL/ min (*see* **Note 16**).

4. Eluting peptides are detected by the coupled high-resolution mass spectrometer, which is operated in positive and data-dependent mode. Here a MS measurement is used to determine the m/z value of the intact peptide and followed by a series of MS/MS scans to record the fragment spectrum of the peptide for sequence determination. The mass range for the MS scan should be adjusted in accordance to the expected m/z-value distribution from the library. Charge state selection for MS/MS acquisition can also differ and even +1 charged species might need to be included. For the presented library a range from 275–1000 *m/z* was chosen.
5. Furthermore, a narrow quadrupole isolation window (e.g., 0.7 *m/z*) ensures cleaner fragment spectra by reducing coisolation to a minimum (*see* **Note 17**).
6. Since these synthetic libraries can contain isobaric peptides, dynamic exclusion should be turned off or reduced to a minimum (e.g., 1 s).

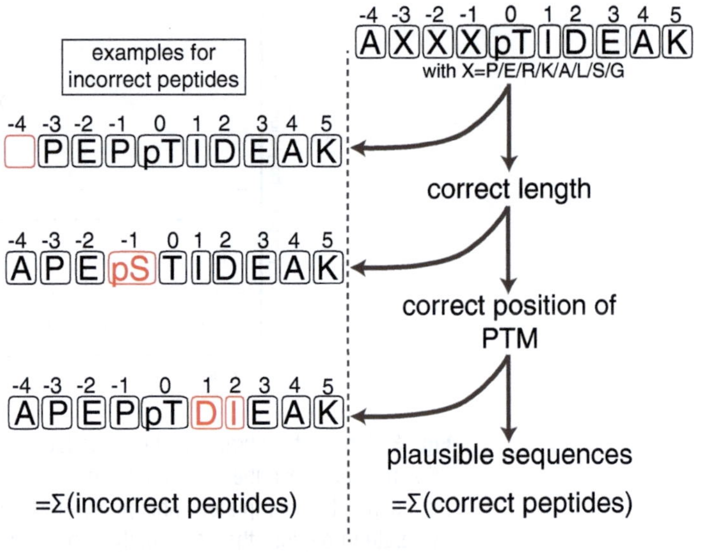

Fig. 4 Graphical representation of the manual data processing steps of the unfiltered database search results to determine empirically correct peptides. Three main steps filter for incorrect peptides of the measurement. In the first step, peptides of incorrect peptide length are excluded. Next, peptides with phosphorylation at an incorrect position are removed. In the final step, the remaining peptides are aligned with the theoretical sequences and excluded if they do not match. The resulting data set are the empirically correct peptides

3.5 MS Data Processing

1. The sequence database in fasta-format needs to be designed as presented previously in Marx et al. [24]. To this end, it is crucial to adapt the search to a tailored database in which every protein corresponds to a single peptide sequence that would theoretically be expected from the synthetic route.
2. The raw data can be analyzed with any database search software available that can handle the required database format. Here we used the pre-mascot script from IsobarQuant [22] for conversion of the raw files to a mascot compatible format (mgf).
3. Next, the spectra are analyzed using Mascot 2.4. (Matrix Science), which is compatible with the library structure presented herein.
4. A decoy database (reverse) was appended to facilitate a target-decoy approach for later false discovery rate (FDR) control in the post-processing step.
5. Search parameters might need adjusting in accordance with library design. Depending on the maximum number of lysines and arginines expected in the library, that is, within a peptide sequence, adjust the number of maximum missed cleavages allowed; mass tolerance should be 10 ppm for MS and 0.05 Da for MS/MS (or <20 ppm). Importantly, the

phosphorylation on the amino acid(s) used in the phosphorylated position should be used as a variable modification and in a similar manner, oxidation on methionine should be accepted if methionine was used for library synthesis (*see* **Note 18**).

6. Export the unfiltered (e.g., no FDR correction) result file (e.g., dat-file in Mascot). Files will need to be converted to txt-files using the post-Mascot script of IsobarQuant and before proceeding with data filtering in R (*see* **Note 19**).
7. The export file should include the following variables in order to process them as described in Subheading 3.6: sequence, modifications,seq_start, and seq_end. All other variables are optional, but might be helpful for further data processing and calculation (e.g., score).

3.6 Data Filtering

1. To start the filtering procedure install the packages tidyverse and Biostrings from the CRAN repository and Bioconductor respectively. Load them into your script by using the *library()* command.

```
library(tidyverse)
library(Biostrings)
```

2. Select the Mascot-output txt-file with *read_tsv()* and import it as a data frame. Do the same for the generated template file containing the theoretically synthesized sequences.

```
Data <- read_tsv("Path\filenameX.txt")
Template_Data <- read_tsv("Path\filenameY.txt")
```

3. To start with the manual filtering procedure, use *filter()* to select peptides with the expected length from the library synthesis (Fig. 4). In the command below, 10-mer peptides are defined as correct peptides (Data_CorrLength). It is helpful to save the incorrect peptide observations in a separate sheet (Data_FalseLength) to analyze potential SPPS issues.

```
Data_CorrLength <- filter(Data, seq_start == 1 & seq_end == 10)
Data_FalseLength <- filter(Data, !(seq_start == 1 & seq_end ==
10))
```

4. Now, determine correct peptide hits by comparing the sequences from Data_CorrLength with the template sequences in Template_Data. Record the correct observations and save them in a new sheet (Data_CorrSeq) (Fig. 4). As before, analyzing peptides with incorrect sequences could be useful in identifying problems during SPPS.

```
Data_CorrSeq <- filter(Data_CorrLength, Data_CorrLength$se-
quence %in% Template_Data$sequence)
Data_IncorrSeq <- filter(Data_CorrLength, !(Data_CorrLength
$sequence %in% Template_Data$sequence))
```

5. Next, the correct position of the phosphorylated amino acid is confirmed (Fig. 4). In untreated samples, peptides with no phosphorylation correspond to spontaneous hydrolysis of the phosphorylation. Peptides with the phosphorylation in the wrong position correspond to incorrect interpretations of the search algorithm. This can be done by using *filter()* again and only selecting observations with the properly placed modifications. In this example it is a phosphorylation on serine or threonine in position five.

```
Data_CorrPhos <- filter(Data_CorrSeq, modifications == "Phos-
pho T5;" | modifications == "Phospho S5;")
Data_IncorrPhos <- filter(Data_CorrSeq, !( modifications ==
"Phospho T5;" | modifications == "Phospho S5;"))
Data_DePhos <- Data_CorrSeq[is.na(Data_CorrSeq$modifica-
tions),]
```

6. The sum of all datapoints excluded in the last three steps (Data_FalseLength, Data_IncorrSeq, Data_IncorrPhos, Data_DePhos) correspond to the empirically incorrect peptides, whereas the sum of Data_CorrPhos corresponds to the empirically correct ones, which are then used for calculation of equal amino acid incorporation frequencies.
7. In the final step, the sequences are split into their individual positions. The *consensusMatrix()* from the Biostrings package computes the consensus matrix of all sequences from a data set on which it is applied. It shows the total number of each amino acid in each position.

```
MatrixA <- consensusMatrix(Data_CorrPhos$sequence)
```

8. The percent incorporation frequency can be derived by dividing the calculated MatrixA by the number of observations from the same dataset used for the MatrixA generation (Data_CorrPhos). The resulting matrix can be used to determine incorporation frequencies in each position of the randomized positions. For biological conclusions please follow Subheadings 3.10 and 3.11.

```
MatrixA_AAFreq <- MatrixA/nrow(Data_CorrPhos)
```

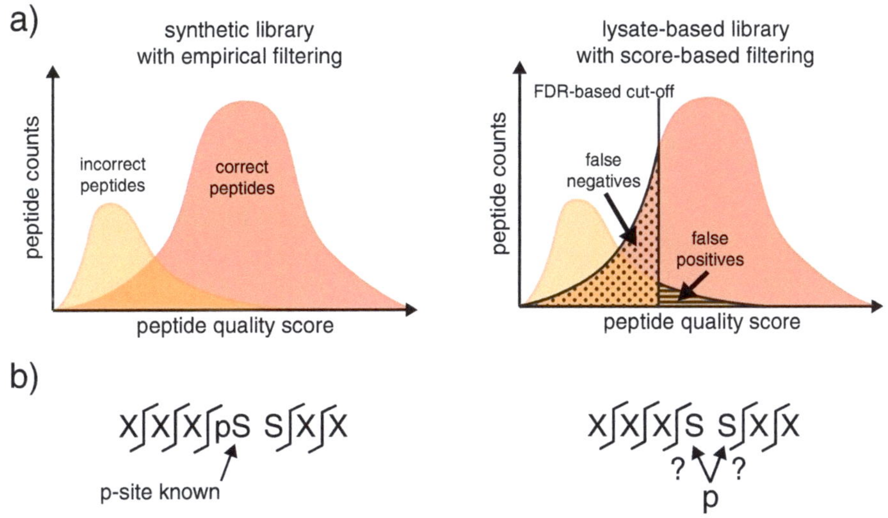

Fig. 5 Comparison of the analysis of synthetic and lysate-based peptide libraries through MS. (**a**) Following the PLDMS approach, correct and incorrect peptides of a synthetic library can be distinguished (left). In a cell lysate-based library an FDR-based cut-off has to be set to distinguish between correct and incorrect hits due to the unknown input (right). However, here the information is biased by including false negatives and excluding false positive peptides. (**b**) Within a synthetic library containing a defined p-site the position of the phosphorylation can be identified with certainty (left), whereas in a cell lysate-based library the determination of the phosphorylated residue is not always possible (right)

Following these steps, only falsely assigned data points that still match the sequence of other theoretically expected sequences are retained. This reduces the number of false positives significantly below an FDR of 5%, which is the normally accepted value for a peptide quality score (Fig. 5a).

Moreover, this approach for library synthesis and filtering overcomes the limitations of false localization rates (FLR) that normally need to be applied to MS data to determine the exact position of a PTM. In this approach, the localization of the PTM is known a priori, ultimately leading to improved motif analysis and refinement (Fig. 5b).

3.7 Performing the Dephosphorylation Assay for MS Analysis

1. Mix 225 μL of the reaction buffer with 60 μL of the prediluted library to a final concentration of 100 μM in a 300 μL in a reaction tube. Incubate for 5 min at 25 °C.
2. The dephosphorylation reaction is started by adding 15 μL of 500 nM phosphatase (25 nM working concentration). Incubate at 25 °C until 30–40% of the library is dephosphorylated. The timepoint can be determined as described in Subheading 3.3 (Fig. 3).
3. The reaction is terminated by adding 13.5 μL of 0.37% HCl to shift the pH from 7.5 to 3, which will inactivate the phosphatase, and then immediately placing the sample on ice (*see* **Note 20**).

4. Spin-down the sample for 5 min at 10,000 × *g* at 4 °C to remove any precipitants. Transfer the supernatant to a new tube. Keep it at 4 °C and continue directly with library purification by liquid chromatography (Subheading 3.8).
5. Depending on the number of treatments to be compared, the dephosphorylation assay described in **step 3** needs to be repeated with all phosphatases to be analyzed.

3.8 Size Exclusion Chromatography to Separate the Library from Recombinant Phosphatase

1. Install a Superdex Peptide 10/300 GL column into an ÄKTA column purification system and equilibrate the column in 0.1 M ammonium acetate buffer (pH 5 at room temperature).
2. Since the quantities of phosphatase used in the dephosphorylation assay will be below the detection limit of the purification system, the expected elution volume needs to be determined by using molecular weight markers. Select a set of marker proteins covering the size range of the phosphatases used and inject them into the purification system for analysis of elution time and volume.
3. Also inject 300 μL of 100 μM library in reaction buffer to determine the peak shape and elution volume of the library by detecting the absorbance at 280 nm. These steps ensure that phosphatase and library are well separated and that the elution volumes of both components are known in subsequent steps.
4. Set up an automated fractionation system using 96-well plates to collect fractions of approx. 200 μL and then load the dephosphorylation assay onto the column. Collect fractions spanning the whole peak corresponding to the peptide library.
5. Pool all peptide library containing fractions and concentrate them by vacuum centrifugation overnight.
6. Dried samples can then be conveniently stored at −20 °C until reconstitution in 0.1% formic acid for MS analysis.

3.9 MS Measurement and Data Processing of Phosphatase-Treated Phosphopeptide Libraries

1. Follow Subheading 3.4 for the measurement of peptide libraries. Measuring in technical duplicates, that is, two injections of the same sample, is recommended.
2. For processing of MS raw data follow Subheading 3.5.
3. For data filtering follow Subheading 3.6. Importantly, due to the phosphatase treatment, dephosphorylated peptides should now be accepted as empirically correct events, and therefore they can be included in the empirically correct peptides (EmpCorr) during the filtering process. In our experience, spontaneous hydrolysis of the phosphorylated residue is a very rare event and therefore can be considered phosphatase-related in this step. Confirm this observation by comparison with the filtering of the reference measurements described under Subheading 3.6.

3.10 Amino Acid Preference Heatmap Generation

1. After filtering for empirically correct sequences as outlined in Fig. 4 and explained in Subheading 3.6, the amino acid preference for phosphatases is analyzed. Continue with the R script started in Subheading 3.6.
2. To this end (Subheading 3.6), the dataset was split between phosphorylated (Data_CorrPhos) and unphosphorylated (Data_DePhos) peptides for each condition. After phosphatase treatment, dephosphorylated peptides can be counted as correct peptides and combined into one data set by using *rbind()*.

```
EmpCorr <- rbind(Data_CorrPhos, Data_DePhos)
```

3. Generate a matrix of the dephosphorylated peptides data set and the sum of all correct peptides as described for phosphorylated peptides.

```
MatrixB <- consensusMatrix(Data_DePhos$sequence)
MatrixC <- consensusMatrix(Data_EmpCorr$sequence)
```

4. Next, divide the number of dephosphorylated peptides by the sum of all empirically correct peptides of the same condition to obtain the global dephosphorylation rate of the measurement. This is done by the *nrow()* function. Compare it to the desired dephosphorylation percentage targeted in Subheading 3.7 obtained with the dephosphorylation assay.

```
DephosRate <- nrow(Data_CorrDePhos)/nrow(EmpCorr)
```

5. Divide MatrixB by MatrixC to obtain the relative fraction of dephosphorylation for each amino acid in each randomized position (*see* **Note 21**). Divide this obtained fraction directly by the overall dephosphorylation rate described in the step above. The resulting values in this matrix will yield the fold-change with which an amino acid in a specific position leads to greater dephosphorylation compared to the average library dephosphorylation (*see* **Note 22**).

```
PreferenceMatrix <- (MatrixB/MatrixC)/DephosRate
```

6. We recommend to depict the resulting matrix in a color-coded heatmap for visualization or in a differential sequence logo, using applications such as IceLogo [25]. This will help in identifying amino acids that lead to greater dephosphorylation and are therefore a driver of phosphatase preference and selectivity.

3.11 Peptide Dephosphorylation Rate

1. As an initial step, it is advisable to minimize the earlier generated data frame EmpCorr by only selecting the necessary columns "sequence" and "modifications" using the function *select ()*.

```
EmpCorr <- select(EmpCorr,"sequence","modifications")
```

2. To calculate to which percentage a peptide sequence was dephosphorylated one has to change the modifications column to numeric values. For the absence of phosphorylation 1 is used, for any phosphorylation (regardless which AA) a 0 is set. First all not identified modifications (NA) have to be set to 1. In the subsequent step, one can use the *ifelse()* function, which sets all values in the modifications column that are not 1, to 0. This selects all phosphorylated residues.

```
EmpCorr$modifications(is.na(EmpCorr$modifications)) <- 1
EmpCorr <- EmpCorr %>% mutate(modifications = ifelse(modifica-
tions!="1",0,1))
```

3. Next, the data is grouped by their sequences and the average value in the modifications column is calculated for each sequence which represents the dephosphorylation rate for each sequence.

```
EmpCorr <- EmpCorr %>% group_by(sequence) %>% mutate(depho-
sphorylation_rate = mean (modifications))
```

4. Set the calculated dephosphorylation rate in relation to the counts (n) of each sequence by using the function *add_count()*.

```
EmpCorr <- add_count(EmpCorr, sequence)
```

5. Next, the modifications column has to be dropped in order to exclude unnecessary replicates of each sequence, which is achieved by the using *select()* followed by the *distinct()*. The argument .keep_all has to be set to TRUE to keep all other columns.

```
EmpCorr <- EmpCorr %>% select(-modifications)
EmpCorr <- distinct(EmpCorr, sequence, .keep_all = TRUE)
```

The resulting data frame contains all detected sequences with their respective dephosphorylation rate (*see* **Note 23**).

4 Notes

1. Adding a *C*-terminal lysine improves mass spectrometry detection. However, impact of the lysine on enzymatic activity or recognition by the enzyme cannot be excluded.
2. Due to the variability of the different peptide properties within the library, a certain percentage might be difficult to dissolve in solution. Ultrasonication and/or addition of low percentages of an organic solvent is recommended to achieve complete solubilization.
3. These values may vary when using a different peptide synthesizer.
4. Each possible peptide sequence from the library has to be represented here as a protein with its separate identifier in the database [24]. The Mascot search engine has been chosen, because it can handle the necessary configuration.
5. The versions mentioned are the ones for the example code in Subheading 3. Using other versions should work as well. However, commands might have to be slightly adjusted.
6. It is important that most of the procedures explained in Subheading 3 are adjusted individually for one's individual purpose before the complete PLDMS approach can be performed for a selected phosphatase-library combination.
7. This parameter can differ greatly across different instruments and measurement settings. A good estimate is the number of identified peptides or scans from a complex proteome measurement on the same system. Especially when comparing multiple conditions in replicates, this can easily result in several days of measurement time, despite relatively short gradients of 1–2 h per sample. Yet, it is of central importance that for all amino acids under consideration, a sufficient number of data points can be acquired for a reproducible and robust statistical analysis. Therefore, one should carefully calculate the theoretical library size and compare it to the expected MS instrumentation time available to derive with the most powerful combination between library size and the depth of MS analysis, that is, incorporation of less amino acids or less randomized positions might lead to more meaningful insights.
8. Avoid leucine and isoleucine in the same isokinetic mixture for the randomized position. Due to the identical mass of the amino acid residues they cannot be distinguished in a regular MS/MS fragment spectrum, thus leading to inconclusive results.

9. We recommend using libraries of fixed length. From our point of view, the advantages in data filtering far outweigh the potential loss of information that could be gathered from variable lengths.
10. If adjustments of the overall incorporation rate of the isokinetic mixture need to be done, we observe that it is constructive to adjust the molarity of the most under- and overrepresented amino acids only. Changing the ratio of all AAs leads to an overall change, which might result in the need for further adjustments.
11. This results in a synthesis ratio of 4 eq.
12. Ensure an air bubble is present in the syringe to enable strong mixing on an orbital shaker after closing the tip of the syringe with a cap.
13. We recommend to always use freshly thawed enzyme for quantitative analysis to avoid changes in enzyme kinetics or dephosphorylation velocity. Furthermore, the concentration might vary drastically with the substrate library and chosen phosphatase. In our experience, 25 nM of the phosphatase is a good starting point.
14. To ensure that point can be accepted as the library dephosphorylation starting point, two things need to be considered: First, minimize the time between phosphatase addition and reaction monitoring to minimize unmonitored library dephosphorylation. Second, verify minimal spontaneous library hydrolysis, which was determined in Subheading 3.6.
15. The absolute amount used for MS-analysis varies greatly based on library complexity. We advise using an amount that leads to a signal intensity on the MS1 level of approx. 500 ng of complex peptide digest.
16. The gradient should be adjusted depending on the physiochemical properties of the library used in order to ensure optimal separation.
17. Peptide fragmentation can also be optimized by testing different fragmentation engines (NCE).
18. In general, no fixed modifications should be used for database search. Unless one introduces them into the peptide synthesis, for example, N-terminal modifications. Furthermore, trypsin only works as a specified enzyme in case of a C-terminal lysine or arginine residue.
19. This way all PSM level data can be accessed. This helps to avoid a potential bias by the search engine reporting.
20. Do not use small molecules or inhibitor cocktails since these might interfere with MS analysis in the next steps. A radical pH

shift is recommended to terminate the reaction. However, one should keep in mind two things, (1) this might have an effect on the peptides and (2) one has to verify that the reaction is terminated.

21. Note that this step normalizes variations in incorporation of amino acids as long as amino acids are sufficiently often incorporated. When the number of amino acids in a position is low, the resulting ratio is probably biased. In our experience at least 500 datapoints for each amino acid in each position should be targeted. However, this depends on the overall complexity of the library. The larger the library, the more data points that should be targeted.
22. It is noteworthy that this also normalizes for variations between 30% and 40% dephosphorylation of the library, as discussed above.
23. Resulting dephosphorylation rates of low peptide counts might be misleading. We recommend considering peptides with more than ten observations ($n \geq 10$) in the data set.

Acknowledgments

This work was funded by an European Research Council (ERC) Consolidator grant (#865119) to M.K., and by the Deutsche Forschungsgemeinschaft (DFG, German Research Foundation) through FOR 2743 and through Germany's Excellence Strategy (BIOSS—EXC 294) to M.K.

References

1. Deribe YL, Pawson T, Dikic I (2010) Post-translational modifications in signal integration. Nat Struct Mol Biol 17:666–672
2. Khoury GA, Baliban RC, Floudas CA (2011) Proteome-wide post-translational modification statistics: frequency analysis and curation of the swiss-prot database. Sci Rep 1:srep00090
3. Ochoa D, Jarnuczak AF, Viéitez C et al (2020) The functional landscape of the human phosphoproteome. Nat Biotechnol 38:365–373
4. Sharma K, D'Souza RCJ, Tyanova S et al (2014) Ultradeep human Phosphoproteome reveals a distinct regulatory nature of Tyr and Ser/Thr-based signaling. Cell Rep 8:1583–1594
5. Beltrao P, Bork P, Krogan NJ, van Noort V (2013) Evolution and functional cross-talk of protein post-translational modifications. Mol Syst Biol 9:714
6. Hardman G, Perkins S, Brownridge PJ et al (2019) Strong anion exchange-mediated phosphoproteomics reveals extensive human non-canonical phosphorylation. EMBO J 38: e100847
7. Endicott JA, Noble MEM, Johnson LN (2012) The structural basis for control of eukaryotic protein kinases. Annu Rev Biochem 81:587–613
8. Brinkworth RI, Breinl RA, Kobe B (2003) Structural basis and prediction of substrate specificity in protein serine/threonine kinases. Proc Natl Acad Sci U S A 100:74–79
9. Oppermann FS, Gnad F, Olsen JV et al (2009) Large-scale proteomics analysis of the human kinome. Mol Cell Proteomics 8:1751–1764
10. Bradley D, Beltrao P (2019) Evolution of protein kinase substrate recognition at the active site. PLoS Biol 17:e3000341

11. Kim M, Shin D-S, Kim J, Lee Y-S (2010) Substrate screening of protein kinases: detection methods and combinatorial peptide libraries. Biopolymers 94:753–762
12. Obata T, Yaffe MB, Leparc GG et al (2000) Peptide and protein library screening defines optimal substrate motifs for AKT/PKB. J Biol Chem 275:36108–36115
13. Till JH, Annan RS, Carr SA, Miller WT (1994) Use of synthetic peptide libraries and phosphopeptide-selective mass spectrometry to probe protein kinase substrate specificity. J Biol Chem 269:7423–7428
14. Zhang M, Yogesha SD, Mayfield JE et al (2013) Viewing serine/threonine protein phosphatases through the eyes of drug designers. FEBS J 280:4739–4760
15. Köhn M (2020) Turn and face the strange: a new view on phosphatases. ACS Cent Sci 6: 467–477
16. Fischer TH, Eiringhaus J, Dybkova N et al (2018) Activation of protein phosphatase 1 by a selective phosphatase disrupting peptide reduces sarcoplasmic reticulum Ca^{2+} leak in human heart failure. Eur J Heart Fail 20: 1673–1685
17. Leonard D, Huang W, Izadmehr S et al (2020) Selective PP2A enhancement through biased Heterotrimer stabilization. Cell 181:688–701.e16
18. Kauko O, O'Connor CM, Kulesskiy E et al (2018) PP2A inhibition is a druggable MEK inhibitor resistance mechanism in KRAS-mutant lung cancer cells. Sci Transl Med 10: eaaq1093
19. Brautigan DL (2013) Protein Ser/ Thr phosphatases - the ugly ducklings of cell signalling. FEBS J 280:324–345
20. Bollen M, Peti W, Ragusa MJ, Beullens M (2010) The extended PP1 toolkit: designed to create specificity. Trends Biochem Sci 35: 450–458
21. Hoermann B, Kokot T, Helm D et al (2020) Dissecting the sequence determinants for dephosphorylation by the catalytic subunits of phosphatases PP1 and PP2A. Nat Commun 11:3583
22. Franken H, Mathieson T, Childs D et al (2015) Thermal proteome profiling for unbiased identification of direct and indirect drug targets using multiplexed quantitative mass spectrometry. Nat Protoc 10:1567–1593
23. Harris JL, Backes BJ, Leonetti F et al (2000) Rapid and general profiling of protease specificity by using combinatorial fluorogenic substrate libraries. Proc Natl Acad Sci U S A 97: 7754–7759
24. Marx H, Lemeer S, Schliep JE et al (2013) A large synthetic peptide and phosphopeptide reference library for mass spectrometry-based proteomics. Nat Biotechnol 31:557–564
25. Colaert N, Helsens K, Martens L et al (2009) Improved visualization of protein consensus sequences by iceLogo. Nat Methods 6:786–787

Chapter 3

FEPS: A Tool for Feature Extraction from Protein Sequence

Hamid Ismail, Clarence White, Hussam AL-Barakati, Robert H. Newman, and Dukka B. KC

Abstract

Machine learning has become one of the most popular choices for developing computational approaches in protein structural bioinformatics. The ability to extract features from protein sequence/structure often becomes one of the crucial steps for the development of machine learning-based approaches. Over the years, various sequence, structural, and physicochemical descriptors have been developed for proteins and these descriptors have been used to predict/solve various bioinformatics problems. Hence, several feature extraction tools have been developed over the years to help researchers to generate numeric features from protein sequences. Most of these tools have some limitations regarding the number of sequences they can handle and the subsequent preprocessing that is required for the generated features before they can be fed to machine learning methods. Here, we present Feature Extraction from Protein Sequences (FEPS), a toolkit for feature extraction. FEPS is a versatile software package for generating various descriptors from protein sequences and can handle several sequences: the number of which is limited only by the computational resources. In addition, the features extracted from FEPS do not require subsequent processing and are ready to be fed to the machine learning techniques as it provides various output formats as well as the ability to concatenate these generated features. FEPS is made freely available via an online web server as well as a stand-alone toolkit. FEPS, a comprehensive toolkit for feature extraction, will help spur the development of machine learning-based models for various bioinformatics problems.

Key words Posttranslational modifications, Feature extraction, Machine learning, Sequence-based features, Protein descriptors

1 Introduction

Over the years, machine learning (ML) has been successfully applied in many bioinformatics problems due, in part, to its ability to examine large volumes of information or training data to discover unique patterns that can be used for prediction/classification. For instance, ML has been used to identify the class and structural characteristics of proteins and DNA, which provides insights into their function in biological systems [1, 2]. ML has proven to be successful in the prediction of the sites of post-translational

Dukka B. KC (ed.), *Computational Methods for Predicting Post-Translational Modification Sites*,
Methods in Molecular Biology, vol. 2499, https://doi.org/10.1007/978-1-0716-2317-6_3,

modifications (PTMs) in proteins [3–9]. To utilize machine learning algorithms for protein sequence analysis, the general standard is to transform protein primary sequence data into numerical values, which is a crucial step for extracting information from biological sequences [10]. Over the past decade, the use of extracted features from the sequences and three-dimensional structures of proteins has become a popular strategy in the field of computational biology. Machine learning-based approaches which heavily rely on the ability to extract meaningful descriptors (features) from protein sequences have proven to be very successful in several computational structural bioinformatics problems such as prediction of protein classes and structures [11], protein–protein interactions [12, 13], subcellular localization of proteins [14], identification of the active sites on enzymes [15, 16], and identification of sites of posttranslational modifications [3–9], among others.

Machine learning-based approaches require numeric features as input for subsequent computational analysis. Subsequently, feature extraction from biological sequences becomes a key step in the process and can be defined as the mechanism to transform biological sequences into numeric vectors representing the information contained in the sequences. The vector obtained from each sequence represents the row of attributes of that sequence. When features are extracted from multiple sequences, the entirety of the data will form a table with the rows representing the attributes of the sequences and the columns representing the variables. Thus, if each row of attributes has a known response or label value, those data can be used in a supervised machine learning framework. The classification machine learning problem requires that the response value be a categorical class name while regression machine learning requires the response value to be numeric. If the data rows have no response values or labels, that data can be used in unsupervised machine learning such as clustering. Dealing with the classification problem using biological data began around four decades ago, but it has recently become an important subdiscipline in computational biology due to the availability of genomic data and the emergence of feature extraction approaches and machine learning algorithms.

For single-domain proteins, the primary amino acid sequence contains all of the information necessary for it to fold into its native tertiary structure, which helps determine its function(s) inside the cell. Therefore, features can be extracted from the primary amino acid sequence of a protein assuming that it contains the information for the protein folding and function. Such features can be then be used in machine learning methods for pattern recognition, knowledge discovery, and protein classification problems, such as the prediction of protein families, protein structures, protein functions, and protein–protein interactions. As a consequence, the development of protein descriptors has been an active area of research in computational biology for several decades. First, we discuss the commonly used protein descriptors. The most extensively used descriptors are based on syntactical information, physicochemical information, or evolutionary information. Thus, we will focus on all of those categories with illustrations and worked examples.

The syntactical-based features are the features that are based on the information of the amino acid composition (AAC) of a protein. The simplest way to extract AAC features is by computing the relative frequencies of each one of the 20 amino acids in a sequence obtaining 20 features. However, there are other variants of AAC including the relative frequencies of any two possible amino acids (dipeptide amino acid composition), any three possible amino acids (tripeptide amino acid composition) [17], and position weight amino acid composition (PWAAC) [18, 19]. AAC was first used in 1983 for grouping proteins into four groups based on their locations in an organism: intracellular enzymes, intracellular nonenzymes, extracellular enzymes, and extracellular nonenzymes [20]. Then, the AAC features were used in machine learning algorithms such as neural networks and support vector machines to predict the subcellular location of proteins [21] and protein family classification [17, 22, 23]. Similarly, AAC features were combined with others to predict the secondary structure of proteins [24, 25]. Features that depend on physicochemical-based information are also used as input for machine learning. Each one of the 20 amino acids has a unique distribution of physicochemical properties, such as hydrophobicity, hydrophilicity, hydrophilicity, and polarity. Currently, there are 566 reported physicochemical properties indexed on the Amino Acid Index Database (AAindex) and available online at "https://www.genome.jp/aaindex/" [26].

Physicochemical properties of amino acids are used in several ways to extract features. The Sequence-Order-Coupling Numbers method and the Quasi Sequence-Order Numbers method are calculated using distance matrices generated from the physicochemical properties. Features from these two methods have been used to predict protein subcellular locations [27] and protein–protein interactions [28]. Likewise, high-quality physicochemical indices (HQI) have been used together with other features to predict posttranscriptional hydroxylation sites on protein sequences [19]. Meanwhile, features from overlapping properties (OP) and Average Cumulative Hydrophobicity (ACH) were used to predict post-transcriptional phosphorylation sites on protein sequences [29]. Some of the physicochemical properties were incorporated in several feature extraction methods, such as Moreau–Broto autocorrelation [30], Moran autocorrelation [31], Geary autocorrelation [32], Composition, Transition and Distribution (CTD) [33, 34], and other feature methods. These features have provided key insights into protein function. For instance, autocorrelation variants have proven to be efficient in the prediction of protein–

protein interactions [28, 35, 36]. Similarly, CTD features have been used to predict sites of protein post-translational modifications (PTM), such as the prediction of methylation and phosphorylation sites on protein sequences [29, 37]. Several studies have shown that the use of evolutionary information in the form of PSSM profiles is much more informative than using the sequence information alone [38, 39]. Therefore, numerous PSSM-based feature descriptors have been used with machine learning to solve a wide range of protein classification problems, such as protein fold recognition [40], protein subcellular localization [41], protein–protein interaction [39], and the prediction of protein structures [42, 43].

There are also hybrid features that combine two or more kinds of information. For example, pseudo amino acid composition (PseAAC) and amphiphilic pseudo amino acid composition (APseAAC) combine syntactical- and physicochemical-based information. These features were developed by Kuo-Chen Chou to deal with one of the primary flaws of AAC, which is that AAC does not take the order of the amino acid sequence into account. PseAAC and APseAAC are combinations of a set of sequence correlation factors using the occurrence frequencies of the 20 amino acid (i.e., the AAC) and amino acid properties [10, 44, 45]. PseAAC was used by many studies in computational proteomics for solving various classification and prediction problems, including prediction of a protein family and subfamily classes [45–48], prediction of the subcellular localization of proteins [49–52], prediction of protein–protein interactions [53–55], and prediction of protein solubility [56, 57].

Other methods are also used for extracting features from protein sequences, such as statistical methods that use probability and probabilistic distributions. Such methods include Shannon entropy, Bayes, and wavelet variance. For instance, prediction of protein structure was first performed using features extracted by combining wavelet variance generated from different physicochemical scales mapped to the protein sequences and SVM [58]. Similarly, Bayes features were used together with SVM to identify protein methylation sites [59] as well as linear B-cell epitopes [60]. Meanwhile, features from protein disordered regions (PDR) are expressed as a probability estimate of each residue in the sequence being disordered. Such features have been used along with other features to predict sites of hydroxylation [19].

Substitution matrices have been used for feature extraction as well. For example, features extracted from substitution matrices were used for protein classification. The process began by extracting motifs from a set of sequences and then the motifs were used to create a learning context table made up of the features' binary values, where 1 and 0 represent the presence and absence of a motif, respectively [61]. Solvent accessible surface (SAS) features were also used as input features for machine learning for predicting binding sites [62] and PTM sites of proteins [29].

Features are usually numeric vectors attempting to score information contained in the protein sequences. Any feature extraction method can be used to extract information that can be efficient for some classification problems. However, there is not a single kind of feature that is suitable for all problems. The general strategy in using features from protein sequences for machine learning is to combine multiple feature types and then to implement feature engineering or feature selection to select only those that contribute the most to the problem. Although regression is possible, most uses of machine learning in proteomics are for classification. This classification may include classifying protein sequences into families and subfamilies or groups, which requires extracting features from the entire sequence. However, the classification may include identification of binding sites, sites of PTMs, prediction of protein structures, and protein–protein interaction sites. Such classifications may require the extraction of features from segments or windows of the sequences. Therefore, it is crucial to pay attention to the feature extraction strategy that works for the classification problem at hand.

The features extracted from protein sequences are usually numeric values that quantify the information. However, feature values can also be discrete values representing categories (e.g., 0 and 1 for "no" and "yes", respectively). For a supervised classification problem, the dataset that is used to train the classification model must include the groups of benchmark sequences that represent the classes. Each sequence will have a class label so that, when features are extracted, that class name will be the value that labels the row of feature values in the class column.

2 Protein Feature Descriptors

A typical sequence dataset for a two-class protein classification problem is shown in Table 1, which includes example protein sequences of two types: DNA binding domain (DBD) and non-DNA binding domain (NBD). Assume that the classes of these sequences are experimentally confirmed.

For such a classification problem, in which a model is trained to classify a protein sequence as one of the possible classes, features can be extracted from the entire sequences. However, this will not be the case when dealing with classifying specific amino acids in the protein sequences. For example, the protein PTMs occur on specific amino acids. In eukaryotic cells, phosphorylation occurs predominantly on the side chains of four amino acids, *serine (S)*, *threonine (T)*, *tyrosine (Y)*, and *histidine (H)*. Meanwhile, hydroxylation may take place on various amino acids, but most frequently on *proline (P)*, *lysine (K)*, *asparagine (N)*, *aspartate (D)*, and *histidine (H)*. However, only some of these three amino acids on a protein sequence are sites for such modifications. Therefore, a machine learning model is trained to identify which of these amino acids is likely to be a PTM site and which one is not. In such cases, extracting features from the entire sequence may not capture any information about a PTM site. For the classification of such sites or small regions of a protein sequence, such as motifs, features are extracted from windows of specific sizes in which the site-of-interest is in the middle. For example, Table 2 shows some window sequences (made from the first sequence in Table 1) that include serine (S) in the middle. Features extracted from these windows can be used to classify serine. The classes in such cases will be positive serine (PTM site) and negative serine (not PTM site).

There might be other strategies depending on the classification problems. Before extracting features from protein sequences, the researcher must adopt the feature extraction strategy that is appropriate for the classification problem.

Table 1
Example protein sequences. *DBD* DNA-binding domain, *NBD* non-DNA-binding domain

Class	Protein sequences
DBD	CKVCGDVASGFHYGVHACEGCKGFFRRSIQQNIQYKRCLKNENCSIVRINRNRCQQCRFK
DBD	CAICGDRSSGKHYGVYSCEGCKGFFKRTIRKDLIYTCRDNKDCLIDKRQRNRCQYCRY
DBD	CKICGDKSSGIHYGVITCEGCKGFFRRSQQSNATYSCPRQKNCLIDRTSRNRCQHCRLQK
NBD	CLVCSDEASGCHYGVLTCGSCKVFFKRAVEGQHNYLCAGRNDCIIDKIRRKNCPACRYRKCL
NBD	CLICGDEASGCHYGVLTCGSCKVFFKRAMEGQHNYLCAGRNDCIVDKIRRKNCPACRLRK
NBD	CPYCRFQKCLSVGMKLEAVRADRMLQXNDKAATCGQNDKILCPPHYCRDVGMSRKC

Table 2
Windows of a sequence (peptide). The window is centered around bold S

Class	Windows
Negative	GDVA**S**GFHY
Positive	FFRR**S**IQQN
Positive	NENC**S**IVRI
Negative	KKCL**S**VGMS

Table 3
A typical feature representation as required by the machine learning methods. The first column is the class (label) and next n-columns are features. L is the number of sequences in Class 1 and K is the number of sequences in class 2

Class	F1	F2	F3	...	Fn
1	$v_{1,1}$	$v_{1,2}$	$v_{1,3}$	...	$v_{1,n}$
1	$v_{2,1}$	$v_{2,2}$	$v_{2,3}$	...	$v_{2,n}$
1	$v_{3,1}$	$v_{3,2}$	$v_{3,3}$	...	v_{3n}
⋮	⋮	⋮	⋮	⋮	⋮
1	$v_{L,1}$	$v_{L,2}$	$v_{L,3}$	...	$v_{L,n}$
2	$v_{1,1}$	$v_{1,2}$	$v_{1,3}$	...	$v_{1,n}$
2	$v_{2,1}$	$v_{2,2}$	$v_{2,3}$	...	$v_{2,n}$
2	$v_{3,1}$	$v_{3,2}$	$v_{3,3}$	...	v_{3n}
⋮	⋮	⋮	⋮	⋮	⋮
2	$v_{K,1}$	$v_{K,2}$	$v_{K,3}$	...	$v_{K,n}$

After extracting features from the sequence, we will generate a table whose columns are the features (variables) and whose rows are the feature values extracted from each sequence. For n features, L number of sequences labeled as Class 1 and K sequences labeled as Class 2, the table of the dataset can be represented by Table 3, where the first column is class labels, F_1, F_2, F_3, ..., F_n are the feature names, $v_{i,\ j}$ are the feature values, L is the number of sequences in Class 1, and K is the number of sequences in Class 2.

In the following sections, we will discuss the most commonly used protein descriptors.

2.1 Amino Acid Composition

The amino acid composition (ACC), which is the simplest feature type, tallies all of the amino acids in the protein but does not take the order of amino acid sequence into account. Because it does not account for the context of the amino acids, the lack of sequence information is often considered a flaw of ACC. However, it can be useful for protein family classification. The AAC is the number of amino acids of each type normalized to the total number of residues.

For the 20 amino acids,

aa = (A, C, D, E, F, G, H, I, K, L, M, N, P, Q, R, S, T, V, W, Y), the ACC is defined as

$$AAC(aa_i) = \left(\sum aa_i\right) \times \frac{100}{N} \tag{1}$$

where $i = 1,2,3,\ldots,20$ are the indexes of amino acids in the above amino acid list; Σaa_i is the number of residues of each type, and N is the total number of amino acids in the sequence.

Since there are 20 amino acids, the number of AAC features is 20 where each feature represents the relative frequency of the corresponding amino acid.

2.2 Dipeptide Composition

The set of dipeptides consists of all possible combinations of two amino acids. For example, the set of possible dipeptides for the first three amino acids (A, C, D) are (AA, AC, AD, CA, CC, CD, DA, DC, DD). Thus, for 20 amino acids, there are $20 \times 20 = 400$ possible dipeptides. This means that there are 400 features for dipeptide composition.

The dipeptide composition (DPC) is the number of dipeptides of each type normalized with the total number of dipeptides in the sequence, where ΣDP_i is the number of each dipeptide of the 400 possible dipeptides; ΣDP is the number of dipeptides in the sequence.

$$\mathrm{DPC}(i) = \left(\sum \mathrm{DP}_i\right) \times \frac{100}{\sum \mathrm{DP}} \quad (2)$$

For example, in the sequence: NALMNACC, the dipeptides are NA, AL, LM, MN, NA, AC, and CC. The frequency of NA, AL, LM, MN, AC, and CC are 2, 1, 1, 1, 1, and 1, respectively. Since in this sample sequence there are only six unique dipeptides that have values, the values of the other 394 features are zeros.

2.3 Tripeptide Composition

As in dipeptide composition, the set of tripeptides consists of all possible combinations of three amino acids. For the 20 amino acids, there are $20 \times 20 \times 20 = 8000$ possible tripeptides, resulting in 8000 features.

The tripeptide composition (TPC) is the number of occurrences of each type of tripeptide normalized to the total number of tripeptides in the sequence, where ΣTP_i is the number of each tripeptide i from among the 8000 possible tripeptides and ΣTP is the total number of tripeptides in the sequence.

$$TPC(i) = \left(\sum TP_i\right) \times \frac{100}{\sum TP} \quad (3)$$

The amino acid order in a tripeptide may add some more information to the feature compared to AAC and DPC. However, the number of features is high, which may cause problems stemming from high dimensionality.

2.4 Position Weight Amino Acid Composition

The position weight amino acid composition (PWAA) is used to extract position information from amino acids surrounding a potential site in a protein sequence. There are 20 PWAA features. Each feature represents the position weight of an amino acid.

aa = (A, C, D, E, F, G, H, I, K, L, M, N, P, Q, R, S, T, V, W, Y).

For a sequence window of length $2n + 1$, in which the potential site is in the center at position $n + 1$ flanked by downstream and upstream amino acids, position weights are given by

$$C_i = \frac{1}{n(n+1)} \sum_{i=1}^{n} x_{ij} \left(\frac{|j|}{n} \right) \quad (4)$$

where n denotes the number of downstream or upstream amino acids from the center of the window, j denotes the position of the amino acid relative to the center ($-n \leq j \leq n$), and $x_{ij} = 1$ if the amino acid, aa_i, is found in position j of the sequence window, otherwise $x_{ij} = 0$.

2.5 One-Hot Encoding

One-hot encoding is suitable in those cases where the protein sequences are divided into windows, as in phosphorylation site prediction, disordered protein region prediction, and motif detection. The one-hot encoding are computed as follows.

1. Order the 20 amino acids and create a 20-digit code for each amino acid as shown in Table 4.
2. Represent each amino acid using the 20-digit code. For instance, for a window size of 10 amino acid residues, the number of features will be $10 \times 20 = 200$ as each residue in the sequence will be replaced by its 20-digit code, and then each digit column will represent a feature.

Table 4
One-hot encoding of 20 amino acids

#	A.A.	20-digit code	#	A.A.	20-digit code
1	A	10,000,000,000,000,000,000	11	M	00000000001000000000
2	C	01000000000000000000	12	N	00000000000100000000
3	D	00100000000000000000	13	P	00000000000010000000
4	E	00010000000000000000	14	Q	00000000000001000000
5	F	00001000000000000000	15	R	00000000000000100000
6	G	00000100000000000000	16	S	00000000000000010000
7	H	00000010000000000000	17	T	00000000000000001000
8	I	00000001000000000000	18	V	00000000000000000100
9	K	00000000100000000000	19	W	00000000000000000010
10	L	00000000010000000000	20	Y	00000000000000000001

2.6 Conjoint Triad Descriptor

To calculate the conjoint triad, the amino acids are clustered into seven groups based on their dipole and the volume of the side chain. The amino acids in the same cluster are likely to substitute for one another since they have similar physicochemical properties.

The scales for dipole are as follows.

1. If the dipole is less than1.0, then the dipole scale is negative (−).
2. If the dipole is between 1.0 and 2.0(+) (i.e., 2.0–3.0), then the scale is 2-plus (++).
3. If the dipole is greater than 3.0, then the scale is 3-plus (+++).

The scales of the side-chain volume scales are as follows.

1. If the volume is less than 50 cubic angstroms ($Å^3$), then the volume scale is negative (−).
2. If the volume is greater than 50 $Å^3$, then the volume scale is positive (+).

Table 5 shows the 20 amino acids clustered into groups based on the dipole and side-chain volume scales.

Cysteine is assigned to a separate class because of its ability to form a disulfide bond with other cysteine residues. The conjoint triad features are calculated as follows.

1. For a protein sequence S, each amino acid is replaced by the corresponding group number. For example, if the sequence is S=NPRESKLCMNPR, then the result of this step leads to S = 425635273425.
2. A 3-amino acid window is used to slide across the whole protein sequence one step at a time from the N-terminus to the C-terminus. For the example sequence, this will result into 425, 256, 563, 635, 352, 527, 273, 734, 342, and 425.

Table 5
Amino acid grouping based on dipole and side-chain volume

#	Group	Dipole scale	Volume scale
1	Ala (A), Gly (G), Val (V)	−	−
2	Ile (I), Lue (L), Phe (F), Pro (P)	−	+
3	Tyr (Y), Met (M), Thre (T), Ser (S)	+	+
4	His (H), Asn (N), Gln (Q), Tpr (W)	++	+
5	Arg (R), Lys (K)	+++	+
6	Asp (D), Glu (E)	+++	+
7	Cys (C)	+	+

Table 6
Conjoint triad features (combinations of 1 to 7 in three-digit positions)

111 - 117	121 - 127	131 - 137	141 - 147	151 - 157	161 - 167	171 - 177
211 - 217	221 - 227	231 - 237	241 - 247	251 - 257	261 - 267	271 - 277
311 - 317	321 - 327	331 - 337	341 - 347	351 - 357	361 - 367	371 - 377
411 - 417	421 - 427	431 - 437	441 - 447	451 - 457	461 - 467	471 - 477
511 - 517	521 - 527	531 - 537	541 - 547	551 - 557	561 - 567	571 - 577
611 - 617	621 - 627	631 - 637	641 - 647	651 - 657	661 - 667	671 - 677
711 - 717	721 - 727	731 - 737	741 - 747	751 - 757	761 - 767	771 - 777

3. The frequencies of these resultant windows are counted. In this case, we have 425(count = 2), 256(count = 1), 563(count = 1), 635(count = 1), 352(count = 1), 527(count = 1), 273 (count = 1), 734(count = 1), and 342(count = 1).
4. The total number of features is the combination of 1, 2, 3, 4, 5, 6, and 7 in three-digit positions 343 ($=7^3$). Thus, the number of conjoint triad features is 343 $[f_1, f_2, f_3, \ldots f_{343}]$, as shown in Table 6. For the example sequence, only 425, 256, 563, 635, 352, 527, 273, 734, and 342 have values, while the values of other features are zero.

Since the lengths of protein sequences may vary, the frequencies are scaled to be in the range between 0.0 and 1.0 as follows.

$$\overset{C}{f}_i = f_i - \frac{min\,(f_1, f_2, f_3, \cdots f_{343})}{max\,(f_1, f_2, f_3, \cdots f_{343})}; \quad (1 < i < 343) \qquad (5)$$

2.7 Overlapping Properties

The amino acid overlapping properties (OP) descriptor was first proposed to capture information from the common physicochemical properties shared by the amino acids in a protein sequence. The amino acids are classified based on ten physicochemical properties into polar, positive, negative, charged, hydrophobic, aliphatic, aromatic, small, tiny, and proline. An amino acid may fall into more than one group due to the overlapping properties. Each group is given a position in a 10-bit code, where each group is represented by a bit, respectively (Table 7). The position of the bit is set to 1 if the amino acid belongs to the corresponding group and 0 if it does not. The number of features extracted using this method depends on the number of residues in the sequence. The OP features are usually used with the window of sequences that are equal in size.

For example, the sequence (EDPA) will have the codes: 00**1**0000000, 00**1**0000000, 0000000**1**00, and 00000000**1**0, respectively. Thus, the resultant feature vector will be (0,0,**1**,0,0,0,0,0,0,0, 0,0,**1**,0,0,0,0,0,0,0, 0,0,0,0,0,0,0,**1**,0,0, 0,0,0,0,0,0,0,0,**1**,0).

Table 7
Overlapping groups of amino acids

Group	Amino acids	Binary code
Polar	N,Q,S,D,E,C,T,K,R,H,Y,W	1,000,000,000
Positive	K,H,R	0100000000
Negative	D,E	0010000000
Charged	K,H,R,D,E	0001000000
Hydrophobic	A,G,C,T,I,V,L,K,H,F,W,Y,M	0000100000
Aliphatic	I,V,L	0000010000
Aromatic	F,Y,W,H	0000001000
Small	P,N,D,T,C,A,G,S,V	0000000100
Tiny	A,S,G,C	0000000010
Proline	P	0000000001

2.8 Sequence-Order-Coupling Numbers

The sequence-order-coupling numbers (SOCN) are calculated using distance matrices, such as the Schneider-Wrede physicochemical distance matrix [63] or the Grantham chemical distance matrices [64]. Those two distance matrices were derived empirically to quantify the pair-wise physicochemical distance between any two of the 20 amino acids. The amino acids that have less distance are chemically similar to each other and are more likely to replace each other via mutation while the amino acids that have more distance are less likely to substitute for each other.

The dth rank of the sequence order coupling number is given by

$$\tau_d = \sum_{i=1}^{N-d} \left(d_{i,i+d}\right)^2, \quad d = 1, 2, 3, \ldots m \tag{6}$$

where $d_{i,\ i\ +\ d}$ is the distance between the two amino acids at position i and $i + d$ while m is the maximum lag-value.

2.9 Quasi Sequence-Order Number

The Quasi sequence-order number (QSO) is an example of a hybrid model that comprises both discrete and sequential models [65–67]. The first 20 features correspond to AAC while features 21 and above reflect the sequence order using four physicochemical properties: hydrophobicity, hydrophilicity, polarity, and side-chain volume. The features are derived from the Schneider–Wrede physicochemical distance matrix [63] and the Grantham chemical distance matrix [64] alluded to above. The first 20 features are given by

$$X_i = \frac{f_i}{\sum_{i=1}^{20} f_i + w\sum_{d=1}^{m} \tau_d} \quad (7)$$

where $i = 1, 2, \ldots, 20$, f_i is the normalized frequency of the amino acid i and w is a weighting factor (the recommended is $w = 0.1$). The features from 21 and above are given by

$$X_i = \frac{w\tau_{d-20}}{\sum_{i=1}^{20} f_i + w\sum_{d=1}^{m} \tau_d}; \quad d = 21, 22, \ldots 20 + m \quad (8)$$

The number of feature elements is determined by the maximum lag-value (m), which is the maximum distance that two residues can be apart.

2.10 Composition, Transition and Distribution

To extract the composition, transition and distribution (CTD) features, [34, 68] the 20 amino acids are categorized into 3 groups based on seven physicochemical properties individually. The seven amino acid properties are (1) hydrophobicity, (2) normalized Van der Waals volume, (3) polarity, (4) polarizability, (5) charge, (6) secondary structures, and (7) solvent accessibility (Table 8). Based on each one of the seven properties, the 20 amino acids are categorized into three groups. The amino acids of a protein sequence are replaced by the group numbers. For example, the sequence MVKELRTA is encoded as 33113122 based on hydrophobicity.

Table 8
Amino acids grouping based on the seven amino acid properties

Properties	Group 1	Group 2	Group 3
Hydrophobicity	*Polar* R,K,E,D,Q,N	*Neutral* G, A, S,T,P,H,Y	*Hydrophobic* C,L,V,I,M,F,W
Normalized van	0–2.78	2.95-4.0	4.03-8.08
der Waals volume	G,A,S,T,P,D	N,V,E,Q,I,L	M,H,K,F,R,Y,W
Polarity	*4.9–6.2* L,I,F,W,C,M,V,Y	*8.0-9.2* P,A,T,G,S	*10.4-13.0* H,Q,R,K,N,E,D
Polarizability	*0–1.08* G,A,S,D,T	*0.128-0.186* C,P,N,V,E,Q,I,L	*0.219-0.409* K,M,H,F,R,Y,W
Charge	*Positive* K,R	*Neutral* A,N,C,Q,G,H,I,L,M,F,P,S,T,W,Y, V	*Negative* D,E
Secondary	*Helix*	*Strand*	*Coil*
Structure	E,A,L,M,Q,K,R,H	V,I,Y,C,W,F,T	G,N,P,S,D
Solvent	*Buried*	*Exposed*	*Intermediate*
accessibil-ity	A,L,F,C,G,I,V,W	R,K,Q,E,N,D	M,S P,T,H,Y

2.10.1 Composition

The composition is defined as the global percent for each encoded class in the sequence based on the property p, where p is any of the abovementioned seven properties. There are 21 features (3 features for each one of the 7 physicochemical properties). The composition is calculated as

$$C_{r,p} = \frac{n_r}{n}; \quad (r = 1, 2, 3) \tag{9}$$

where n_r is the number of groups coded with r in the sequence and n is the number of amino acids in the sequence.

2.10.2 Transition

The transition is defined as the frequency with which a code (r) is followed by another code (s). As there are three group codes (e.g., 1, 2, and 3), the possible combinations are (1,2), (1,3), and (2,3). The number of features is 21 (3 for each one of the seven physicochemical properties). The transition probability that the code r is followed by the code s is given as:

$$T_{rs} = \frac{n_{rs} + n_{sr}}{N - 1} \tag{10}$$

where n_{rs} is the number of the sequence (r,s) and n_{sr} is the number of the sequence (s,r) for the above possible combinations.

2.10.3 Distribution

The distribution is defined as the distribution of each encoded group (1, 2, and 3) in the sequence for the first, 25%, 50%, 75%, and 100% of distributions for a particular property. The number of feature elements for the distribution is 105 (15 for each one of the seven physicochemical properties). The residue position is calculated by

$$R = Frequency\ of\ codes\ (1\ or\ 2\ or\ 3) \times D \tag{11}$$

where D is 25%, 50%, 75%, or 100%. The distribution is then calculated by dividing R with the length of the sequence and multiplying by 100.

2.11 Average Cumulative Hydrophobicity

The average cumulative hydrophobicity (ACH) is often used for prediction of PTMs, where windows of sequences are used [69, 70]. ACH is computed by averaging the cumulative hydrophobicity indices around a site in the sliding windows as shown in Table 9.

2.12 High-Quality Physicochemical Indices

High-quality physicochemical indices (HQI) are suitable for extracting features from windows. HQIs are the indices of eight physicochemical properties that represent the eight groups identified by clustering the 544 correlated amino acid properties in the AAindex1 database [26]. Each amino acid in the sequence will have eight features representing the eight high-quality physicochemical property indices. The properties are (1) propensity

Table 9
Example of how ACH is calculated

Size	Window	Cumulative hydrophobicity	Avg
9	KAGVSPHED	−1.15 + 0.62 + 0.48 + 1.08–0.18 + 0.12–0.4-0.74-0.9	−0.12
7	AGVSPHE	0.62 + 0.48 + 1.08–0.18 + 0.12–0.4-0.74	0.14
5	GVSPH	0.48 + 1.08–0.18 + 0.12–0.4	0.22
3	VSP	1.08–0.18 + 0.12	0.34

(BLAM930101) (HQI1), (2) information value for accessibility (BIOV880101) (HQI2), (3) normalized frequency of alpha-helix (MAXF760101) (HQI3), (4) volumes including the crystallographic waters (TSAJ990101) (HQI4), (5) amino acid composition of MEM of multispanning proteins (NAKH920108) (HQI5), (6) composition of amino acids in intracellular proteins (CEDJ970104) (HQI6), (7) conformational preference for all beta-strand (LIFS790101) (HQI7), and (8) optimized relative partition energies (MIYS990104) (HQI8). The number of HQI features for a sequence window of size n is $8 \times (n - 1)$.

2.13 Composition of k-Spaced Amino Acid Pairs

The Composition of k-spaced amino-acid pairs (CKSAAP) descriptor was originally developed by Chen et al. [71]. Essentially, for a given protein sequence all the adjacent pairs of Amino Acids (AAs) (dipeptides) in the sequence are counted. Since there are 400 possible AA pairs (*AA*, *AC*, *AD*, ..., *YY*), a feature vector of that size is used to represent occurrences of these pairs in the window. To accommodate for the short-range interactions between AAs, rather than only interactions between immediately adjacent AAs, CKSAAP also considers k-spaced pairs of AAs, that is, pairs that are separated by k other AAs. For our purpose, we generally use $k = 0, 1, \dots 5$, where for $k = 0$ the pairs reduce to dipeptides. For each value of k, there are 400 corresponding features.

2.14 Autocorrelation

The autocorrelation, which is one of the spatial statistics, is used to detect nonrandomness in time series data for spatial dependency. Rather than the correlation, which measures the relationship between two variables, autocorrelation is used to find the correlation between points in the same variable. The application of the autocorrelation as a protein descriptor arises from the fact that a protein sequence can be considered as a space in which each residue represents a point. Thus, the autocorrelation between any two amino acids on the sequence can be computed based on one of their physicochemical properties. The autocorrelation between two residues can be either positive or negative. The autocorrelation between any two amino acid residues will quantify the level of dependency or association of those two residues on the protein sequence.

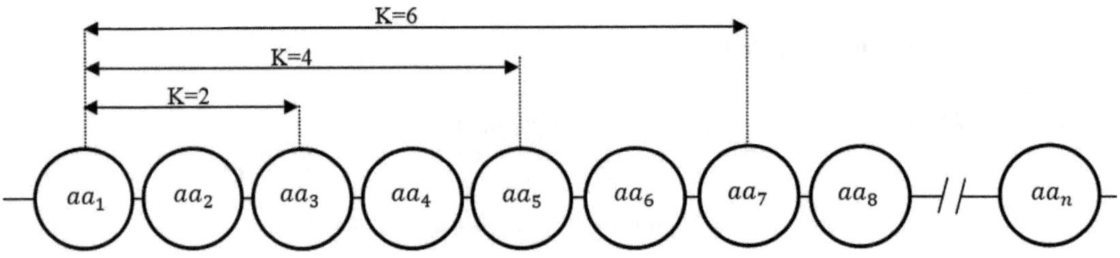

Fig. 1 Illustration depicting the lag between two amino acid residues on a sequence

The following steps are used to compute the autocorrelation features from a protein sequence:

1. Choose a physicochemical property of interest from the Amino Acid Index Database (e.g., hydrophilicity).
2. Standardize the physicochemical property values of the twenty amino acids by subtracting the mean of each value and dividing by the standard deviation of the property.

$$\acute{P}_i = \frac{P_i - \overline{P}}{S} \tag{12}$$

where $\acute{P}_i$ is the standardized value of the property for the amino acid i; the mean is given by $\overline{P} = \frac{\sum_{i=1}^{20} P_i}{20}$; and the standard deviation is calculated as $S = \sqrt{\frac{1}{20} \sum_{j=1}^{20} (P_i - \overline{P})^2}$.

3. Choose a suitable lag (interval between the residues) along the protein sequence as shown in Fig. 1. The lag can be any integer less than the length of the sequence (e.g., $k = 6$).
4. The autocorrelations between the residues of a protein sequence based on the chosen lag are computed according to the following equation.

$$r_k = \frac{\sum_{i=1}^{n-k} \left(\acute{P}_i - \overline{P}\right)\left(\acute{P}_{i+k} - \overline{\acute{P}}\right)}{\sum_{i}^{n} \left(\acute{P}_i - \overline{\acute{P}}\right)^2} \tag{13}$$

where r_k is the autocorrelation coefficient between residue i (aa_i) and residue $i + k$ (aa_{i+k}); n is the length of the protein sequence.

The autocorrelation features are computed by incrementing by $i + k$ until n-k is reached.

The following autocorrelation variants were proposed as protein descriptors.

2.14.1 Moran Autocorrelation

Moran autocorrelation is like the Pearson correlation coefficient in that it ranges between +1 (strong positive autocorrelation) and −1 (strong negative autocorrelation). An autocorrelation coefficient close to zero indicates a weak dependency.

$$r_k = \frac{\frac{1}{n-k}\sum_{i=1}^{n-k}\left(\acute{P}_i - \overline{\acute{P}}\right)\left(\acute{P}_{i+k} - \overline{\acute{P}}\right)}{\frac{1}{n}\sum_{i=1}^{n}\left(\acute{P}_i - \overline{\acute{P}}\right)^2} \quad (14)$$

2.14.2 Geary Autocorrelation

The Geary autocorrelation ranges between 0 and 2, where a value less than 1 and close to zero indicates strong positive autocorrelation while a value greater than 1 and close to 2 indicates strong negative autocorrelation and a value close to 1 indicates very weak autocorrelation.

$$r_k = \frac{\frac{1}{2(n-k)}\sum_{i=1}^{n-k}\left(\acute{P}_i - \acute{P}_{i+k}\right)^2}{\frac{1}{n-1}\sum_{i=1}^{n}\left(\acute{P}_i - \overline{\acute{P}}\right)^2} \quad (15)$$

2.14.3 Moreau–Broto autocorrelation

The Moreau–Broto autocorrelation formula, which does not depend on the Pearson correlation, can be described as follows.

$$r_k = \sum_{i=1}^{n-k}\acute{P}_i\acute{P}_{i+k} \quad (16)$$

where $\acute{P}_i$ and $\acute{P}_{i+k}$ represent the standardized value of the property for the amino acid, i, and is the standardized value of the property for the amino acid, $i + k$, respectively. The normalized version of Moreau–Broto autocorrelation is computed by dividing Eq. 16 by (n-k):

$$r_k = \frac{\sum_{i=1}^{n-k}\acute{P}_i\acute{P}_{i+k}}{n-k} \quad (17)$$

2.15 Pseudo Amino Acid Composition

The pseudo amino acid composition (PseAAC) is another example of a hybrid descriptor that combines both physicochemical properties and amino acid composition and takes the order of the amino acids into account. The first 20 features are the AAC features while the others describe the sequence order features computed using physicochemical properties usually hydrophobicity (H1), hydrophilicity (H2), and side-chain mass (M) of the amino acids. The value of property P of the 20 amino acids must be standardized, as described above.

For example, given a protein sequence S with L amino acid residues,

$$S = A_1A_2A_3A_4A_5 \ldots A_{L.}$$

The features can be extracted as follows.

$$F = (F_1,\ F_2, F_3,\ F_4,\ F_5 \ldots, F_{20}, F_{20+1}, F_{20+2}, \ldots, F_{20+2\lambda}).$$

The first 20 features are AAC features, which are the frequencies of the 20 amino acids in the sequence S and the remaining 2λ features are the correlation factors describing hydrophobicity and hydrophilicity distribution patterns along a protein sequence.

The PseAAC features are calculated using the following steps.

1. Determine which physicochemical properties to use and standardize their values as shown in Eq. 5. The proposed properties are hydrophobicity (H1), hydrophilicity (H2), and side-chain mass (M) but can be any other properties and any number.
2. Determine the maximum number of tiers ($\lambda < n$). Correlation factors are given by coupling the most contiguous residues whose contiguity condition varies according to the considered tier. Figure 2 is a schematic showing the first-tier, second-tier, and third-tier sequence order correlation modes along a protein sequence. For the first tier, the correlation will be between the direct adjacent residues; for the second tier, the correlation will be between residues with 2 lags in between; for the third tier, the correlation will be between the residues with 3 lags in between; and so forth until the maximum number of tiers is

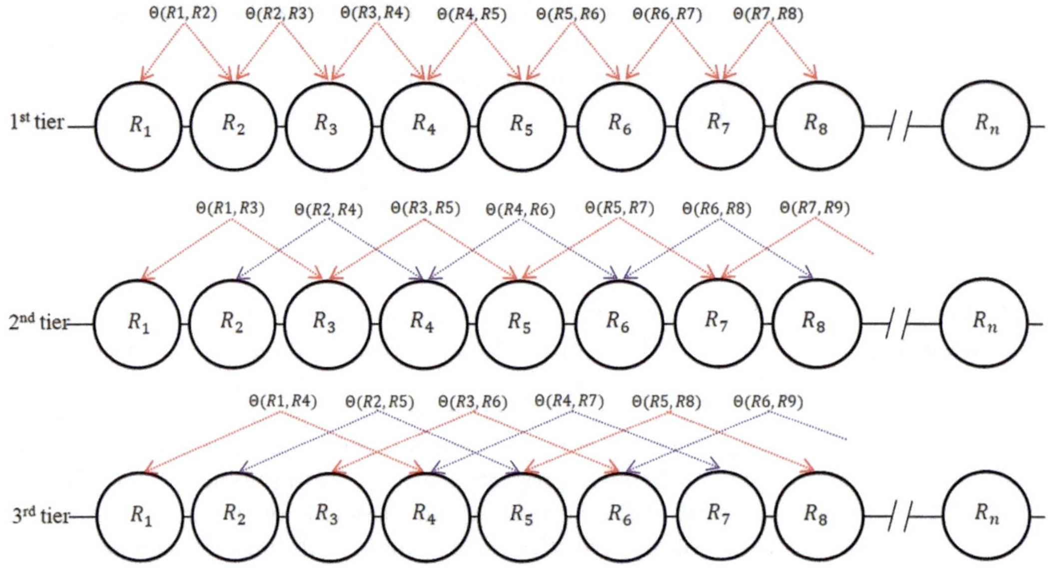

Fig. 2 First-tier, second-tier, and third-tier order correlation

reached. The maximum number of tiers is denoted by λ, which is a parameter to be specified and it must be less than the length of the protein sequence ($\lambda < n$).

3. The function for the correlation between any two residues is defined as an average for the set of the amino acid properties. The following function describes the three amino acid property values, hydrophobicity, hydrophilicity, and side chain mass.

$$\Theta(R_i, R_j) = \frac{1}{3} \times \left[[H_1(R_i) - H_1(R_i)]^2 + [H_2(R_i) - H_2(R_i)]^2 + [M(R_i) - M(R_i)]^2 \right] \tag{18}$$

The above definition of the correlation function can be extended to a set of n amino acid properties as follows.

$$\Theta(R_i, R_j) = \frac{1}{n} \sum_{k=1}^{n} [H_k(R_i) - H_k(R_i)]^2 \tag{19}$$

where $H_k(R_i)$ is the k-th property in the amino acid property set for amino acid R_i.

Thus, if we need to use only one physicochemical property, we can use the following.

$$\Theta(R_i, R_j) = [H_1(R_i) - H_1(R_i)]^2 \tag{20}$$

4. Next, compute the sequence order-correlated factor (θ) for each tier as follows.

$$\theta_k = \frac{1}{N-k} \sum_{i=1}^{N-k} \Theta(R_i, R_{i+k}) \tag{21}$$

where k is the tier order ($k = 1, 2, 3, \ldots, \lambda$).

For example, for sequence order correlated factor for the first, second, third, and λth tier are as follows.

$$\theta_1 = \frac{1}{N-1} \sum_{i=1}^{N-1} \Theta(R_i, R_{i+1})$$

$$\theta_2 = \frac{1}{N-2} \sum_{i=1}^{N-2} \Theta(R_i, R_{i+2})$$

$$\theta_3 = \frac{1}{N-3} \sum_{i=1}^{N-3} \Theta(R_i, R_{i+3})$$

$$\theta_\lambda = \frac{1}{N-\lambda} \sum_{i=1}^{N-\lambda} \Theta(R_i, R_{i+\lambda})$$

5. Finally, compute the PseAAC features. Let f_a be the normalized frequency of an amino acid in the protein sequence, where $(1 < a < 20)$. There will be a set of 20+ λ PseAAC features that can be defined as follows.

$$X_a = \frac{f_a}{\sum_{r=1}^{20} f_r + \omega \sum_{j=1}^{\lambda} \theta_j}; \quad (1 < a < 20) \tag{22}$$

$$X_a = \frac{\omega \theta_{a-20}}{\sum_{r=1}^{20} f_r + \omega \sum_{j=1}^{\lambda} \theta_j}; \quad (21 < a < 20 + \lambda) \tag{23}$$

where ω is the weighting factor for the sequence-order effect and is usually set to $\omega = 0.05$, as suggested by Kuo-Chen Chou.

2.16 Amphiphilic Pseudo Amino Acid Composition

Amphiphilic pseudo amino acid composition (APseAAC), also called type II pseudo amino acid composition, is similar to the PseAAC descriptors. In APseAAC, the correlation functions are defined using only hydrophobicity $H_1(i)$ and hydrophilicity $H_2(j)$. The APseAAC features are computed as follows.

1. Normalize the values of hydrophobicity $H_1(i)$ and hydrophilicity $H_2(j)$.
2. The correlation functions are defined as follows.

$$H^1_{i,j} = H_1(i)\, H_1(i) \tag{24}$$

$$H^2_{i,j} = H_2(i)\, H_2(i) \tag{25}$$

3. From the above functions and with the aid of Fig. 3, which shows the first and second ranks for the APseAAC correlation modes, compute the sequence order factors as follows.

$$\tau_1 = \frac{1}{N-1} \sum_{i=1}^{N-1} H^1_{i,i+1} \tag{26}$$

$$\tau_2 = \frac{1}{N-1} \sum_{i=1}^{N-1} H^2_{i,i+1} \tag{27}$$

$$\tau_3 = \frac{1}{N-1} \sum_{i=1}^{N-2} H^1_{i,i+2} \tag{28}$$

$$\tau_4 = \frac{1}{N-1} \sum_{i=1}^{N-2} H^2_{i,i+2} \tag{29}$$

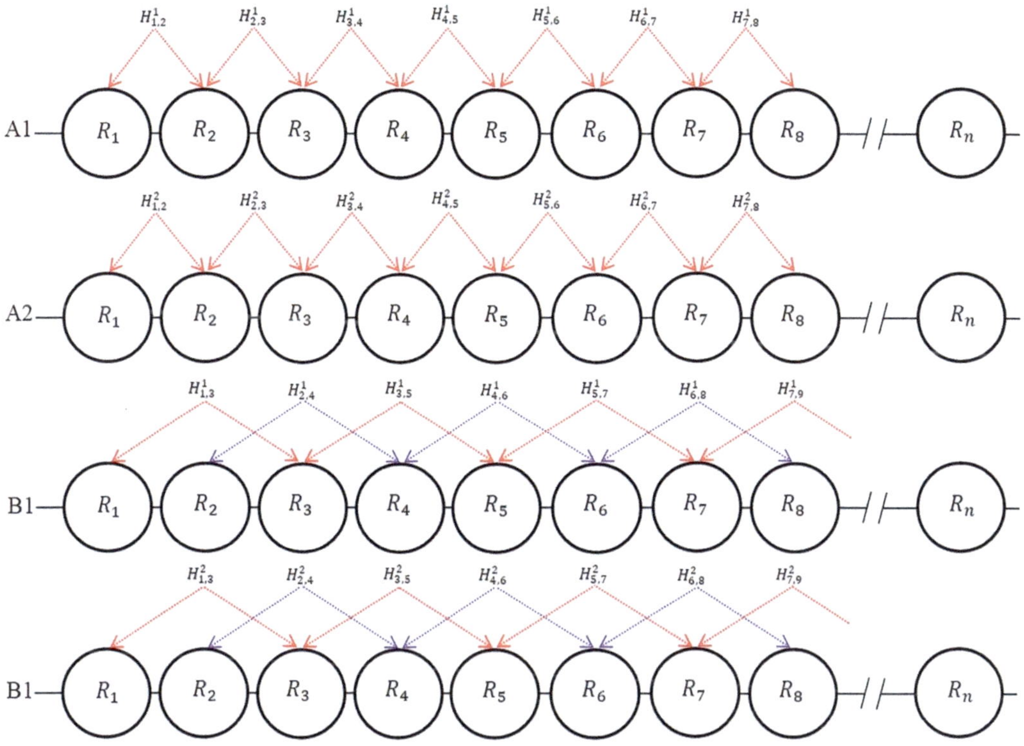

Fig. 3 Schematic diagram showing first rank (A1 and A2) and second rank (B1 and B2)

$$\tau_{2\lambda-1} = \frac{1}{N-1} \sum_{i=1}^{N-\lambda} H^1_{i,i+\lambda} \quad (30)$$

$$\tau_{2\lambda} = \frac{1}{N-1} \sum_{i=1}^{N-\lambda} H^2_{i,i+\lambda} \quad (31)$$

4. Finally, compute the APseAAC) features using the following formulas.

$$X_a = \frac{f_a}{\sum_{r=1}^{20} f_r + \omega \sum_{j=1}^{2\lambda} \tau_j}; \quad (1 < a < 20) \quad (32)$$

$$X_a = \frac{\omega \tau_u}{\sum_{r=1}^{20} f_r + \omega \sum_{j=1}^{2\lambda} \tau_j}; \quad (21 < a < 20 + \lambda) \quad (33)$$

where ω is the weighting factor for the sequence-order effect and is usually set to $\omega = 0.05$, as suggested by Kuo-Chen Chou.

2.17 Profile-Based Derived by Position-Specific Scoring Matrix

A position-specific scoring matrix (PSSM) was originally used by the NCBI position-specific iterated BLAST (PSI-BLAST) algorithm to detect distantly related proteins [72]. A PSSM represents a protein sequence as a matrix whose elements are the scores describing the conservation of amino acids in a protein sequence. Thus, the first step for extracting features from a PSSM is to transform each protein sequence into a PSSM by utilizing the NCBI PSI-BLAST utility program. There are over 20 PSSM-based descriptors that can be used with machine learning [73]. However, all PSSM-based features are transformed from the PSSM score.

The PSI-BLAST utility program is available to download from the NCBI website. A FASTA-formatted protein sequence, which is needed to transform in a PSSM, is used as a query for the PSI-BLAST program that uses the NCBI BLAST to search iteratively for evolutionarily related sequences on the nonredundant (nr) protein database. The hits of the related sequences are then aligned using multiple sequence alignment. The probabilities of the position-specific substitution are computed for each amino acid in the protein sequence, creating a $L \times 20$ PSSM matrix, where L is the protein sequence length. Figure 4 shows the PSI-BLAST process and the final PSSM for the sample sequence (KMNKQMKNLNM).

Once a PSSM is generated for a given sequence, there are numerous feature extraction algorithms available to generate a spectrum of feature extractions. We will discuss two methods as examples: PSSM composition and PSSM auto covariance transformation.

2.17.1 PSSM Composition

The PSSM is an $L \times 20$ matrix, as shown in Fig. 5. The column headers of the PSSM are the 20 amino acids while the row headers are the amino acid residues of the protein sequence. The elements of the PSSM are position-specific substitution probabilities, but they are transformed by logarithmic transformation. The PSSM composition [38, 74] is computed by summing the amino acid rows in the PSSM. The red arrow in Fig. 5 indicates the direction of summing. The rows of each amino acid, as indicated by the same color, are summed together, generating a vector of length 20. Therefore, there are 400 PSSM composition features (20 features for each amino acid type of the protein sequence).

2.17.2 PSSM Autocovariance Transformation

Autocovariance transformation is similar to autocorrelation in that it measures the correlation of a property along the protein sequence. The PSSM autocovariance transformation [38, 74] transforms a PSSM into a fixed-length vector. For instance, let us represent the PSSM as follows.

$$PSSM = (C_1, C_2, C_3, \cdots, C_{20})$$

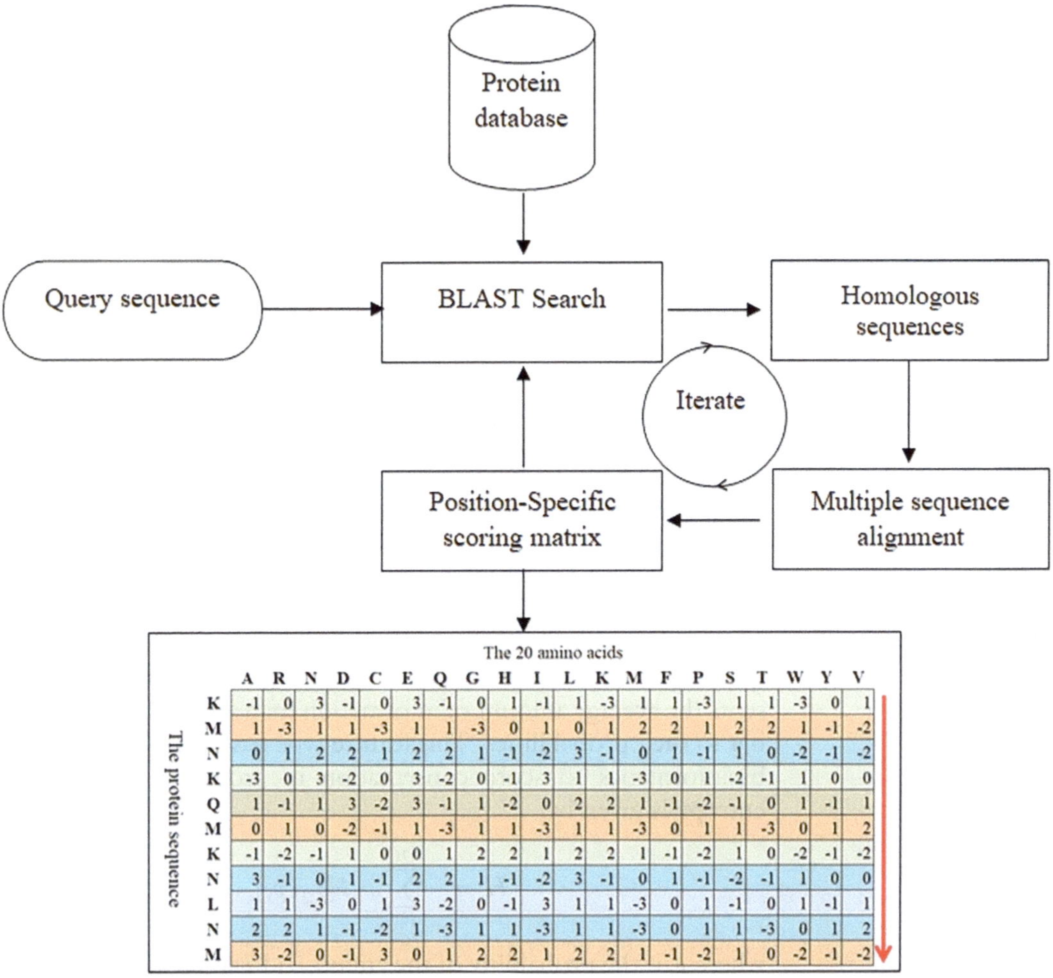

Fig. 4 The use of Basic Local Alignment Search Tool (BLAST) to generate the Position-specific scoring matrix (PSSM)

The 20 amino acids

The protein sequence	A	R	N	D	C	E	Q	G	H	I	L	K	M	F	P	S	T	W	Y	V
K	-1	0	3	-1	0	3	-1	0	1	-1	1	-3	1	1	-3	1	1	-3	0	1
M	1	-3	1	1	-3	1	1	-3	0	1	0	1	2	2	1	2	2	1	-1	-2
N	0	1	2	2	1	2	2	1	-1	-2	3	-1	0	1	-1	1	0	-2	-1	-2
K	-3	0	3	-2	0	3	-2	0	-1	3	1	1	-3	0	1	-2	-1	1	0	0
Q	1	-1	1	3	-2	3	-1	1	-2	0	2	2	1	-1	-2	-1	0	1	-1	1
M	0	1	0	-2	-1	1	-3	1	1	-3	1	1	-3	0	1	1	-3	0	1	2
K	-1	-2	-1	1	0	0	1	2	2	1	2	2	1	-1	-2	1	0	-2	-1	-2
N	3	-1	0	1	-1	2	2	1	-1	-2	3	-1	0	1	-1	-2	-1	1	0	0
L	1	1	-3	0	1	3	-2	0	-1	3	1	1	-3	0	1	-1	0	1	-1	1
N	2	2	1	-1	-2	1	-3	1	1	-3	1	1	-3	0	1	1	-3	0	1	2
M	3	-2	0	-1	3	0	1	2	2	1	2	2	1	-1	-2	1	0	-2	-1	-2

Fig. 5 A position-specific scoring matrix (PSSM) for the sequence KMNKQMKNLNM

where C_j (i = 1, 2, 3, ..., 20) are the columns of the PSSM.

Let us also denote a column C_i as follows.

$$C_j = (c_{1,j}\ , c_{2,j}\ , c_{3,j}\ , \cdots, c_{L,j}); \text{where } (1 < j < 20)$$

where L is the length of the protein sequence and $c_{i,\ j}$ is the PSSM score in the position i corresponding to amino acid j. The PSSM autocovariance is then defined as follows.

$$PSSM - AC_{\ j,g} = \frac{1}{L-g} \sum_{i=1}^{L-g} \left(c_{i,j} - \frac{1}{L} \sum_{i=1}^{L} c_{i,j} \right) \times \left(c_{i+g,j} - \frac{1}{L} \sum_{i=1}^{L} c_{i,j} \right) \quad (34)$$

where g = 1, 2, ..., G and G is a positive number indicating the transformation parameter. The PSSM autocovariance transformation generated 20 × G features.

2.18 Shannon Entropy

Shannon entropy (H) is known in information theory as a measure of randomness and diversity of a set of objects distributed into a space. It was defined by Shannon as a unique function that represents the average amount of information for a set of objects according to their probabilities. This feature type has been widely used in bioinformatics to score conservation in protein. The Shannon entropy is given by

$$H = -\sum_{i=1}^{20} p_i \, log_2(p_i) \quad (35)$$

where p_i is the probability of the amino acid i and it is computed as the total number of amino acids i divided by the length of the sequence. Entropy ranges between zero, where only one type of residue in the entire sequence is found, and 4.3219, where all types of amino acids have equal occurrence in the sequence.

2.19 Relative Entropy

The relative entropy (RE) of two distributions p_i and p_0, is also known as the Kullback–Leibler distance, which is defined as the sum of entropies over 20 positions. The p_0 is the background distribution that takes place in a random state where all residues have equal chance of *occurrence* (1/20), *indicating that the sequence was not affected by a* constraint. The RE was used in previous studies to identify the conserved position and to find a motif among the patterns composed from frequent residues.

$$RE = \sum_{i=1}^{20} p_i \, log_2\left(\frac{p_i}{p_0}\right) \quad (36)$$

RE is always nonnegative and becomes zero if and only if $p_i = p_0$. The RE is represented by one feature for a protein sequence.

2.20 Information Gain

The information gain (IG) is computed by subtracting *the relative entropy (RE)* from entropy (H). It measures the transform of information pattern from the background state to the state due to *a consequence of* constraint such as homology.

$$IG = H - RE \tag{37}$$

3 Current Feature Extraction Tools from Protein Sequence/Structure

Over the years, researchers have developed several feature encoding methods and descriptors for protein sequences (and structures). In addition, since feature extraction requires programming and mathematical skills, to facilitate the development of ML-based approaches, various stand-alone, as well as web servers have been developed to facilitate feature extraction for researchers. These tools include PROFEAT [66], Psuedoamino Acid Composition (PseAAC) [75], PseAAC-Builder [76], propy [65], PseAAC-General [77], protr/ProtrWeb [78], Rcpi [12], PseKRAAC [79], iFeature [80], and Seq2Feature [81] (Tables 10 and 11). However, in addition to extracting these features, the ability to integrate (combine) multiple features together so that they be readily input to the machine learning apparatus and extract features from a large number of proteins sequences become crucial. To the best of our knowledge, the existing tools have limitations such as the number of protein sequences used as an input for feature extraction, the lack of flexibility to integrate (combine) different numbers of descriptors, and a large amount of time taken for feature extraction which

Table 10
Several feature extraction tools reviewed in this chapter. WS stands for web server, NA for not available, and SA for stand-alone

Method	Year	Availability	References
PROFEAT	2006	WS	[66]
PseAAC	2008	NA	[75]
PseAAC-Builder	2012	SA	[76]
Propy	2013	SA	[65]
PseAAC-General	2014	SA	[77]
protr/ProtrWeb	2015	SA/WS	[78]
Rcpi	2015	SA	[12]
PseKRAAC	2017	WS	[79]
iFeature	2018	SA/WS	[80]
Seq2Feature	2019	SA/WS	[81]

Table 11
Comparison between the existing feature extraction tools and FEPS in regard to the types of features that can be extracted (There were several features shown in this table with functional properties: Y for "YES" and N for "NO")

Types	PseKRAAC	iFeature	Seq2Feature	FEPS
Physicochemical Properties	4	544	16	544
Type I PseAAC	N	Y	N	Y
Type II PseAAC	N	Y	N	Y
Amino acid composition	N	Y	Y	Y
di-Peptide composition	N	Y	N	Y
tri-Peptide composition	N	Y	N	Y
Normalized Moreau–Broto autocorrelation	N	Y	N	Y
Moran autocorrelation	N	Y	N	Y
Geary autocorrelation	N	Y	N	Y
Composition–Transition–Distribution (CTD)	N	Y	N	Y
Quasi-sequence order	N	Y	N	Y
PSSM	N	N	N	Y
k-spaced amino acid pairs (KSAAP)	N	N	N	Y
PSSM	N	Y	N	N
BLOSUM62	N	Y	N	N
Z-scale	N	Y	N	N
Secondary structure types	N	Y	N	N
Disorder types	N	Y	N	N
Accessible surface area	N	Y	N	N
main-chain torsional angles	N	Y	N	N
Pseudo *K*-tuple reduced amino acids composition	Y	Y	N	N
Average property	N	N	Y	N
Change in properties upon amino acid substitutions	N	N	Y	N
Substitution matrices	N	N	Y	N
Pairwise contact potentials	N	N	Y	N
User-defined	Y	Y	Y	Y
Graphical User Interface (GUI)	Y	Y	Y	Y
Online updates	N	N	N	Y

renders these tools not to be widely used by biologists. To overcome some of these limitations and to provide a tool for fast, integrated feature extraction and readily usable to biologists from protein sequences, we developed the Feature Extraction from Protein Sequences (FEPS) [82], which is capable to handle a large number of sequence (limited by the computational resources only) and it provides ML-ready data sets with both features and labels. FEPS is available both as a web server and a stand-alone program designed to quickly extract many features from input protein sequence data. To promote flexibility in model development, FEPS can be customized based on end-user preferences to select a specific descriptor. Likewise, it accepts an unlimited number of sequences as input and can integrate all features (limited only by the computational resources). FEPS can be used for generating a dataset of protein features ready to be passed as an input to machine learning apparatus. First, we briefly review existing tools/packages for extracting features. We then discuss various types of numerical descriptors (features). Finally, we discuss in detail about the new FEPS toolkit.

In the following, we briefly summarize some of the existing tools for extracting features from protein sequence/structure.

PROFEAT [66] is a web server developed by Chen et al. [66]. This method is used to extract structural and physicochemical features. It can extract 51 types of features, leading to a total length of 1447.

PseAAC [75] is a computational method established by Shen et al.. This method was developed to generate different types of the pseudo amino acids (PseAA). The web server allows the user to specify any of 63 correlation types of PseAA. The lengths of the descriptor depend on the value of lambda, which is used to show correlation along the protein sequences.

PseAAC-Builder [76], which is a stand-alone software developed by Du et al., is very similar to PseAAC [75] in that it is designed to generate distinctive varieties of a PseAAs. One of the primary advantages of PseAAC-Builder over the original PseAAC method is that it allows more than 500 protein sequences as input. However, this also increases the time needed for generating the output files. The stand-alone software allows users to choose any of 544 physicochemical properties with massive groupings to mine correlation types of PseAAs from sequences.

Propy [65] is a stand-alone software package developed by Liang et al. It was originally used to calculate five groups of features. These features represent either physiochemical, structural properties of proteins or peptide fragments. The total length of features using default parameters is 9487.

PseAAC-General [77] is a stand-alone software package developed by Du et al. This method is an updated version of the PseAAC-Builder [76] method that extended features, such as gene ontology mode, functional domain mode, and sequential evolution mode. The computational efficiency of this method is also faster than that of PseAAC-Builder [77].

Protr and ProtrWeb [78] are stand-alone and web-server versions, respectively, of the feature extraction tool developed by Xiao et al. This method can be used to calculate eight groups of features. These groups express and represent twenty-two descriptors. Both the stand-alone software package and web server allow users to choose any of 544 physicochemical properties and to select several specific features.

Rcpi [12] is a stand-alone software package developed by Cao et al. It is designed to extract both physicochemical and structural descriptors. It can also mine biological sequences and interaction descriptors such as protein–protein interaction and compound–protein interactions [12].

PseKRAAC [79] is a web server developed by Zuo et al. that uses different types of pseudo-K-tuples to reduce amino acid composition. One of the primary advantages of this method is that it reduced the computational time and escaped from the overfitting problems. The web server allows the user to specify the parameter and the number of selected types for sixteen groups of reduced amino acids [79].

iFeature [80] is a stand-alone software and web server developed by Chen et al. iFeature can be used to calculate fifty-three groups of features. In addition, iFeature also consists of twelve varieties of features via selection, clustering, and reduction algorithms. Both the stand-alone software and the web server allow users to choose any of the physiochemical properties, with the flexibility to select several specific features [80].

Seq2Feature [81] is a stand-alone software package and web server developed by Nikam et al. This method was used to calculate 252 descriptors for proteins and 41 descriptors for DNA. The primary features used for protein/DNA sequences are physicochemical, energetic, and conformational properties, contact potentials properties, and mutation matrices [81]. Table 10 summarizes the feature extraction tools. A comparison of FEPS and other existing feature extraction tools is provided in Table 11.

4 Description of Feature Extraction from Protein Sequences (FEPS) Toolkit

The Feature Extraction from Protein Sequences (FEPS) server is available at http://www.hamiddi.com/feps/feps/. A link (Fig. 6) is provided on the main FEPS page to download example FASTA protein sequences of five groups (Group1.fasta, Group2.fasta, Group3.fasta, Group4.fasta, and Group4.fasta). The following are the steps to generate features from protein sequences using FEPS.

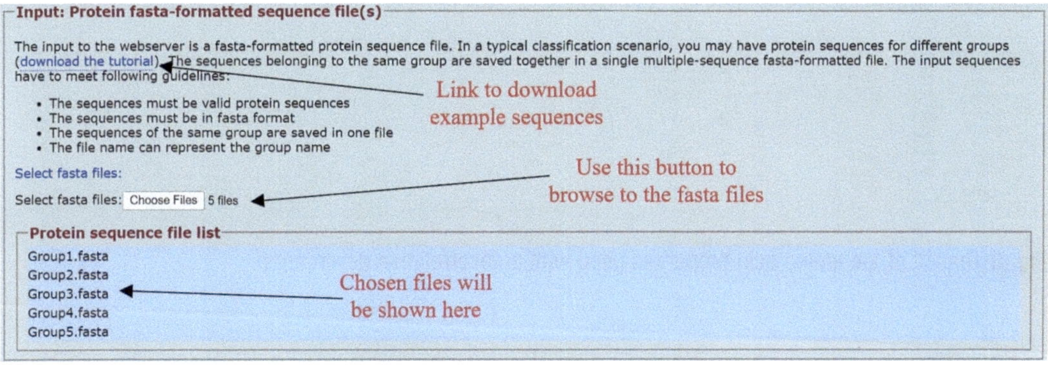

Fig. 6 Choosing the FASTA-formatted protein sequence files for uploading

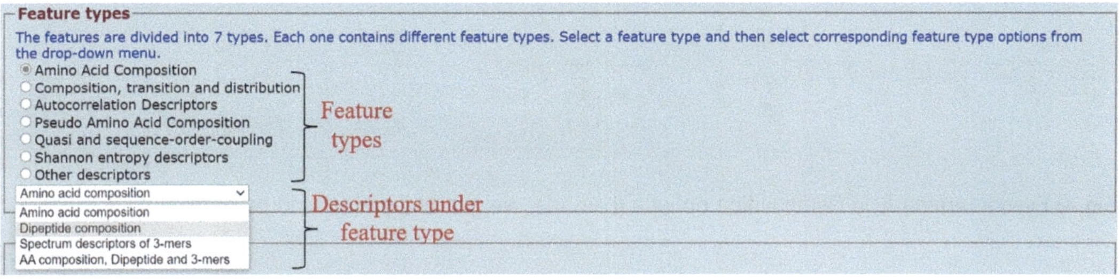

Fig. 7 Selecting one of the feature types and a protein descriptor

4.1 Choosing the Input Fasta Files

In a typical case users may need features either for supervised machine learning or unsupervised machine learning. For the supervised machine learning, the sequences are provided as on the above example sequence files; each file contains the protein sequences (in FASTA format) of a classified group. The file name must be unique, with no white space, and it is recommended to be short and informative as it will be used for coding the class label of the sequences. For the unsupervised machine, the FASTA protein sequences can be in a single file.

4.2 Selecting the Feature Type and Options

After choosing input fasta files, a feature type can be selected from the list by checking the right radio button (Fig. 7). The features are classified into seven types: Amino Acid Composition, Composition, transition and distribution, Autocorrelation Descriptors, Pseudo Amino Acid Composition, Quasi and sequence-order-coupling, Shannon entropy descriptors, and other descriptors. Each feature type may include several feature descriptors. When a type is selected, the dropdown list is populated by a list of the descriptors under that feature type.

The current number of feature choices provided by FEPS is 48. However, three of the autocorrelation descriptors (those with amino acid properties) can be computed with one of eight amino properties: hydrophobicity, average flexibility, polarizability, free

Fig. 8 The list of the amino acid properties used with autocorrelation descriptors

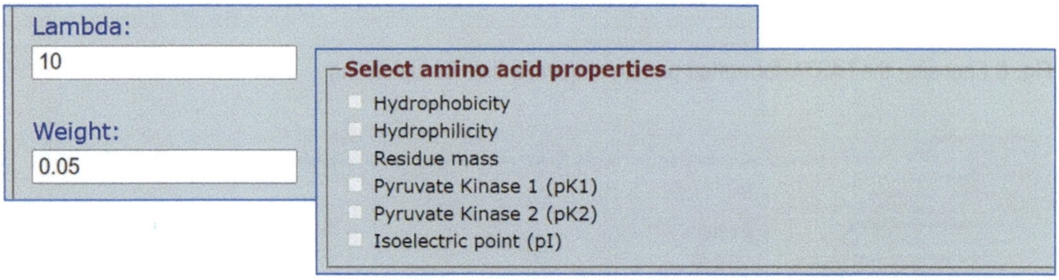

Fig. 9 Pseudo Amino Acid Composition options (Lambda, weight, and amino acid properties)

energy, residue ASA (Accessible Surface Area), steric property, and mutability (Fig. 8). The other autocorrelation descriptors are computed with the eight amino acid properties altogether.

All pseudo amino acid composition (PsAAC) descriptors require lambda and weight options (Fig. 9). Those parameters can either be adjusted or left unchanged by accepting the default values. Three of PsAAC descriptors (those with single or multiple amino acid properties) also require one or more of the amino acid properties (Fig. 9). When selecting more than two amino acid properties, the computation of the feature may take a longer time.

From PsAAC descriptors, only "Type I Pseudo amino acid composition with physicochemical property ID" requires selecting one of the 544 amino acid physicochemical properties (Fig. 10).

The Quasi and sequence-order-coupling descriptors require using a suitable maximum lag. Moreover, those with "select a matrix" require selecting one of the two distance matrices: Grantham distance matrix or Schneider-Wrede distance matrix (Fig. 11).

4.3 The Feature File Format

The available feature file formats include comma-separated value (CSV), SVM-light file, Weka format, and tab-delimited text format (Fig. 12).

4.4 Downloading Feature and Description Files

After choosing the protein fasta sequence files and selecting a feature type and a descriptor, and adjusting the required parameters, the request can be submitted. When the features have been computed, the output page will be displayed showing the download links and feature statistics (Fig. 13).

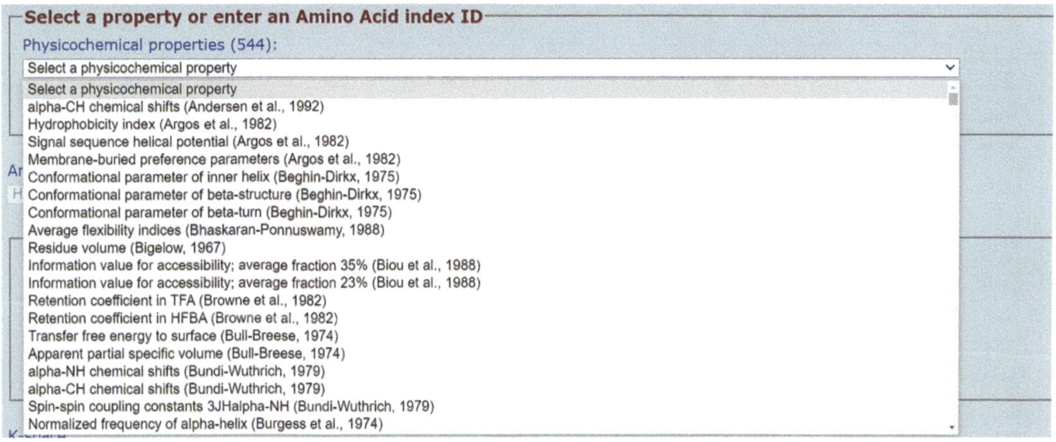

Fig. 10 The list of the 544 physicochemical properties (AAIndex)

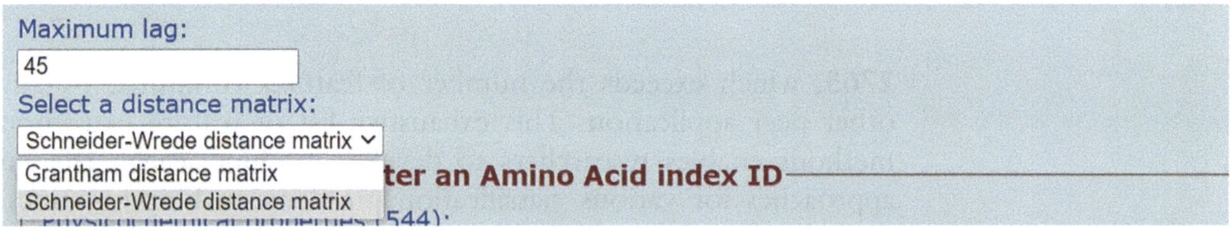

Fig. 11 Quasi and sequence-order-coupling descriptor options (Maximum lag and distance matrices)

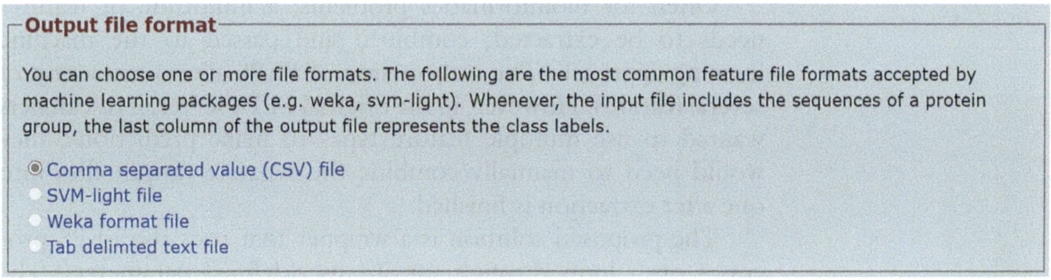

Fig. 12 The list of the feature file formats that can be generated by FEPS

5 EPS_CFS: A Wrapper in Python for FEPS

FEPS uses published feature extraction methods of proteins from single or multiple-FASTA formatted files. In addition, FEPS also provides users with the ability to redefine some of the features by choosing one of the 544 physicochemical properties or to enter any set of user-defined amino acid indices, thereby increasing feature choices. The FEPS server includes 48 published feature extraction methods, six of which can use any of the 544 physicochemical properties. The total number of features calculated by FEPS is

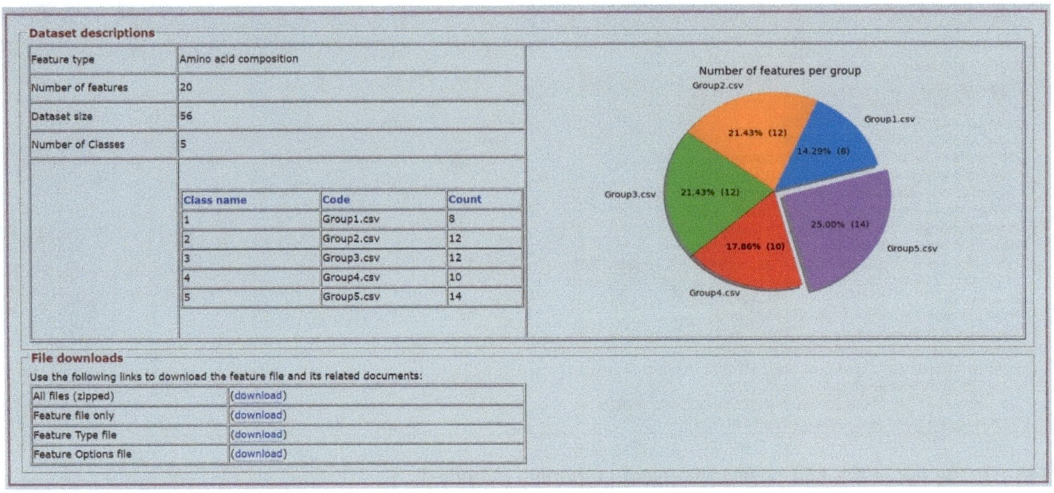

Fig. 13 The feature download page (Feature file, feature type file, and a file for the options used)

2765, which exceeds the number of features computed by any other peer application. This exhaustive list of feature extraction methods enables researchers to develop machine learning-based approaches for various classification problems in bioinformatics. FEPS has been successfully applied for the prediction and classification of nuclear receptors [83], prediction of phosphorylation sites [29], and prediction of hydroxylation sites [19].

Often for bioinformatics problems, a multitude of features needs to be extracted, combined and passed to the machine learning approach. The web version of FEPS allows extraction of several features. However, in the original FEPS server, if researchers wanted to use multiple feature types to make predictions, they would need to manually combine the separate output files into one after extraction is finished.

The proposed solution is a wrapper that runs the FEPS processes on a loop through various user-defined parameters. The results in the ability to extract all features programmed into FEPS and place those features in a single feature vector file. For this, we developed the FEPS_CFS (FEPS Complete Feature Set) a python-based wrapper for FEPS toolkit. In this wrapper, we use FEPS code previously developed by H. Ismail and colleagues in the KC lab [83]. To make it easier for researchers to use multiple feature types during the development of prediction algorithms, the developed wrapper allows concurrent runs of FEPS and places the output into a single CSV file. This gives users the ability to perform various analyses (including feature selection) over the extracted entire FEPS feature set. Importantly, this allows users to determine feature importance using the entire features set, which eliminates the need to try different feature sets to address a particular problem.

5.1 Protocol for Feature Extraction Using FEPS_CFS wrapper

1. FEPS_CFS can be retrieved from https://github.com/dukkakc/FEPS_CFS.
2. Once the download is complete, open a terminal and navigate to the repos root directory and run the following command.

```
$ python setup.py bdist_wheel
```

3. Then run the following command.

```
$ pip install dist/FEPS-0.1.0-py3-none-any.whl
```

In this case, the wheel file is in *dist* directory, essentially do

```
pip install /path/to/wheelfile.whl
```

4. To make things easier, the input file (infile/Group1.fasta), outfile (output directory) along with a sample script file called run_feps.py is provided. To extract features for a sample file provided, we just need to run the following command in the repos root directory.

```
python run_feps.py
```

This script will generate the features for sequence file that is in infile to the outfile. The main output file will be *.fasta_full_features.csv that has all the extracted features.

USAGE:

Once the FEPS library is installed, you can now import FEPS as any other library by opening a Python terminal or file and import it using the following.

1. import FEPS.FEPS_CFS as feps,

   Next, two new directories are created: One for input (e.g., 'infile') and another for output (e.g., 'outfile') (if already no there) and place the input fasta files inside the 'infile' folder (outfile will contain the output of FEPS).
2. After this step, the FEPS_CFS can be run using the following command:

```
$feps.feps('infile/','outfile/')#update in and out file
here
```

5.2 FEPS_CFS Details

Table 12 shows the various features contained in FEPS_CFS. The first feature k-spaced Amino Acid Pairs is set using the tag "AAP" and requires the user to set either an integer value or a range of

Table 12
Features included in FEPS_CFS along with position (when using the "ALL" flag), # of features in each set, options/subfeatures, and references

Feature	Vector Position	# Features	Options	Reference
k-spaced Amino Acid Pairs (CKSAAP)	1–2400	2400	$k = 0,1,2,3,4,5$	[71]
Conjoint Triad (CT)	2401–2912	512		[84]
Entropy (ENT)	2913–2915	3	Entropy Relative entropy Gain	[85]
Composition, Transition, and Distribution	2916–3062	147	Composition descriptors based on charge of Amino Acid Descriptors (AADs) Composition descriptors based on hydrophobicity of AADs Composition descriptors based on normalized Van der Waals Volume (VDWV) of AADs Composition descriptors based on polarity of AADs Composition descriptors based on polarizability of AADs Composition descriptors based on secondary structure of AADs Composition descriptors based on solvent accessibility of AADs Distribution descriptors based on charge of AADs Distribution descriptors based on hydrophobicity of AADs Distribution descriptors based on normalized VDWV of AADs Distribution descriptors based on polarity of AADs Distribution descriptors based on polarizability of AADs Distribution descriptors based on secondary structure of AADs Distribution descriptors based on solvent accessibility of AADs Transition descriptors based on charge of AADs Transition descriptors based on hydrophobicity of AADs Transition descriptors based on normalized VDWV of AADs Transition descriptors based on polarity of AADs	[65]

(continued)

Table 12 (continued)

Feature	Vector Position	# Features	Options	Reference
			Transition descriptors based on polarizability of AADs Transition descriptors based on 2ndary structure of AADs Transition descriptors based on solvent accessibility of AADs	
Amino Acid Composition (AAC)	3063–11,462	8400	Dipeptide composition Spectrum descriptors of 3-mers	[65]
Auto Correlation Descriptors (ACD)	11,463–12,183	721	Geary autocorrelation Moran autocorrelation descriptors Normalized Moreau–Broto autocorrelation	[65]
Psuedo Amino Acid Composition (PseAAC)	12,184–12,229	46	Type I pseudo amino acid composition Type II pseudo amino acid composition	[65]
Sequence Order Coupling	12,230–12,551	322	Quasi sequence order descriptors Quasi sequence order descriptors (matrix) Sequence order coupling numbers Sequence order coupling numbers (matrix) Sequence order correlation factor (Hydrophobic and Hydrophilic)	(65)

values for the variable, *k*. The second feature, Conjoint Triad, is set using the tag "CT". The third feature is Entropy and is set using the tag "ENT". This requires the user to provide the options for entropy, relative entropy, and/or information gain, which is the difference between the entropy and relative entropy. The next feature is CTD, which is set using the tag "AAD" (for Amino Acid Descriptors), of which the user can select the various descriptors provided (see table below). The next feature set, Amino Acid Composition, is set using the tag "AAC" and requires the user to provide the options for Amino Acid Composition, Dipeptide Composition, and/or Spectrum of 3-mers. The next feature is Auto Correlation Descriptors and can be selected with the tag "ACD". This requires the user to select options for Geary autocorrelation, Moran–Broto autocorrelation, and/or Normalized Moran–Broto autocorrelation. The next feature is psuedo amino acid composition, which can be set using the tag "PAAC". The PAAC setting requires the user to provide options for Type I and/or Type II PAAC. The last feature set is Sequence Order Coupling, which can be set using the tag "SOC". This requires the user to provide options for quasi-sequence order descriptors, quasi-sequence order descriptors (matrix), sequence order coupling numbers, sequence order coupling numbers (matrix), and/or sequence order correlation factor (Hydrophobic and Hydrophilic). Table 12 lists the features included in FEPS_CFS along with position.

6 Conclusion

Features are one of the most critical components of machine learning-based approaches. There are several tools available online for feature extraction from protein sequences. Most of the existing methods have limitations as they do not provide ML-ready data set and also the number of descriptors is limited. Feature Extraction from Protein Sequences (FEPS) was developed to overcome those limitations and to provide most of the published feature descriptors. The workflow of FEPS is simple and straightforward, consisting of choosing sequence data files (in FASTA format), selecting a feature type and a descriptor, and adjusting options if required. The features are can be generated in CSV, SVM-light, WEKA, and comma delimited file format. Additionally, we also provide a python-based wrapper (FEPS_CFS) that allows a seamless combination of several types of descriptors.

References

1. Larranaga P, Calvo B, Santana R, Bielza C, Galdiano J, Inza I, Lozano JA, Armananzas R, Santafé G, Pérez A (2006) Machine learning in bioinformatics. Brief Bioinform 7(1):86–112
2. Libbrecht MW, Noble WS (2015) Machine learning applications in genetics and genomics. Nat Rev Genet 16(6):321–332
3. Kim JH, Lee J, Oh B, Kimm K, Koh I (2004) Prediction of phosphorylation sites using SVMs. Bioinformatics 20(17):3179–3184
4. Xie Y, Luo X, Li Y, Chen L, Ma W, Huang J, Cui J, Zhao Y, Xue Y, Zuo Z (2018) DeepNitro: prediction of protein nitration and nitrosylation sites by deep learning. Genomics Proteomics Bioinformatics 16(4):294–306
5. Yu B, Yu Z, Chen C, Ma A, Liu B, Tian B, Ma Q (2020) DNNAce: prediction of prokaryote lysine acetylation sites through deep neural networks with multi-information fusion. Chemom Intell Lab Syst 200:103999
6. Ismail HD, Jones A, Kim JH, Newman RH, Kc DB (2016) RF-Phos: a novel general Phosphorylation site prediction tool based on random Forest. In: BioMed research international 2016
7. AL-barakati HJ, EW MC, Hicks LM, Poole LB, Newman RH (2018) SVM-SulfoSite: A support vector machine based predictor for sulfenylation sites. Sci Rep 8(1):11288
8. Ismail HD, Newman RH (2016) RF-Hydroxysite: a random forest based predictor for hydroxylation sites. Mol BioSyst 12(8): 2427–2435
9. AL-barakati HJ, Saigo H, Newman RH (2019) RF-GlutarySite: a random forest based predictor for glutarylation sites. Mol Omics 15(3): 189–204
10. Chou K-C (2011) Some remarks on protein attribute prediction and pseudo amino acid composition. J Theor Biol 273(1):236–247
11. Chou PY, Fasman GD (1978) Prediction of the secondary structure of proteins from their amino acid sequence. Adv Enzymol Relat Areas Mol Biol 47:45–148. https://doi.org/10.1002/9780470122921.ch2

12. Cao D-S, Xiao N, Xu Q-S, Chen AF (2015) Rcpi: R/Bioconductor package to generate various descriptors of proteins, compounds and their interactions. Bioinformatics 31(2):279–281
13. Shen J, Zhang J, Luo X, Zhu W, Yu K, Chen K, Li Y, Jiang H (2007) Predicting protein–protein interactions based only on sequences information. Proc Natl Acad Sci 104(11):4337–4341
14. Chou K-C, Shen H-B (2008) Cell-PLoc: a package of Web servers for predicting subcellular localization of proteins in various organisms. Nat Protoc 3(2):153
15. Barkan DT, Hostetter DR, Mahrus S, Pieper U, Wells JA, Craik CS, Sali A (2010) Prediction of protease substrates using sequence and structure features. Bioinformatics 26(14):1714–1722
16. Röttig M, Rausch C, Kohlbacher O (2010) Combining structure and sequence information allows automated prediction of substrate specificities within enzyme families. PLoS Comput Biol 6(1):e1000636
17. Ismail HD, Saigo H, KC DB (2018) RF-NR: Random Forest Based Approach for Improved Classification of Nuclear Receptors. IEEE/ACM Trans Comput Biol Bioinform 15(6):1844–1852. https://doi.org/10.1109/TCBB.2017.2773063
18. Shi SP, Chen X, Xu HD, Qiu JD (2015) PredHydroxy: computational prediction of protein hydroxylation site locations based on the primary structure. Mol BioSyst 11(3):819–825. https://doi.org/10.1039/c4mb00646a
19. Ismail HD, Newman RH, Kc DB (2016) RF-Hydroxysite: a random forest based predictor for hydroxylation sites. Mol BioSyst 12(8):2427–2435. https://doi.org/10.1039/c6mb00179c
20. Nishikawa K, Kubota Y, Ooi T (1983) Classification of proteins into groups based on amino acid composition and other characters. II. Grouping into four types. J Biochem 94(3):997–1007. https://doi.org/10.1093/oxfordjournals.jbchem.a134443
21. Reinhardt A, Hubbard T (1998) Using neural networks for prediction of the subcellular location of proteins. Nucleic Acids Res 26(9):2230–2236. https://doi.org/10.1093/nar/26.9.2230
22. Bhasin M, Raghava GP (2004) Classification of nuclear receptors based on amino acid composition and dipeptide composition. J Biol Chem 279(22):23262–23266. https://doi.org/10.1074/jbc.M401932200
23. Huang WL, Tung CW, Huang HL, Ho SY (2009) Predicting protein subnuclear localization using GO-amino-acid composition features. Biosystems 98(2):73–79. https://doi.org/10.1016/j.biosystems.2009.06.007
24. Rost B, Sander C (1994) Combining evolutionary information and neural networks to predict protein secondary structure. Proteins 19(1):55–72. https://doi.org/10.1002/prot.340190108
25. Rost B, Sander C (1993) Prediction of protein secondary structure at better than 70% accuracy. J Mol Biol 232(2):584–599. https://doi.org/10.1006/jmbi.1993.1413
26. Kawashima S, Kanehisa M (2000) AAindex: amino acid index database. Nucleic Acids Res 28(1):374. https://doi.org/10.1093/nar/28.1.374
27. Chou KC (2000) Prediction of protein subcellular locations by incorporating quasi-sequence-order effect. Biochem Biophys Res Commun 278(2):477–483. https://doi.org/10.1006/bbrc.2000.3815
28. Li J, Ma X, Li X, Gu J (2020) PPAI: a web server for predicting protein-aptamer interactions. BMC Bioinformatics 21(1):236. https://doi.org/10.1186/s12859-020-03574-7
29. Ismail HD, Jones A, Kim JH, Newman RH, Kc DB (2016) RF-Phos: a novel general phosphorylation site prediction tool based on random forest. Biomed Res Int 2016:3281590. https://doi.org/10.1155/2016/3281590
30. Moreau G, Broto P (1980) The autocorrelation of a topological structure: a new molecular descriptor. Nouv J Chim 4(6):359–360
31. Horne DS (1988) Prediction of protein helix content from an autocorrelation analysis of sequence hydrophobicities. Biopolymers 27(3):451–477. https://doi.org/10.1002/bip.360270308
32. Sokal RR, Thomson BA (2006) Population structure inferred by local spatial autocorrelation: an example from an Amerindian tribal population. Am J Phys Anthropol 129(1):121–131. https://doi.org/10.1002/ajpa.20250
33. Dubchak I, Muchnik I, Mayor C, Dralyuk I, Kim SH (1999) Recognition of a protein fold in the context of the Structural Classification of Proteins (SCOP) classification. Proteins 35(4):401–407
34. Dubchak I, Muchnik I, Holbrook SR, Kim SH (1995) Prediction of protein folding class using global description of amino acid sequence. Proc Natl Acad Sci U S A 92(19):8700–8704. https://doi.org/10.1073/pnas.92.19.8700

35. Ding Y, Tang J, Guo F (2016) Predicting protein-protein interactions via multivariate mutual information of protein sequences. BMC Bioinformatics 17(1):398. https://doi.org/10.1186/s12859-016-1253-9
36. Huang DS, Zhang L, Han K, Deng S, Yang K, Zhang H (2014) Prediction of protein-protein interactions based on protein-protein correlation using least squares regression. Curr Protein Pept Sci 15(6):553–560. https://doi.org/10.2174/1389203715666140724084019
37. Hou R, Wu J, Xu L, Zou Q, Wu YJ (2020) Computational prediction of protein arginine methylation based on composition-transition-distribution features. ACS Omega 5(42): 27470–27479. https://doi.org/10.1021/acsomega.0c03972
38. Zou L, Nan C, Hu F (2013) Accurate prediction of bacterial type IV secreted effectors using amino acid composition and PSSM profiles. Bioinformatics 29(24):3135–3142. https://doi.org/10.1093/bioinformatics/btt554
39. An Y, Wang J, Li C, Leier A, Marquez-Lago T, Wilksch J, Zhang Y, Webb GI, Song J, Lithgow T (2018) Comprehensive assessment and performance improvement of effector protein predictors for bacterial secretion systems III, IV and VI. Brief Bioinform 19(1):148–161. https://doi.org/10.1093/bib/bbw100
40. Lobley A, Sadowski MI, Jones DT (2009) pGenTHREADER and pDomTHREADER: new methods for improved protein fold recognition and superfamily discrimination. Bioinformatics 25(14):1761–1767. https://doi.org/10.1093/bioinformatics/btp302
41. Xie D, Li A, Wang M, Fan Z, Feng H (2005) LOCSVMPSI: a web server for subcellular localization of eukaryotic proteins using SVM and profile of PSI-BLAST. Nucleic Acids Res 33(Web Server issue):W105–W110. https://doi.org/10.1093/nar/gki359
42. Guo Y, Wu J, Ma H, Wang S, Huang J (2021) Comprehensive study on enhancing low-quality position-specific scoring matrix with deep learning for accurate protein structure property prediction: using bagging multiple sequence alignment learning. J Comput Biol 28(4):346–361. https://doi.org/10.1089/cmb.2020.0416
43. Chang DT, Huang HY, Syu YT, Wu CP (2008) Real value prediction of protein solvent accessibility using enhanced PSSM features. BMC Bioinformatics 9(Suppl 12):S12. https://doi.org/10.1186/1471-2105-9-S12-S12
44. Chou KC (2001) Prediction of protein cellular attributes using pseudo-amino acid composition. Proteins 43(3):246–255. https://doi.org/10.1002/prot.1035
45. Chou KC (2005) Using amphiphilic pseudo amino acid composition to predict enzyme subfamily classes. Bioinformatics 21(1): 10–19. https://doi.org/10.1093/bioinformatics/bth466
46. Gu Q, Ding YS, Zhang TL (2010) Prediction of G-protein-coupled receptor classes in low homology using Chou's pseudo amino acid composition with approximate entropy and hydrophobicity patterns. Protein Pept Lett 17(5):559–567. https://doi.org/10.2174/092986610791112693
47. Mohabatkar H (2010) Prediction of cyclin proteins using Chou's pseudo amino acid composition. Protein Pept Lett 17(10):1207–1214. https://doi.org/10.2174/092986610792231564
48. Lin H, Ding H (2011) Predicting ion channels and their types by the dipeptide mode of pseudo amino acid composition. J Theor Biol 269(1):64–69. https://doi.org/10.1016/j.jtbi.2010.10.019
49. Liu T, Zheng X, Wang C, Wang J (2010) Prediction of subcellular location of apoptosis proteins using pseudo amino acid composition: an approach from auto covariance transformation. Protein Pept Lett 17(10):1263–1269. https://doi.org/10.2174/092986610792231528
50. Ma J, Gu H (2010) A novel method for predicting protein subcellular localization based on pseudo amino acid composition. BMB Rep 43(10):670–676. https://doi.org/10.5483/BMBRep.2010.43.10.670
51. Zakeri P, Moshiri B, Sadeghi M (2011) Prediction of protein submitochondria locations based on data fusion of various features of sequences. J Theor Biol 269(1):208–216. https://doi.org/10.1016/j.jtbi.2010.10.026
52. Shi SP, Qiu JD, Sun XY, Huang JH, Huang SY, Suo SB, Liang RP, Zhang L (2011) Identify submitochondria and subchloroplast locations with pseudo amino acid composition: approach from the strategy of discrete wavelet transform feature extraction. Biochim Biophys Acta 3: 424–430. https://doi.org/10.1016/j.bbamcr.2011.01.011
53. Zhao XW, Ma ZQ, Yin MH (2012) Predicting protein-protein interactions by combing various sequence- derived features into the general form of Chou's Pseudo amino acid composition. Protein Pept Lett 19(5):492–500. https://doi.org/10.2174/092986612800191080
54. Xiao X, Min JL, Wang P, Chou KC (2013) Predict drug-protein interaction in cellular networking. Curr Top Med Chem 13(14): 1707–1712. https://doi.org/10.2174/15680266113139990121

55. Jia J, Liu Z, Xiao X, Liu B, Chou KC (2016) Identification of protein-protein binding sites by incorporating the physicochemical properties and stationary wavelet transforms into pseudo amino acid composition. J Biomol Struct Dyn 34(9):1946–1961. https://doi.org/10.1080/07391102.2015.1095116
56. Xiaohui N, Nana L, Jingbo X, Dingyan C, Yuehua P, Yang X, Weiquan W, Dongming W, Zengzhen W (2013) Using the concept of Chou's pseudo amino acid composition to predict protein solubility: an approach with entropies in information theory. J Theor Biol 332: 211–217. https://doi.org/10.1016/j.jtbi.2013.03.010
57. Niu XH, Hu XH, Shi F, Xia JB (2012) Predicting protein solubility by the general form of Chou's pseudo amino acid composition: approached from chaos game representation and fractal dimension. Protein Pept Lett 19(9):940–948. https://doi.org/10.2174/092986612802084492
58. Gupta R, Mittal A, Singh K (2008) A time-series-based feature extraction approach for prediction of protein structural class. EURASIP J Bioinform Syst Biol 2008:235451. https://doi.org/10.1155/2008/235451
59. Shao J, Xu D, Tsai SN, Wang Y, Ngai SM (2009) Computational identification of protein methylation sites through bi-profile Bayes feature extraction. PLoS One 4(3):e4920. https://doi.org/10.1371/journal.pone.0004920
60. Wee LJ, Simarmata D, Kam YW, Ng LF, Tong JC (2010) SVM-based prediction of linear B-cell epitopes using Bayes Feature Extraction. BMC Genomics 11(Suppl 4):S21. https://doi.org/10.1186/1471-2164-11-S4-S21
61. Saidi R, Maddouri M, Mephu Nguifo E (2010) Protein sequences classification by means of feature extraction with substitution matrices. BMC Bioinformatics 11:175. https://doi.org/10.1186/1471-2105-11-175
62. Krivak R, Hoksza D (2018) P2Rank: machine learning based tool for rapid and accurate prediction of ligand binding sites from protein structure. J Cheminform 10(1):39. https://doi.org/10.1186/s13321-018-0285-8
63. Schneider G, Wrede P (1994) The rational design of amino acid sequences by artificial neural networks and simulated molecular evolution: de novo design of an idealized leader peptidase cleavage site. Biophys J 66(2 Pt 1): 335–344. https://doi.org/10.1016/s0006-3495(94)80782-9
64. Grantham R (1974) Amino acid difference formula to help explain protein evolution. Science 185(4154):862–864
65. Cao DS, Xu QS, Liang YZ (2013) propy: a tool to generate various modes of Chou's PseAAC. Bioinformatics 29(7):960–962. https://doi.org/10.1093/bioinformatics/btt072
66. Li ZR, Lin HH, Han LY, Jiang L, Chen X, Chen YZ (2006) PROFEAT: a web server for computing structural and physicochemical features of proteins and peptides from amino acid sequence. Nucleic Acids Res 34(Web Server issue):W32–W37. https://doi.org/10.1093/nar/gkl305
67. Ong SA, Lin HH, Chen YZ, Li ZR, Cao Z (2007) Efficacy of different protein descriptors in predicting protein functional families. BMC Bioinformatics 8:300. https://doi.org/10.1186/1471-2105-8-300
68. Govindan G, Nair AS (2011) Composition, Transition and Distribution (CTD)—a dynamic feature for predictions based on hierarchical structure of cellular sorting. In: India Conference (INDICON), 2011 Annual IEEE, 16–18 Dec, 2011. IEEE, London, pp 1–6. https://doi.org/10.1109/INDCON.2011.6139332
69. Dou Y, Yao B, Zhang C (2014) PhosphoSVM: prediction of phosphorylation sites by integrating various protein sequence attributes with a support vector machine. Amino Acids 46(6): 1459–1469. https://doi.org/10.1007/s00726-014-1711-5
70. Dou Y, Wang J, Yang J, Zhang C (2012) L1pred: a sequence-based prediction tool for catalytic residues in enzymes with the L1-logreg classifier. PLoS One 7(4):e35666. https://doi.org/10.1371/journal.pone.0035666
71. Chen K, Kurgan LA, Ruan J (2007) Prediction of flexible/rigid regions from protein sequences using k-spaced amino acid pairs. BMC Struct Biol 7(1):25. https://doi.org/10.1186/1472-6807-7-25
72. Altschul SF, Koonin EV (1998) Iterated profile searches with PSI-BLAST—a tool for discovery in protein databases. Trends Biochem Sci 23(11):444–447. https://doi.org/10.1016/s0968-0004(98)01298-5
73. Wang J, Yang B, Revote J, Leier A, Marquez-Lago TT, Webb G, Song J, Chou KC, Lithgow T (2017) POSSUM: a bioinformatics toolkit for generating numerical sequence feature descriptors based on PSSM profiles. Bioinformatics 33(17):2756–2758. https://doi.org/10.1093/bioinformatics/btx302
74. Dong Q, Zhou S, Guan J (2009) A new taxonomy-based protein fold recognition approach based on autocross-covariance transformation. Bioinformatics 25(20):2655–2662.

https://doi.org/10.1093/bioinformatics/btp500

75. Shen HB, Chou KC (2008) PseAAC: a flexible web server for generating various kinds of protein pseudo amino acid composition. Anal Biochem 373(2):386–388. https://doi.org/10.1016/j.ab.2007.10.012
76. Du P, Wang X, Xu C, Gao Y (2012) PseAAC-builder: a cross-platform stand-alone program for generating various special Chou's pseudo-amino acid compositions. Anal Biochem 425(2):117–119. https://doi.org/10.1016/j.ab.2012.03.015
77. Du P, Gu S, Jiao Y (2014) PseAAC-General: fast building various modes of general form of Chou's pseudo-amino acid composition for large-scale protein datasets. Int J Mol Sci 15(3):3495–3506. https://doi.org/10.3390/ijms15033495
78. Xiao N, Cao DS, Zhu MF, Xu QS (2015) protr/ProtrWeb: R package and web server for generating various numerical representation schemes of protein sequences. Bioinformatics 31(11):1857–1859. https://doi.org/10.1093/bioinformatics/btv042
79. Zuo Y, Li Y, Chen Y, Li G, Yan Z, Yang L (2017) PseKRAAC: a flexible web server for generating pseudo K-tuple reduced amino acids composition. Bioinformatics 33(1):122–124. https://doi.org/10.1093/bioinformatics/btw564
80. Chen Z, Zhao P, Li F, Leier A, Marquez-Lago TT, Wang Y, Webb GI, Smith AI, Daly RJ, Chou K-C, Song J (2018) iFeature: a Python package and web server for features extraction and selection from protein and peptide sequences. Bioinformatics 34(14):2499–2502. https://doi.org/10.1093/bioinformatics/bty140
81. Nikam R, Gromiha MM (2019) Seq2Feature: a comprehensive web-based feature extraction tool. Bioinformatics 35(22):4797–4799. https://doi.org/10.1093/bioinformatics/btz432
82. Ismail HD, Saigo H, DB KC (2018) RF-NR: random forest based approach for improved classification of nuclear receptors. IEEE/ACM Trans Comput Biol Bioinform 15(6):1844–1852
83. Ismail HD, Saigo H, DB KC (2015) RF-NR: random forest based approach for improved classification of nuclear receptors. IEEE/ACM Trans Comput BIol Bioinfom 15(6):1844–1852. (Also appeared in GIW/INCoB 2015, Tokyo, Japan)
84. Yin Z, Tan J (2012) New encoding schemes for prediction of protein phosphorylation sites. In: 2012 IEEE 6th international conference on systems biology (ISB), 18–20 Aug. 2012. IEEE, London, pp 56–62. https://doi.org/10.1109/ISB.2012.6314113
85. Jiao X, Wang B, Su J, Chen W, Wang C (2006) Protein design based on the relative entropy. Phys Rev E Stat Nonlinear Soft Matter Phys 73:061903. https://doi.org/10.1103/PhysRevE.73.061903

Chapter 4

A Pretrained ELECTRA Model for Kinase-Specific Phosphorylation Site Prediction

Lei Jiang, Duolin Wang, and Dong Xu

Abstract

Phosphorylation plays a vital role in signal transduction and cell cycle. Identifying and understanding phosphorylation through machine-learning methods has a long history. However, existing methods only learn representations of a protein sequence segment from a labeled dataset itself, which could result in biased or incomplete features, especially for kinase-specific phosphorylation site prediction in which training data are typically sparse. To learn a comprehensive contextual representation of a protein sequence segment for kinase-specific phosphorylation site prediction, we pretrained our model from over 24 million unlabeled sequence fragments using ELECTRA (Efficiently Learning an Encoder that Classifies Token Replacements Accurately). The pretrained model was applied to kinase-specific site prediction of kinases CDK, PKA, CK2, MAPK, and PKC. The pretrained ELECTRA model achieves 9.02% improvement over BERT and 11.10% improvement over MusiteDeep in the area under the precision-recall curve on the benchmark data.

Key words Deep leaning, Pretraining, Transformer, ELECTRA, Phosphorylation, Kinase-specific phosphorylation site prediction

1 Introduction

Posttranslational modification (PTM), including protein phosphorylation, is a crucial mechanism to increase the functional diversity of the proteome [1, 2]. In protein phosphorylation, an amino acid residue is phosphorylated by a protein kinase by adding a covalently bound phosphate group. Phosphorylation plays a vital role in intracellular signal transduction and cell cycle [3–6]. Approximately one-third of mammalian proteins may be phosphorylated [7], and the phosphorylation on serine (S) and threonine (T) is most studied. Experimental determination of phosphorylation sites and associated kinases is time-consuming and costly, and hence computational prediction plays a critical role in phosphorylation studies.

Dukka B. KC (ed.), *Computational Methods for Predicting Post-Translational Modification Sites*, Methods in Molecular Biology, vol. 2499, https://doi.org/10.1007/978-1-0716-2317-6_4,

Predicting protein phosphorylation sites by machine-learning methods has a long history. Protein phosphorylation site prediction can be roughly divided into two tasks: (1) general phosphorylation site prediction, which predicts sites that can be phosphorylated, and (2) kinase-specific phosphorylation site prediction, which predicts sites that can be phosphorylated by a specific kinase. Multiple features of a protein sequence have been explored by different machine-learning-based methods, such as amino acid substitution matrix used in GPS-YNO2 [8], GPS 2.0 [9], and GPS-SUMO [10], disorder scores, amino acid frequencies used in Musite [11, 12], and physicochemical properties adopted by AMS 4.0 [13] and phosphorylation site prediction in plants [14]. These methods are successful for general phosphorylation site prediction, but the error rates for kinase-specific phosphorylation site prediction are often high due to limited training data. There are only ~6000 labeled positive data for many kinases, and each kinase typically has tens to hundreds of its phosphorylation sites for training. It could easily cause overfitting when applying a deep and complex neural network on such a small dataset, limiting the model's generalization and resulting in poor performance.

Is it possible to enrich the information in the small dataset of this problem by pretraining an unsupervised representation model on a large dataset? In Natural Language Processing (NLP), the rapid increase of data and unsupervised learning capacity with Masked Language Modelling (MLM) have led to major advances in representation learning. In particular, it is possible to learn and extract information and latent features from a large-scale corpus and use the learned model to transfer the feature embeddings to the small dataset for downstream NLP tasks. This approach greatly improves the complexity and capacity of the model while effectively avoiding the small-training-sample issue. Protein sequences are similar to sentences in NLP in syntax and semantics [15–18]. Different representation learning methods were employed for modeling representations of biological properties of protein sequences for downstream protein sequence analyses. BioVec [19], using the same negative sampling method from like Word2Vec [20], represents a protein sequence with a single dense n-dimensional vector by presenting each 3-gram of a protein sequence as a vector of size 100, under the idea that the meaning of a word is characterized by its context, that is, neighboring words. However, BioVec only uses one single vector representation for each word and fails to capture the position information to distinguish different semantic meanings for the same word in other contexts. Later, BERT [21] made substantial progress for representations of sequential data. BERT generates embeddings with multiple vector representations for the same word with different semantic meanings based on different contexts. The BERT model was applied for protein sequence representations, mapping semantically similar objects to nearby vectors in representation space [22]. However, MLM like BERT is computationally expensive to train because the pretraining task is only defined over the small masked-out subset. Recently, another pretraining method ELECTRA (Efficiently Learning an Encoder that Classifies Token Replacements Accurately) [23], defined over all the input tokens, has been successfully applied to several NLP tasks, such as sentiment analysis [24], question and answering [25], event temporal reasoning [26], and biomedical semantic indexing and question answering [27].

In this study, we applied ELECTRA to pretrain a fragment representation model for phosphorylation site prediction. To our best knowledge, this is the first attempt to use ELECTRA for biological sequence analyses. This paper presents and compares BERT and ELECTRA models, which are pretrained on a large unlabeled dataset for kinase phosphorylation site prediction. We evaluated their performance by comparing them with the state-of-the-art deep learning model for kinase-specific phosphorylation site prediction—MusiteDeep [28], which we developed previously. MusiteDeep uses a multilayer convolutional neural network (CNN) [29] with a two-dimensional attention mechanism to learn representations from small labeled raw data.

2 Materials and Methods

2.1 Dataset

For pretraining the ELECTRA protein fragment representation model, the protein sequences were collected from the latest 2021 UniProt/Swiss-Prot [30]. We extracted a peptide of 33 residues centered at the potential phosphorylation site S or T together with 16 residues at each side from protein sequences, which is the same window size as MusiteDeep (*see* **Note 1**). Both general and kinase-specific phosphorylation site protein sequences were obtained from the MusiteDeep paper. Table 1 summarizes the collected phosphorylation data used in this study. For general phosphorylation site prediction, phosphorylation sites on S or T of *Homo sapiens* annotated by UniProt/Swiss-Prot were used as positive data. In contrast, the sites with the same target amino acids but were not annotated as phosphorylation sites from the same proteins were regarded as the negative data.

For kinase-specific phosphorylation site prediction, we only consider kinase families CDK, CK2, MAPK, PKA, and PKC, each of which has more than 100 known substrate phosphorylation sites in our collected dataset. For each kinase family, only sites annotated by the specific kinase family were used as positive data, whereas all other residues of the same types (S or T) in the same substrates were used as negative data. Since serine/threonine-specific kinase typically can phosphorylate both S and T residues [31], we combined phosphoserine and phosphothreonine sites in the data collection

Table 1
Phosphorylation data collected in this study

Fragments	Data source	# of fragments	
Training	Uniprot	24,293,085	
Homo sapiens	Data source	# of positive sites	# of negative sites
Training	Swiss-Prot	31,883	622,075
Testing	Swiss-Prot	2074	60,880
Kinase family	Data source	# of positive sites	# of negative sites
CDK training	Swiss-Prot/RegPhos	269	12,657
CDK testing	Swiss-Prot/RegPhos	46	3221
PKA training	Swiss-Prot/RegPhos	274	16,794
PKA testing	Swiss-Prot/RegPhos	80	3527
CK2 training	Swiss-Prot/RegPhos	252	7509
CK2 testing	Swiss-Prot/RegPhos	51	2178
MAPK training	Swiss-Prot/RegPhos	327	13,603
MAPK testing	Swiss-Prot/RegPhos	72	2969
PKC training	Swiss-Prot/RegPhos	368	15,933
PKC testing	Swiss-Prot/RegPhos	88	3846

and trained one model for both S and T sites. For the general phosphorylation site prediction, the split of the training and testing data was made by the time stamp. The testing data was extracted from the protein sequences generated after 2008 and the training data was extracted from the remaining protein sequences. Protein sequences in the training set that have more than 50% of sequence similarity with the testing set were removed. For the kinase-specific phosphorylation site prediction, the training and testing data were split by a ratio of 80% to 20%.

2.2 Methods

The framework of our kinase-specific phosphorylation site prediction model implemented with ELECTRA is summarized in Fig. 1. Like most representation learning models, ELECTRA has two training stages—pretraining and fine-tuning. The pretraining models the representations of the fragment sequences through replaced tokens detection performed by the discriminator on a sizeable unlabeled dataset, as shown in the left part of Fig. 1. The generator is only used to train the discriminator and is discarded after pretraining. The fine-tuning adjusts the weights of the ELECTRA model for downstream tasks on a small labeled dataset. Due to the limited kinase-specific phosphorylation site data, the model was first fine-tuned on the general phosphorylation site data to

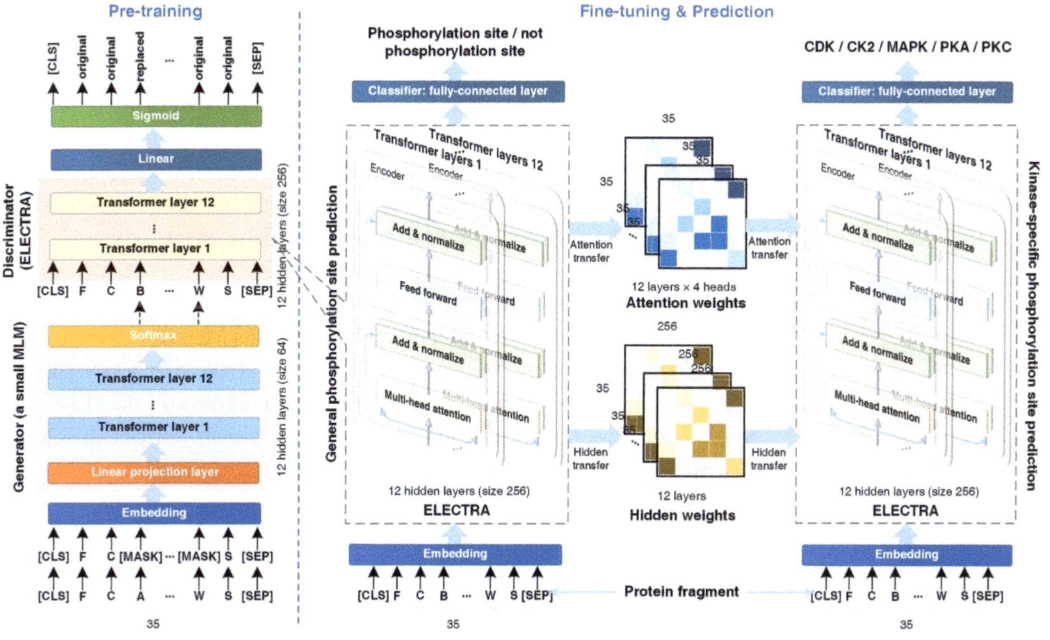

Fig. 1 The framework of the kinase-specific phosphorylation site prediction model. The left part represents pretraining. The ELECTRA discriminator is trained to model the representations for protein fragments by detecting the replaced tokens from the corrupt input generated by the generator on a sizeable unlabeled dataset. The generator is discarded after pretraining. The right part represents fine-tuning, where the ELECTRA model is trained on the labeled general phosphorylation site dataset, and the learned weights are transferred to the kinase-specific phosphorylation site prediction model for other kinase-specific training tasks

predict if the site could be phosphorylated or not, and then fine-tuned on the kinase-specific phosphorylation site data, with the attention weights and feature map weights transferred from the general phosphorylation site prediction model, as shown in the right part of Fig. 1.

2.2.1 Pretraining ELECTRA Model

In the pretraining stage, ELECTRA consists of two Transformer models—a generator G and a discriminator D. The basic idea of the pretraining task in ELECTRA is to train the discriminator to detect the replaced tokens in the corrupted input sequence created by the generator to learn context-dependent biological properties of protein sequence fragments from a large-scale task-independent and unlabeled protein dataset. This knowledge can then be embedded into a smaller task-specific and labeled dataset in the downstream tasks of kinase phosphorylation site prediction, drastically cutting down the amount of labeled data needed since the pretrained model has already captured the underlying patterns related to classification. We chose the ELECTRA-Small model (12 layers, 256 hidden layer size, and 14 million hyperparameters) with 128-dimension embedding (*see* **Note 2**).

The major advantage of MLM is that it uses a huge amount of unlabeled fragment data, which is shown in Table 1. After extracting the fragments, we have over 24 million fragment samples (*see* **Note 3**). The vocabulary size is 25, including 20 common amino acids, and using a dash token ("-") as a pseudo amino acid for the padding positions at protein termini (*see* **Note 4**). Special tokens [MASK], [CLS], [SEP], and [UNK] are used, where [MASK] is for the masked-out token, [CLS] represents the start token, [SEP] represents the end token, and [UNK] is for unknown amino acids. In the embedding layer, each input fragment has 15% of its amino acids randomly replaced with special [MASK] tokens, and then is encoded into a token embeddings matrix with a size of [35, 128] and position embeddings matrix with a size of [35, 128] (*see* **Note 5**). The token embeddings and position embeddings are summed together as input tokens $x = [x_1, \ldots, x_n]$ $(n = 35)$ and fed into the generator.

The generator is typically a small MLM. We used a generator of 1/4 size of the discriminator with 12 layers and a hidden layer size of 64 (*see* **Note 6**). The generator is trained with maximum likelihood to predict the original value of the masked-out tokens based on the context provided by the other, nonmasked amino acids in a fragment input. For a given masked token x^{masked}, the generator outputs a probability for predicting its original identity with a softmax layer. The generator as an MLM approach only learns from the 15% masked tokens per fragment because the loss is only calculated over the masked tokens. The loss function of the generator is defined as follows.

$$L_{MLM}(x, \theta_G) = E\left(\sum_{i \in m} - \log p_G\left(x_i | x^{masked}\right)\right)$$

By replacing the masked-out tokens with generator output prediction, a corrupted example $x^{corrupt}$ is created and fed into the discriminator to predict which positions are masked. To improve the efficiency of the pretraining, the generator and discriminator share the embeddings (both the token and positional embeddings) by adding a linear layer to the generator to project the embeddings of size [35, 128] into the generator hidden layer size [35, 64] representations.

The ELECTRA discriminator is a Transformer network with 12 layers and a hidden layer size of 256. For each token in $x^{corrupt}$, the discriminator is trained to predict whether it matches the original input x or not with a sigmoid output layer. The discriminator's loss is calculated over all the input tokens. The loss function of the discriminator is defined as

$$L_{Disc}(x, \theta_D) = E\left(\sum_{t=1}^{n} -1\left(x_t^{corrupt} = x_t\right) \log D(x^{corrupt}, t) - 1\left(x_t^{corrupt} \neq x_t\right) \log\left(1 - D(x^{corrupt}, t)\right)\right)$$

where t denotes the position of amino acid in the fragment, $t \in \{1, n\}$.

The generator and discriminator are trained to minimize the combined loss

$$\min_{\theta_G, \theta_D} \sum_{x \in \chi} L_{MLM}(x, \theta_G) + \lambda L_{Disc}(x, \theta_D)$$

over a huge corpus χ of raw protein fragments.

After pretraining, the generator was discarded and only the discriminator was fine-tuned for downstream tasks.

2.2.2 Fine-Tune ELECTRA Model for General Phosphorylation Site Prediction

The phosphorylation site prediction can be formulated as a binary classification problem, and each potential phosphorylation site can be classified as either a phosphorylation site or a nonphosphorylation site. During the pretraining, the ELECTRA learns how to model the protein fragment representation. Then, the fine-tuning procedure learns how to classify phosphorylation sites using that information with the labeled data.

We implemented two different fine-tuning techniques: In the first one, we froze the weights of the pretrained ELECTRA model, attached a multilayer convolutional neural network (CNN) to the framework, and only updated the weights of the attached CNN layers during the training. In this way, the pretrained ELECTRA model was only used to generate the representations of the input fragments. In the second one, we built a fully connected layer on top of the pretrained ELECTRA model and fine-tuned the entire network with 12 layers of the discriminator. In this way, the error was back-propagated through the whole architecture and the weights of the discriminator were updated based on the phosphorylation dataset. We evaluated the performance of the two fine-tuning approaches and found training the entire architecture yielded better results. The final ELECTRA models were fine-tuned on the entire structure and then compared with BERT and MusiteDeep.

2.2.3 Fine-Tune ELECTRA Model for Kinase-Specific Phosphorylation Site Prediction

For kinase-specific phosphorylation prediction, the positive sites range from 1.7% to 3.1% of the negative sites. The number of positive samples of each category is less than 500, which may be challenging to train a highly accurate generalized model. To solve the small sample issue, we fine-tuned an ELECTRA model for each kinase-specific site prediction based on the ELECTRA model for general phosphorylation site prediction by sharing the model weights. In this way, the kinase-specific models learn from the general feature representations and the overfitting problem is relieved.

3 Results

We used a learning rate of 2e-4, a generator size of 25%, 50%, or 100% of the discriminator size, and a batch size of 1024. We trained the ELECTRA model for 200K steps and 1M steps, and the BERT model without next sentence prediction for 500K steps on a v100 GPU 32 GB VRAM machine, using MusiteDeep as a baseline with the same dataset, as shown in Table 1. MusiteDeep does not require pretraining thus is not pretrained on the unlabeled 24M datasets.

We fine-tuned all layers of the ELECTRA and BERT models with the same hyperparameters as pretraining except for a learning rate of 1e-5 for 500 epochs (~28 h on a single v100 GPU of 32GB VRAM) on the general phosphorylation site prediction data. Because the generator is thrown out after pretraining, the number of hyperparameters of the ELECTRA-small model is the same as the BERT-small model, and the fine-tuning cost is also close. Unlike the general phosphorylation site classification model, each of the kinase-specific phosphorylation site prediction models only took less than 10 min to fine-tune.

Although pretraining is computationally expensive, the pretrained fragment representation model can be reused for different downstream tasks performed on the same category of fragments, and the fine-tuning and prediction are fast. In addition, learning and extracting features automatically from raw data saves a lot of time in feature engineering, and it will avoid bias either by humans or from the limited data.

3.1 Evaluation of the ELECTRA Model for Kinase-Specific Phosphorylation Site Prediction

After fine-tuning, we had one model for the general phosphorylation site (S/T) prediction, and five models for kinase-specific site prediction as CDK, PKA, CK2, MAPK, and PKC for ELECTRA, BERT, and MusiteDeep. For the general phosphorylation site prediction, the performances of all the methods were evaluated on an independent testing set, as shown in Table 1. For the kinase-specific phosphorylation site prediction, the performances were evaluated by 20% of the whole kinase-specific phosphorylation site dataset in Table 1, which was not used in training. Since our dataset is highly imbalanced, accuracy is not the best metric for evaluation because it can be misleading. A model can get a high accuracy by predicting the majority class, but it fails to capture the minority class, resulting in trivial and meaningless results. Instead, we evaluate the models by the area under the Receiver Operating Characteristics (ROC) curve (AUC), and the area under the precision-recall curve (AUPR) as it shows a tradeoff between precision and recall for different thresholds. The ROC and the precision-recall curves of the general phosphorylation site prediction and the kinase-specific phosphorylation site prediction for the ELECTRA, BERT, and MusiteDeep are shown in Fig. 2a, b. For the S/T

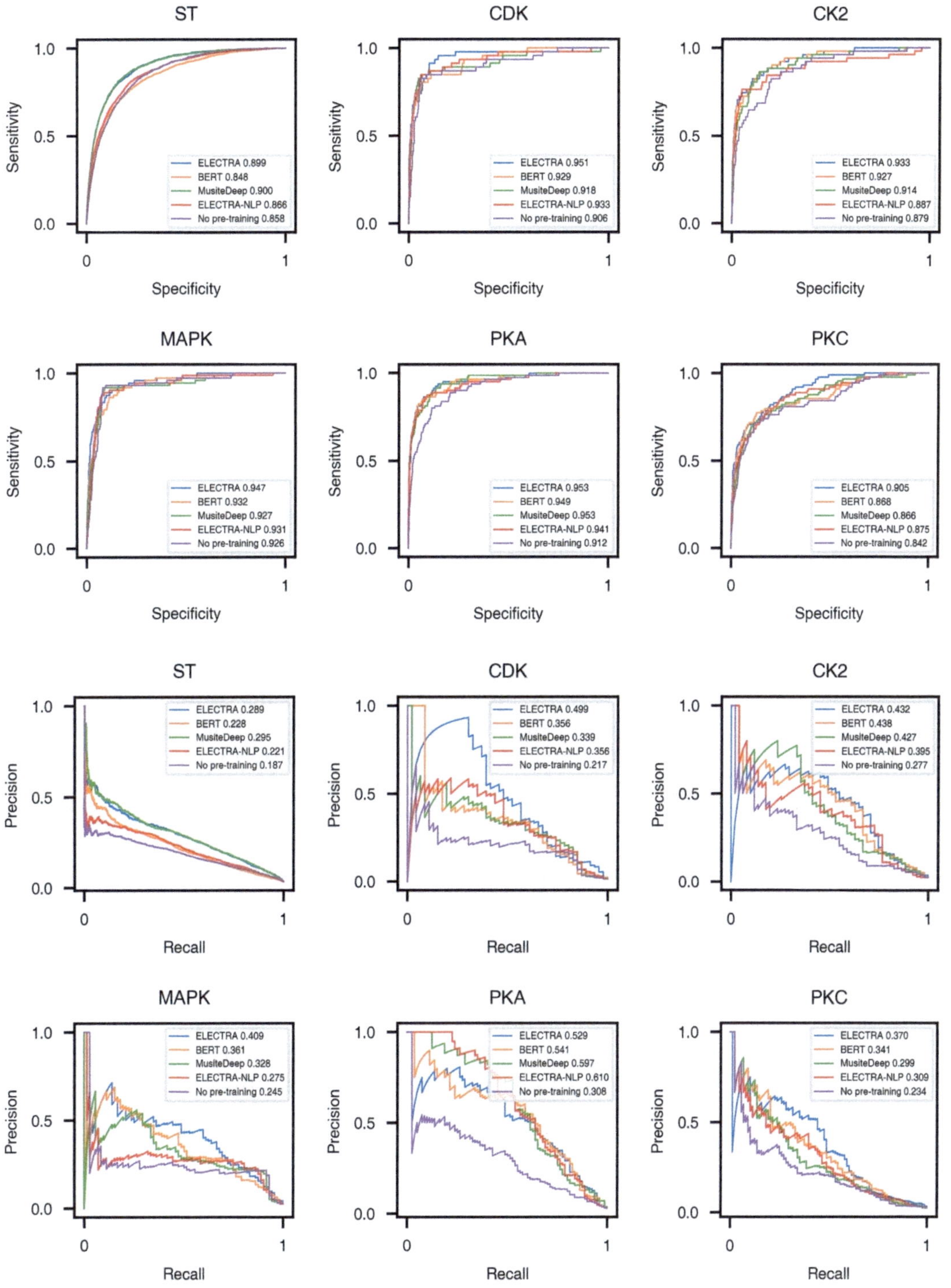

Fig. 2 ROC curves and precision-recall curves for S/T phosphorylation site prediction, and five models for kinase-specific site prediction as CDK, PKA, CK2, MAPK, and PKC

general phosphorylation site prediction, MusieDeep achieved the best AUC with 0.9 and AUPR 0.295, ELECTRA obtained the second AUC with 0.899 and AUPR 0.289, and BERT got AUC of 0.848 and AUPR 0.228 on the testing dataset. For the kinase-specific phosphorylation site, ELECTRA achieved the best AUC of CK2 (0.933), MAPK (0.947), PKA (0.953), and PKC (0.907), the best AUPR of CDK (0.499) and MAPK (0.409), and MusiteDeep got the best AUPR of PKA (0.597). Both the highest AUC (0.938) and AUPR (0.448) were achieved by ELECTRA with the AUC 1.83% higher than BERT (0.921) and 2.2% higher than Musite-Deep (0.918), and the AUPR 9.02% higher than BERT (0.407) and 11.1% higher than MusiteDeep (0.398). The result demonstrated the vital role of the pretrained ELECTRA protein fragments representation model in small-sample learning.

To explore the effect of pretraining on protein fragments for the ELECTRA model, we performed an ablation experiment by fine-tuning a small pretrained NLP ELECTRA model (ELECTRA-NLP) by mapping amino acids to "stop words" (stop words are a set of commonly used words with very little information in a language, such as I, you, she, etc.) and fine-tuning an ELECTRA model initiated with random weights (no pretraining). The fine-tuning loss curves for general phosphorylation site prediction of the three models are plotted in Fig. 3. The loss curve of the pretrained model of ELECTRA on protein fragments converges much faster and deeper than the ELECTRA-NLP model and ELECTRA model without pretraining. The ELECTRA model pretrained on protein fragments significantly outperformed the ELECTRA-NLP model or the ELECTRA model without pretraining by over 13.1% and 42.8%, respectively, in the area under the precision-recall curve in the kinase-specific phosphorylation site prediction. The ELECTRA-NLP model and ELECTRA model without pretraining could not achieve a good generalization and robustness compared with the pretrained ELECTRA models on protein fragments on general phosphorylation site prediction or kinase-specific site prediction, which demonstrates the important role of pretraining on protein fragments. However, an interesting finding is that the ELECTRA-NLP model performs better than the random model on both AUC and AUPR.

3.2 Efficiency Analysis

To better understand the efficiency of ELECTRA models with different configurations and in comparison with MLM like BERT, we trained and evaluated them in both pretraining and fine-tuning stages. First, we pretrained a small BERT model without next sentence prediction for 500K steps, namely BERT-500K, to measure the efficiency of the MLM model. Then, we pretrained and fine-tuned small ELECTRA models with different generators varying in sizes of 25%, 50%, and 100% of the discriminator size for 200 K steps, namely, ELECTRA-0.25G-200K, ELECTRA-0.5G-

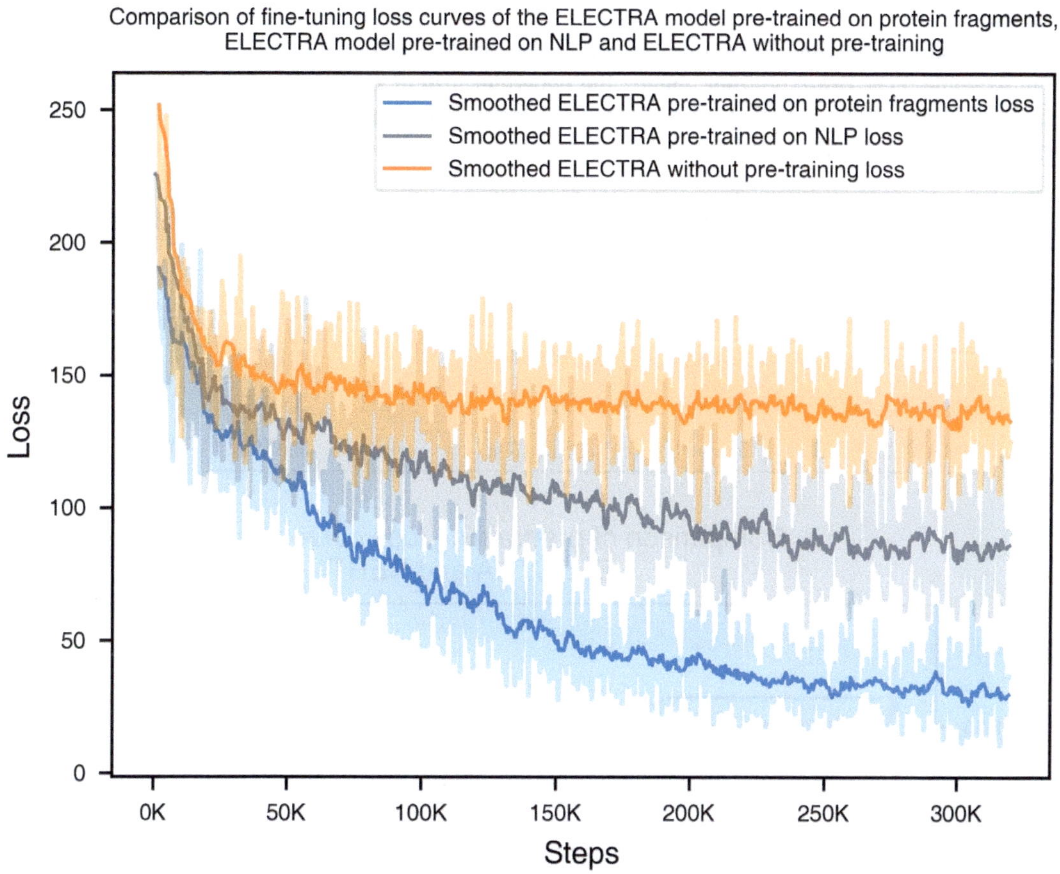

Fig. 3 Comparison of fine-tuning loss curves of the ELECTRA model pretrained on protein fragments, and the ELECTRA model pretrained on NLP and ELECTRA without pretraining

200K, and ELECTRA-1G-200K, and a generator with 25% of the discriminator size for 1 M steps namely ELECTRA-0.25G-1M to find an optimized size of generator to train the discriminator.

During pretraining, floating-point operations (FLOPs) were chosen to measure the compute usage. The results are shown in Table 2. BERT-500 K only has MLM loss because it does not have a discriminator. The training loss curves for ELECTRA and BERT models are plotted in Fig. 4. The curves are smoothed using Exponential Moving Averages (EMA) with a factor of 0.01, and the loss curve of BERT-500 K is projected on the loss curve of ELECTRA-0.25G-1M with a dashed line by taking the same starting point. Compared to the loss curve of BERT-500 K, the loss of ELECTRA-0.25G-1M converges much faster and deeper, and ELECTRA-0.25G-1M should achieve decent results after 200K steps. However, the MLM loss of BERT-500K decreases very slowly after about 20K, and the loss of the 200K is almost the same after training for 300K more steps, which might suggest the model was converged and there is little room to improve further.

Table 2
Comparison of small models in pretraining: G represents Generator while D represents Discriminator. For example, ELECTRA-0.5G-200K means the generator size of ELECTRA is 0.5 (50%) of the discriminator and the model is trained for 200 K steps. D loss means discriminator loss. The BERT model was trained without next sentence prediction and it has no discriminator loss since it is MLM

Model	Train FLOPs	Training Time	Params	MLM Loss	D Loss	Total
BERT-500K	7.02E+17 (2.3×)	2D 3 h	14M	2.055		
ELECTRA-0.25G-200K	**3.02E+17 (1×)**	1D 1 h	14M	2.19	0.289	16.652
ELECTRA-0.5G-200K	3.55E+17 (1.2×)	1D 5 h	14M	2.157	0.292	16.779
ELECTRA-1G-200K	5.63E+17 (1.9×)	3D 7 h	14M	**2.013**	0.287	**16.380**
ELECTRA-0.25G-1M	1.51E+18 (5×)	5D 18 h	14M	2.689	**0.278**	16.600

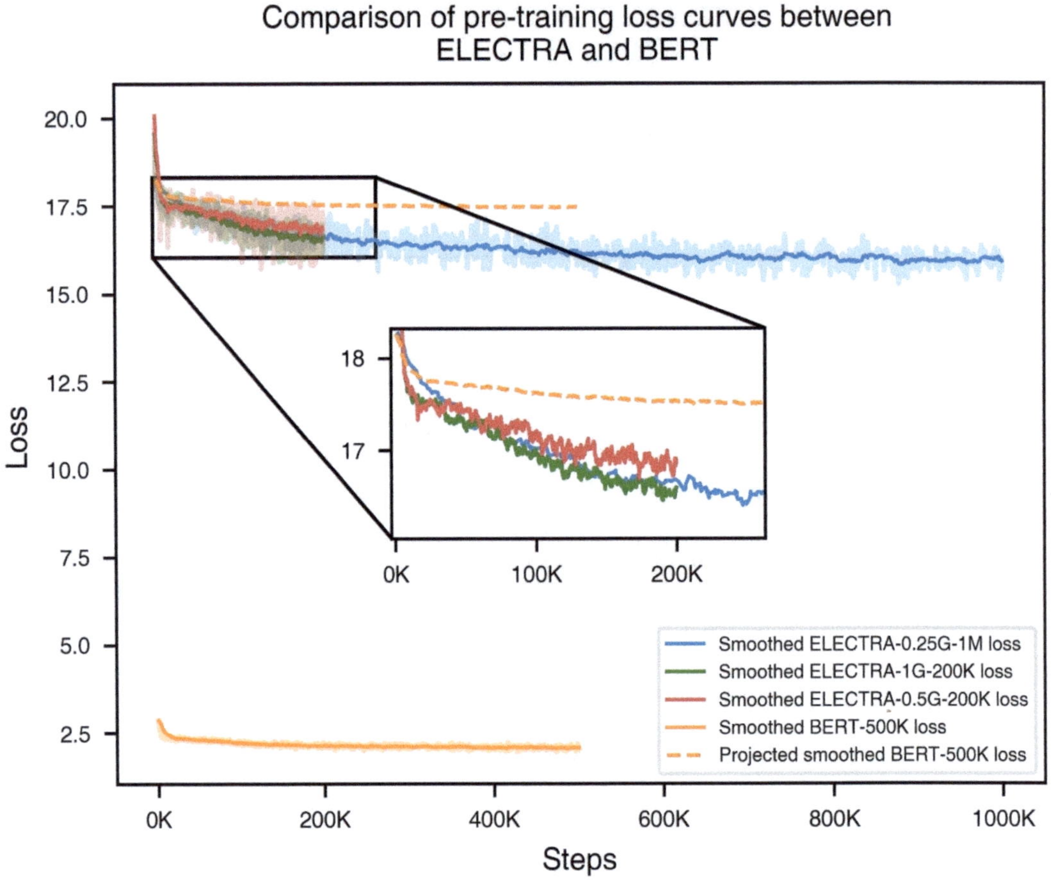

Fig. 4 Comparison of pretraining loss curves between ELECTRA and BERT. The loss curves are smoothed using Exponential Moving Averages (EMA) with a factor of 0.01, and the transparent curves are the corresponding original loss curves for each model of the same color. BERT has a much smaller loss because it only has the MLM loss, while ELECTRA has a combined loss of MLM loss and discriminator loss. The loss curve of BERT-500K is projected on the loss curve of ELECTRA-0.25G-1M by taking the same starting point with a dashed line to compare their tendency

For ELECTRA models with different generator sizes, interestingly, the loss of ELECTRA-0.25G-1M before 200K (3.02E+17 FLOPs) converges faster and is also more stable than that of ELECTRA-0.5G-200K and is close to ELECTRA-1G-200K. At the same time, the computational cost is just slightly higher than that of ELECTRA-1G-200K (5.63E+17 FLOPs).

In the fine-tuning stage, the architecture of the ELECTRA small discriminator is the same as the small BERT model. We evaluated the models by comparing their AUC and AUPR scores, as shown in Table 3. ELECTRA-0.25G-200K achieves a higher average AUC and AUPR (0.931/0.456) than BERT-500K (0.921/0.407) in kinase-specific phosphorylation site prediction with less than half of its pretraining FLOPs. This result supports the assumption that ELECTRA is more computationally efficient than the BERT model. For all ELECTRA models trained for 200 K steps, ELECTRA-0.25G-200K achieves the highest average AUC and AUPR with the lowest cost, ELECTRA-1G-200K (0.9186/ 0.3952) obtains a slightly lower average AUC than ELECTRA-0.5G-200K in kinase-specific phosphorylation site prediction while the pretraining FLOPs are also 1.6× higher than the latter model. ELECTRA-0.25G-1M, with the highest MLM loss and lowest discriminator loss in pretraining, gets the best AUC (0.899) and AUPR (0.289) in S/T general phosphorylation site prediction and average AUC (0.9382) and average AUPR (0.4478) in kinase-specific phosphorylation site prediction. The fine-tuning results of ELECTRA models are consistent with the pretraining results, suggesting that more computational cost does not always mean better performance. While a bigger generator might learn faster during the first few epochs, the learning capability of the discriminator decreases when the generator becomes too strong. ELECTRA-0.25G-1M beats all other models with the lowest discriminator loss, and hence, we use 25% as the generator size.

3.3 Effect of Pretrained ELECTRA Model on Small-Sample Data Learning

To show how the general feature representations from the pretrained ELECTRA may be transferred and help the small-sample data learning for the kinase-specific phosphorylation site prediction, we applied PCA (Principal component analysis) [32] and t-SNE (t-Distributed Stochastic Neighbor Embedding) [33] to project the representations of the ELECTRA model to their corresponding two-dimensional space. Specifically, the 256 dimensions of the embeddings of the last hidden layer were reduced to 50 through PCA and then decreased further to 2 using t-SNE to visualize pretrained representation and untrained one-hot representation for comparison in Fig. 5. The protein fragments from the same kinase family are generally grouped into clusters in the reduced dimension representation of the pretrained ELECTRA model. Since CDK and MAPK are from the CMGC group and PKA and PKC are from the AGC group, they tend to overlap with the fragments in the same group due to similar biological

Table 3
Comparison of small models in fine-tuning

Kinase	ST		CDK		CK2		MAPK		PKA		PKC	
Measure	AUC	AUPR	AUC	AUPR	AUC	AUPR	AUC	AUPR	AUC	AUPR	AUC	AUPR
MusiteDeep	**0.900**	**0.295**	0.918	0.339	0.914	0.427	0.927	0.328	0.953	0.597	0.866	0.299
BERT-500K	0.848	0.228	0.929	0.356	0.927	0.438	0.932	0.361	0.949	0.541	0.868	0.341
ELECTRA-0.25G-200K	0.855	0.218	**0.959**	0.464	0.917	**0.482**	0.945	0.407	0.942	0.549	0.894	**0.378**
ELECTRA-0.5G-200K	0.862	0.244	0.956	0.436	0.891	0.236	0.941	0.356	0.948	0.508	0.875	0.355
ELECTRA-1G-200K	0.870	0.226	0.907	0.312	0.917	0.476	0.939	0.362	**0.956**	0.474	0.874	0.352
ELECTRA-0.25G-1M	0.899	0.289	0.951	**0.499**	**0.933**	0.432	**0.947**	**0.409**	0.953	0.529	**0.907**	0.370
ELECTRA-NLP	0.866	0.221	0.933	0.356	0.887	0.395	0.931	0.275	0.941	**0.610**	0.875	0.309
ELECTRA-no-pretraining	0.858	0.187	0.906	0.217	0.879	0.277	0.926	0.245	0.912	0.308	0.842	0.234

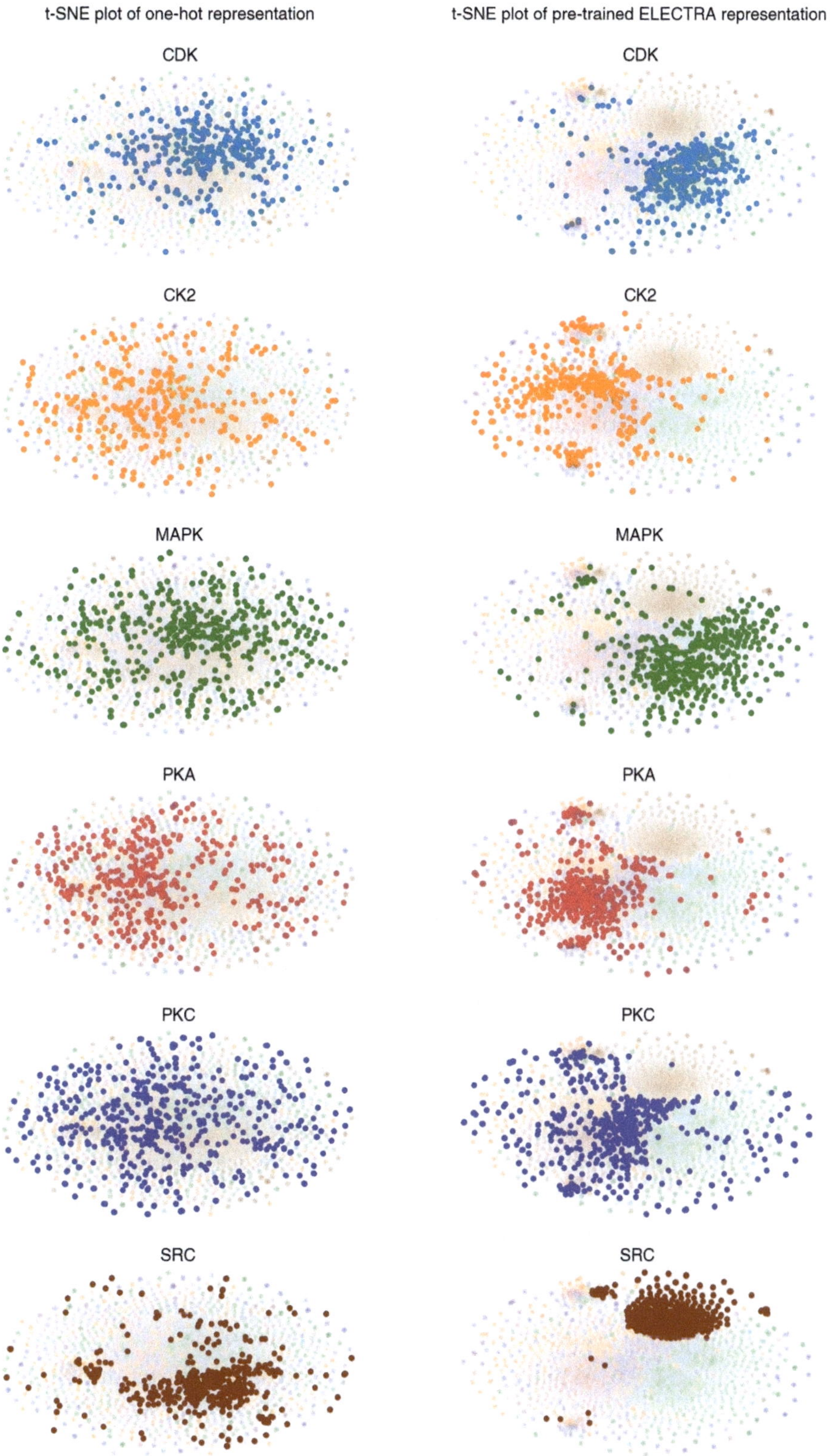

Fig. 5 Comparison of the t-SNE plots of untrained one-hot representation and the pretrained representation of protein fragments

properties. Through visualizing the representations generated by the ELECTRA pretraining, we show that the raw protein fragments can be transformed into a biologically meaningful representation; in particular, even without the kinase labels, it can classify unknown kinase families to some extent, and hence, help the small-sample data learning for the kinase-specific phosphorylation site prediction.

3.4 Tool

The pretrained ELECTRA model for kinase-specific phosphorylation site prediction is provided as an open-source tool available at https://github.com/cirisjl/ELECTRA-for-Kinase-specific-Phosphorylation-Site-Prediction. The input of the tool is a file containing protein sequences in the FASTA format, and the output of the tool is a file that includes the probability scores for each candidate site for each kinase. Residues with probability scores higher than a threshold (0.5 used as the default) are predicted as the substrates for the kinase under consideration. The ELECTRA model was trained on one GPU, and hence it needs one GPU to run the prediction. The software environment requirements are Python 3, TensorFlow-GPU 1.15. CUDA 10, NumPy, Pandas, scikit-learn, and SciPy. The default batch size is 128 for prediction, which can be adjusted according to the VRAM of the user's local machine.

4 Notes

1. The window size of 33 is the longest among various tools—Netphos 3.1 and MusiteDeep use up to 33, Musite uses 27, and GPS 2.0 uses 15.
2. A larger model is more computationally expensive to train and the improvement might not be significant given the relatively small corpus.
3. A large-scale task-independent and unlabeled protein fragment dataset is essential for the pretrained model to unsupervised learn a comprehensive context-dependent biological properties representation of protein sequence segments.
4. A dash token ('-') works better than using a separate padding label in our case.
5. After adding [CLS] to the start of each sequence and [SEP] to the end of each sequence with an original length of 33, we have a maximum input sequence size of 35.
6. While a bigger generator might learn faster during the first few epochs, the learning capability of the discriminator decreases when the generator becomes too strong, making the task too challenging for the discriminator to learn effectively. Also, a larger generator is much more computationally expensive.

5 Discussion

In this paper, we present a pretrained ELECTRA model for both general and kinase-specific phosphorylation site prediction. The ELECTRA is pretrained on a large task-independent raw protein fragment dataset in an unsupervised fashion and fine-tuned on small-sample labeled datasets also with raw protein fragments. The ELECTRA model has been demonstrated to significantly outperform the BERT MLM model and the MusiteDeep model in the kinase-specific phosphorylation site prediction by over 9.02% and 11.1% relative improvement, respectively in the area under the precision-recall curve. We believe the superior performance is due to the following three reasons. (1) The ELECTRA pretraining model learns the contextual information and complex latent features from the vast protein fragment dataset, and the knowledge is transferred to capture the underlying sequence patterns related to phosphorylation and enrich the information in the small dataset. (2) The attention mechanism of the Transformer model inside the ELECTRA helps to learn from the imbalanced dataset and improves the performance. (3) ELECTRA learns more efficiently than MLM because it adopts a more sample-efficient detection strategy of replaced tokens in the pretraining procedure, wherein the detection is conducted over all input tokens rather than just the small subset that was masked out.

This work has explored different deep-learning architectures for protein fragment representation, including BERT and ELECTRA with different hyperparameters and configurations. We measured the computational costs in terms of FLOPs and demonstrated the following. (1) Both BERT and ELECTRA have shown very powerful capability in extracting complex features from massive unlabeled protein databases. (2) ELECTRA learns more efficiently than MLM models such as BERT. (3) The computational cost and learning speed usually increase as the size of the ELECTRA generator increases, but when the size of the generator and training steps reach a certain point, the efficiency for the discriminator to learn decreases. We also demonstrated the importance of the domain-specific pretraining on protein fragments using the ELECTRA model for both general and kinase-specific phosphorylation site prediction by comparing between the ELECTRA model pretrained on NLP and the ELECTRA model without pretraining. The overall performance is protein fragments specific pretrained ELECTRA model > ELECTRA-NLP model > ELECTRA model without pretraining.

By visualizing the fragment representations generated by the pretrained ELECTRA and the untrained one-hot encoding method, the pretrained ELECTRA model demonstrated its ability to extract different latent biologically interpretable patterns underlying a specific kinase family to tell where it belongs to. The pretrained ELECTRA was proved effective and robust in general and kinase-specific phosphorylation feature representations through unsupervised learning. To our best knowledge, this work represents the first attempt to apply ELECTRA for biological sequence analyses. Like different representation models trained for different languages in NLP tasks, the biological sequence representation models could vary in terms of granularity and scope to reduce the size of the model, the computational cost of training, or the noise contained in the information. Our fragment representation model for phosphorylation site prediction provides a good example for other similar studies. The pretrained ELECTRA model is also a general representation model that could be used for other PTM prediction models with problem-specific data.

Although the deep representation model from NLP has improved classification performance and become a promising approach, there are still many differences and significant challenges for applying it to other domains like biology. The original replaced token detection strategy might be modified to adapt bioinformatics applications. Recent research shows that (nonneural) alignment-based features substantially outperform features learned via unsupervised pretraining LSTM on structural tasks [34], indicating that there is plenty of room for utilizing prior knowledge of protein sequence data such as multiple sequence alignment to improve pretraining tasks and architectures, especially for the small-sample problem. In future work, we will explore a semisupervised learning approach by incorporating the Multiple Sequence Alignment (MSA) information to provide the distance information about amino acids and train the discriminator to predict the similarity between the replaced tokens. We will also upgrade the architecture to make the model more interpretable to capture the underlying biological mechanisms.

Acknowledgments

This work was supported by the National Institutes of Health's National Institute of General Medical Sciences (R35-GM126985). In addition, this work used the high-performance computing infrastructure provided by Research Computing Support Services at the University of Missouri, as well as the Extreme Science and Engineering Discovery Environment (XSEDE), which is supported by National Science Foundation grant number ACI-1548562.

References

1. Metcalf SR (2019) An overview of post-translational modifications on PIF1 family helicases. Indiana University
2. Prabakaran S, Lippens G, Steen H, Gunawardena J (2012) Post-translational modification: nature's escape from genetic imprisonment and the basis for dynamic information encoding. Wiley Interdiscip Rev Syst Biol Med 4(6): 565–583
3. Cohen P (2002) The origins of protein phosphorylation. Nat Cell Biol 4(5):E127–E130
4. Cohen PT (2002) Protein phosphatase 1–targeted in many directions. J Cell Sci 115(2): 241–256
5. Hubbard MJ, Cohen P (1993) On target with a new mechanism for the regulation of protein phosphorylation. Trends Biochem Sci 18(5): 172–177
6. Pawson T, Scott JD (1997) Signaling through scaffold, anchoring, and adaptor proteins. Science 278(5346):2075–2080
7. Sefton BM (1998) Overview of protein phosphorylation. Curr Protoc Cell Biol. Chapter 14:Unit 14.1.
8. Liu Z, Cao J, Ma Q, Gao X, Ren J, Xue Y (2011) GPS-YNO2: computational prediction of tyrosine nitration sites in proteins. Mol BioSyst 7(4):1197–1204
9. Xue Y, Ren J, Gao X, Jin C, Wen L, Yao X (2008) GPS 2.0, a tool to predict kinase-specific phosphorylation sites in hierarchy. Mol Cell Proteomics 7(9):1598–1608
10. Zhao Q, Xie Y, Zheng Y, Jiang S, Liu W, Mu W, Liu Z, Zhao Y, Xue Y, Ren J (2014) GPS-SUMO: a tool for the prediction of sumoylation sites and SUMO-interaction motifs. Nucleic Acids Res 42(W1): W325–W330
11. Gao J, Thelen JJ, Dunker AK, Xu D (2010) Musite, a tool for global prediction of general and kinase-specific phosphorylation sites. Mol Cell Proteomics 9(12):2586–2600
12. Yao Q, Gao J, Bollinger C, Thelen J, Xu D (2012) Predicting and analyzing protein phosphorylation sites in plants using musite. Front Plant Sci 3:186
13. Plewczynski D, Basu S, Saha I (2012) AMS 4.0: consensus prediction of post-translational modifications in protein sequences. Amino Acids 43(2):573–582
14. Yao Q, Schulze WX, Xu D (2015) Phosphorylation site prediction in plants. In: Plant Phosphoproteomics. Springer, New York, pp 217–228
15. Yandell MD, Majoros WH (2002) Genomics and natural language processing. Nat Rev Genet 3(8):601–610
16. Searls DB (2002) The language of genes. Nature 420(6912):211–217
17. Motomura K, Fujita T, Tsutsumi M, Kikuzato S, Nakamura M, Otaki JM (2012) Word decoding of protein amino acid sequences with availability analysis: a linguistic approach. PLoS One 7(11):e50039
18. Cai Y, Lux MW, Adam L, Peccoud J (2009) Modeling structure-function relationships in synthetic DNA sequences using attribute grammars. PLoS Comput Biol 5(10): e1000529
19. Asgari E, Mofrad MR (2015) Continuous distributed representation of biological sequences for deep proteomics and genomics. PLoS One 10(11):e0141287
20. Goldberg Y, Levy O (2014) word2vec Explained: deriving Mikolov et al. 's negative-sampling word-embedding method. arXiv:preprint arXiv:14023722
21. Devlin J, Chang M-W, Lee K, Toutanova K (2018) Bert: pre-training of deep bidirectional transformers for language understanding. arXiv:preprint arXiv:181004805
22. Rives A, Meier J, Sercu T, Goyal S, Lin Z, Liu J, Guo D, Ott M, Zitnick CL, Ma J (2021) Biological structure and function emerge from scaling unsupervised learning to 250 million protein sequences. Proc Natl Acad Sci U S A 118(15):e2016239118
23. Clark K, Luong M-T, Le QV, Manning CD (2020) Electra: pre-training text encoders as discriminators rather than generators. arXiv: preprint arXiv:200310555
24. Cortiz D (2021) Exploring transformers in emotion recognition: a comparison of BERT, DistillBERT, RoBERTa, XLNet and ELECTRA. arXiv:preprint arXiv:210402041
25. Xiong W, Li XL, Iyer S, Du J, Lewis P, Wang WY, Mehdad Y, Yih W-T, Riedel S, Kiela D (2020) Answering complex open-domain questions with multi-hop dense retrieval. arXiv:preprint arXiv:200912756
26. Han R, Ren X, Peng N (2020) DEER: a data efficient language model for event temporal reasoning. arXiv:preprint arXiv:201215283
27. Nentidis A, Krithara A, Bougiatiotis K, Krallinger M, Rodriguez-Penagos C, Villegas M, Paliouras G (2020) Overview of BioASQ 2020: the eighth BioASQ challenge on large-scale biomedical semantic indexing

and question answering. In: International conference of the cross-language evaluation forum for European languages. Springer, New York, pp 194–214
28. Wang D, Zeng S, Xu C, Qiu W, Liang Y, Joshi T, Xu D (2017) MusiteDeep: a deep-learning framework for general and kinase-specific phosphorylation site prediction. Bioinformatics 33(24):3909–3916
29. Kalchbrenner N, Grefenstette E, Blunsom P (2014) A convolutional neural network for modelling sentences. arXiv:preprint arXiv:14042188
30. UniProt (2021) The universal protein knowledgebase in 2021. Nucleic Acids Res 49(D1): D480–D489
31. Shi Y (2009) Serine/threonine phosphatases: mechanism through structure. Cell 139(3): 468–484
32. Wold S, Esbensen K, Geladi P (1987) Principal component analysis. Chemom Intell Lab Syst 2(1–3):37–52
33. Van der Maaten L, Hinton G (2008) Visualizing data using t-SNE. J Mach Learn Res 9(11)
34. Rao R, Bhattacharya N, Thomas N, Duan Y, Chen X, Canny J, Abbeel P, Song YS (2019) Evaluating protein transfer learning with tape. Adv Neural Inf Proces Syst 32:9689

Chapter 5

iProtGly-SS: A Tool to Accurately Predict Protein Glycation Site Using Structural-Based Features

Iman Dehzangi, Alok Sharma, and Swakkhar Shatabda

Abstract

Posttranslational modification (PTM) is an important biological mechanism to promote functional diversity among the proteins. So far, a wide range of PTMs has been identified. Among them, glycation is considered as one of the most important PTMs. Glycation is associated with different neurological disorders including Parkinson and Alzheimer. It is also shown to be responsible for different diseases, including vascular complications of diabetes mellitus. Despite all the efforts have been made so far, the prediction performance of glycation sites using computational methods remains limited. Here we present a newly developed machine learning tool called iProtGly-SS that utilizes sequential and structural information as well as Support Vector Machine (SVM) classifier to enhance lysine glycation site prediction accuracy. The performance of iProtGly-SS was investigated using the three most popular benchmarks used for this task. Our results demonstrate that iProtGly-SS is able to achieve 81.61%, 93.62%, and 92.95% prediction accuracies on these benchmarks, which are significantly better than those results reported in the previous studies. iProtGly-SS is implemented as a web-based tool which is publicly available at http://brl.uiu.ac.bd/iprotgly-ss/.

Key words Posttranslational modification, Protein glycation, Feature selection, Support vector machine, Evolutionary features, Structural features

1 Introduction

Posttranslational modification (PTM) of proteins is an important mechanism to increase proteomic diversity by covalently attaching small macromolecules to selected amino acid residues after the translation process. PTM plays critical roles in biological reactions ranging from amino acid degradation, cell signaling, regulating the folding process of proteins or their interaction with other molecules to disease causing interactions such as in Parkinson, and many cancers. Therefore, the identification of PTM sites provides detailed insights into the function of proteins and their biological interactions [1]. Proteins undergo a wide range of PTMs. Among them, glycation is among the most important ones. Glycation is defined as

Dukka B. KC (ed.), *Computational Methods for Predicting Post-Translational Modification Sites*,
Methods in Molecular Biology, vol. 2499, https://doi.org/10.1007/978-1-0716-2317-6_5,

the covalent bonding of sugar molecules with proteins [1, 2]. In this process, sugar molecules including glucose, fructose or their other derivatives interact with lysine (K) and arginine (R) residues to form Amadori ketoamine. In the next step, Amadori ketoamine undergoes degradation to form advanced glycation end products (AGEs). The glycation and its related by-products have been shown to be responsible for many diseases, including neurodegenerative disorders, such as Alzheimer and Parkinson [3, 4]. It is also shown to be responsible for complications in vascular complications of diabetes mellitus. Glycation is also shown related in the aging process [5].

Glycation can be detected using experimental methods [6]. Among the experimental methods, Liquid chromatography with mass spectrometry is considered the most successful approach to detect this and other PTMs. However, this method is complicated, expensive, and time-consuming [7, 8]. Hence, it is unable to meet the exponential growth of protein databanks. Therefore, the implementation of fast and accurate computational methods is considered as a feasible alternative to experimental methods to detect glycation and other PTMs [9]. Among different computational methods, machine learning based approaches have emerged as the most effective and accurate methods to tackle this problem [10].

During the past decade, several machine learning-based methods have been proposed to tackle this problem [2, 11, 12]. Among them, GlyNN [13], PreGly [14], and Gly-PseAAC [15] are the most notable ones which demonstrated the best results for this task. GlyNN was proposed in [13]. To build this model, they used the composition of the amino acids around a given lysine as the input features. Consequently, they used artificial neural network (ANN) as their employed classification to predict lysine glycation sites. Later on, [14] proposed a new method called PreGly to solve this problem. To build this model, they used several sequence-based features including amino acid occurrence frequency, amino acid factors, and the composition of k-spaced amino acid pairs (CKSAAP). They also used support vector machine (SVM) as their classification technique. More recently, [15] used propensity-based features and SVM as the classifier to proposed Gly-PseAAC. Similar to GlyNN and PreGly, Gly-PseAAC also relied solely on sequence-based features to solve this problem. Despite aching promising results, the lysine glycation prediction accuracy remains limited.

In a recent study, we have developed iProtGly-SS, a machine learning based model that uses the combination of structural and sequential based information extracted from the protein sequence to accurately predict lysine glycation sites. We also used SVM as the classification technique to build this model. We investigated the effectiveness of iProtGly-SS on the three most popular benchmarks used for this task. Our results showed that this model is able to significantly outperform previous studies found in the literature to solve this problem. iProtGly-SS achieves 81.61%, 93.62%, and 92.95% accuracies on the employed benchmarks, which in all cases are over 10% better than the results reported in previous studies. It is important to note that iProtGly-SS is the first machine learning-based method to achieve over 90% prediction accuracy for this task on two of the benchmarks used in this study [15].

It is also worth mentioning that iProtGly-SS performance and its achieved results are still better or comparable to more recent machine learning models proposed to predict lysine glycation sites on the employed benchmarks [16–21].

2 Materials and Algorithm

In this section, the datasets used for our experiments and the method used to develop iProtGly-SS are explained in detail.

2.1 Datasets

For the experimental analysis, we used the three most popular datasets that are widely used for the lysine glycation prediction task. The first dataset (Dataset A) was collected from CPLM (http://cplm.biocuckoo.org/) [10]. It consists of 72 proteins containing 323 glycation sites. We then used CD-Hit to remove proteins with over 40% sequence similarity. The resulting dataset contains significantly more negative samples compared to positive samples. To avoid bias toward negative samples, we balanced the dataset using random sampling with the ratio of 2:1 (two negative samples for every positive sample). As a result, the final dataset contained 223 positive and 446 negative samples.

The second dataset (Dataset B) was introduced in [13]. This dataset was generated manually by inspecting a large number of proteins that undergone glycation. The resulting dataset consists of 20 proteins with 89 glycation sites and 126 nonglycation sites.

The third dataset (Dataset C) was introduced in [14], which was manually collected from Uniprot [22, 23]. This dataset was originally used to compare the performance of PreGly with other methods. This dataset contains 82 glycation sites and 117 nonglycation sites.

2.2 Algorithm

To build iProtGly-SS we first studied the impact of several features to investigate their effectiveness and select those with the most important discriminatory information for the lysine glycation site prediction task. Here we studied nine feature groups based on four properties: propensity, composition, secondary structure, and physicochemical properties of the amino acids. We also studied different window sizes to extract our features from the neighboring amino acids. Among them, using a window size of 35 (17 upstream + 17 downstream + 1 central lysine) demonstrated the best results and used to build our model.

Propensity was first used in [15] to build Gly-PseAAC to predict lysine glycation sites. It was shown in this study that propensity can provide important local information about the interaction of neighboring amino acids [24]. Therefore, it is considered as a sequence-based feature and investigated in this study.

Another sequence-based feature investigated in this study was the composition of amino acids. The composition of the amino acids provides information about the occurrence of the neighboring amino acids to a lysine. Considering 20 standard amino acids plus an unknown amino acid "X," we extracted 21 features based on the composition property.

On the other hand, we used predicted secondary structure using SPIDER 3.0 to extract structure-based features [25, 26]. Structural information provides information about the local structure of the proteins and how amino acids interact in 3D structure [27–29]. Structural information has been widely used to predict different types of PTMs, including SUMOylating, malonylation, pupylation, phosphoglycerylation, succinylation, glycosylation, and other similar problems and obtained promising results [30–42].

Finally, we also extracted features based on different physicochemical properties of the amino acids. These features can provide important information about different attributes of the amino acids and how they interact. We studied six different physicochemical attributes of the amino acids that were used in [43], namely, polarity, polarizability, normalized van der valse volume, secondary structure tendency, charge, and solvent accessibility. As a result, we extract six feature groups based on these properties.

We also studied a wide range of classifiers which among them, Support Vector Machine (SVM), attained the best results and therefore used as the primary classifier to build iProtGly-SS. SVM is a nonparametric classification technique that aims at finding the maximum marginal hyperplane to minimize the prediction error [44].

We then investigated the impact of each feature group to select those with the highest discriminatory information for lysine glycation prediction task [14]. To do this, we design an incremental subset feature group selection approach, named INFUSE: INcremental Feature groUp Selection. INFUSE is designed similar to incremental feature selection [45]. In this model, we first investigate the impact of each feature group, separately. We then choose the one with the best performance. In the next step, we add the remaining feature groups to the one selected in the first step and repeat the experiment to identify the best performance. Consequently, we then choose the combination of two with the best

Table 1
Results achieved using iProtGly-SS on our employed datasets compared to PreGly, Gly-PseAAC, and GlyNN

Dataset	Method	Accuracy (%)	Sensitivity (%)	Specificity (%)	AUC	MCC
Dataset A	*iProtGly-SS*[a]	*81.61 ± 4.7*	*92.38 ± 1.1*	*60.09 ± 2.3*	*0.8676 ± 0.04*	*0.5619 ± 0.13*
	iProtGly-SS[b]	80.72	93.05	56.05	0.8816	0.5474
	Gly-PseAAC[a]	68.69	57.48	74.30	0.7199	0.3166
	Gly-PseAAC[b]	68.91	58.74	73.99	0.7258	0.3198
Dataset B	*iProtGly-SS*[a]	*93.62 ± 4.1*	*93.68 ± 2.8*	*93.42 ± 1.6*	*0.9773 ± 0.02*	*0.8783 ± 0.03*
	iProtGly-SS[b]	94.15	95.79	92.11	0.9841	0.8815
	GlyNN[c]	79.50	78.65	80.15	0.77	0.58
	PreGly[a]	85.51	71.06	95.85	–	0.70
	Gly-PseAAC[a]	68.12	56.06	80.17	0.7705	0.38
Dataset C	*iProtGly-SS*[a]	*92.95 ± 6.0*	*95.73 ± 1.2*	*89.02 ± 3.8*	*0.9672 ± 0.03*	*0.8543 ± 0.12*
	iProtGly-SS[b]	92.96	95.73	89.02	0.9635	0.8545
	PreGly[a]	79.92	64.20	91.57	–	0.5849

[a] Tenfold cross-validation
[b] Leave-one-out cross-validation
[c] Threefold cross-validation

performance. We repeat this process for all nine feature groups to determine the best combination of feature groups for the lysine glycation site prediction task.

Using INFUSE, we identify that we can obtain the best results by using amino acid composition, secondary structure, and polarity. Hence, we build iProtGly-SS using the combination of these three feature groups and SVM as our classification method.

Table 1 demonstrates the results achieved using iProtGly-SS on our employed datasets compared to PreGly, Gly-PseAAC, and GlyNN using threefold, tenfold, and leave-one-out cross-validation as our validation methods [13–15]. We compared our results using different evaluation measurements such as Accuracy, Area Under the Curve (AUC), Sensitivity, Specificity, and Matthews Correlation Coefficient (MCC).

As shown in Table 1, iProtGly-SS is able to significantly outperform PreGly, GlyNN, and Gly-PseAAC on all three datasets with respect to all the evaluation measurement. In particular, iProtGly-SS achieves, 81.61%, 93.62%, and 92.95% prediction accuracies which are 12.92%, 8.11%, and 13.03% better than those reported by previous studies in Datasets A, B, and C, respectively. It is also important to note that iProtGly-SS is able to achieve significantly better sensitivity than those reported in other studies [13–15]. This highlights the effectiveness of this method to predict lysine glycation sites compared to other methods.

#	AA	SS	ASA	Phi	Psi	Theta(i-1=>i+1)	Tau(i-2=>i+1)	P(C)	P(E)	P(H)
1	M	C	137.7	-101.4	137.8	118.7	-148.8	0.988	0.011	0.001
2	A	C	83.3	-82.7	136.1	113.9	-129.6	0.978	0.021	0.001
3	T	E	45.7	-110.3	139.5	122.6	-158.2	0.364	0.615	0.002
4	K	E	81.8	-112.6	134.5	120.7	-155.5	0.422	0.545	0.005
5	A	E	7.9	-118.7	141.0	126.6	-151.9	0.190	0.799	0.012
6	V	E	14.7	-125.6	137.0	127.1	-158.1	0.034	0.965	0.005
7	C	E	4.7	-119.6	139.2	127.2	-161.1	0.060	0.909	0.040
8	V	E	15.5	-116.0	127.4	118.7	-158.8	0.051	0.946	0.007
9	L	E	14.2	-105.9	133.7	119.2	-156.1	0.098	0.892	0.004
10	K	C	96.4	-120.6	136.1	124.6	-167.9	0.493	0.491	0.004
11	G	C	16.9	-147.6	174.8	129.3	171.8	0.655	0.365	0.005
12	D	C	110.7	-105.5	71.1	109.4	-105.3	0.806	0.184	0.013
13	G	C	43.4	-167.8	-168.7	136.7	32.2	0.943	0.053	0.008
14	P	C	105.4	-76.2	134.1	112.4	-50.9	0.913	0.075	0.012
15	V	E	22.8	-112.3	133.3	120.8	162.9	0.306	0.676	0.005
16	Q	C	110.5	-121.7	138.8	125.6	-165.3	0.530	0.462	0.002

Fig. 1 Partial SPD3 matrix sample, which is produced as the output of SPIDER 3.0. iProtGly-SS uses predicted secondary structure (P(C), P(E), P(E)) to extract structural based features

2.3 Webserver

1. iProtGly-SS is implemented as an online tool which is publicly available at http://brl.uiu.ac.bd/iprotgly-ss/ (*see* **Note 1**). The input to the webserver is in the SPD3 format which is the output of SPIDER 3.0 (*see* **Note 2**). The partial input SPD3 sample is shown in Fig. 1. It is important to note that iProtGly-SS can work with any other predicted secondary structure output file with the same output format as SPIDER 3.0 [26]. For example, the output of SPIDER 2.0, an earlier version of SPIDER 3.0 (*see* **Note 3**), which is also used for predicting protein local structure, is in the similar format [27, 28]. Hence, it can also be used as an input to iProtGly-SS (*see* **Note 4**). However, for such a case, the performance will not be guaranteed to be exactly the same as what was reported since the local structure prediction accuracy of different tools might be different, which directly impacts the iProtGly-SS performance.
2. iProtGly-SS uses the secondary structure predicted values (P (C), P(E), P(H)) from SPD3 to extract related features as it was mentioned earlier in Subheading 2.
3. Figure 2 demonstrates iProtGly-SS web-based tool and its output for a sample sequence (Q72FV4). As shown in this figure, user can upload the SPD3 for a given protein and submit it. The results will be then shown at the bottom of the page. For each lysine (shown as 'K') in the input protein sequence, iProtGly-SS predict if it undergoes glycation or not. it also prints the lysine and its neighboring amino acids with respect to the window size that is adopted for this model (which is equal to 35), the position of the lysine in the protein sequence, and if it is a glycation site or not. iProtGly-SS is fast, very easy to use, and its generated output is straightforward and easy to interpret.

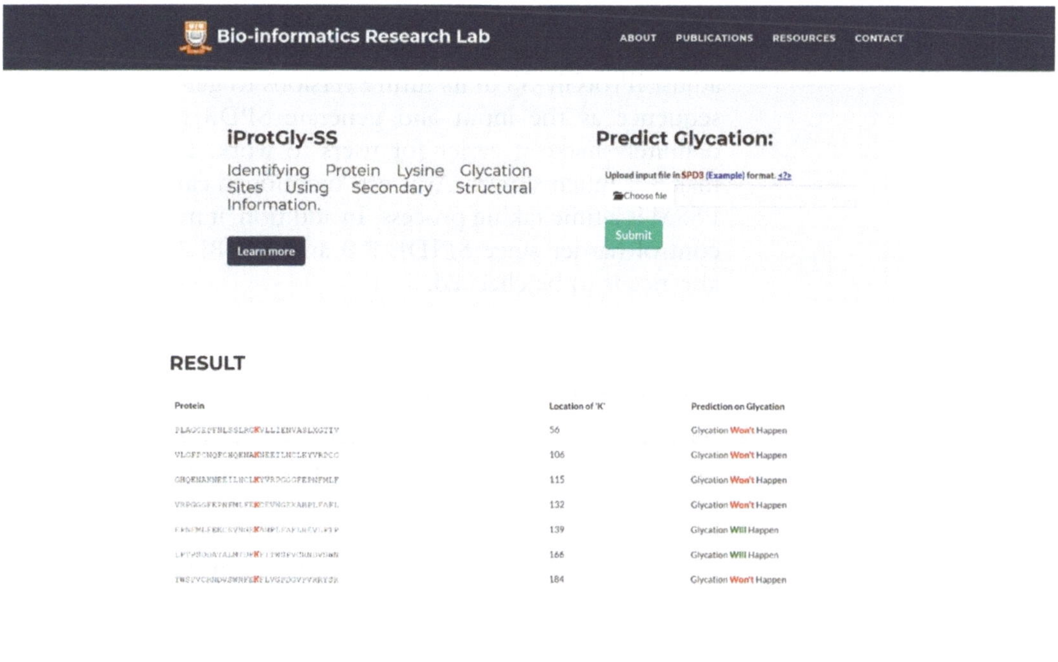

Fig. 2 iProtGly-SS and its prediction result for the sample sequence "Q72FV4"

4. To generate SPD3, SPIDER 3.0 requires Position Specific Scoring Matrix (PSSM), which is generated as the output of PSI-BLAST. Generating PSSM is a time taking process (depend on the available computational resources). Hence, to maintain the simplicity of the usage, iProtGly-SS uses SPD3 directly as an input. It also makes it faster. iProtGly-SS produces the output in less than 5 min, depending on the length of the input protein. It also makes it independent from those tools (SPIDER 3.0 and PSI-BLAST). iProtGly-SS is written in python, which also makes it easier to adjust and maintain the version control. User can write scripts to automatically upload the SPD3 files to generate the prediction output for more than one input protein sequence.

3 Notes

1. iProtGly-SS is just available as a web-based tool. Hence, to generate the glycation prediction sites, proteins are required to be submitted online and one by one. We aim to implement future versions of iProtGly-SS as a standalone tool as well. It will make it easier to use it iteratively for more than one protein in an offline mode.

2. The query sequence should be in SPD3 format. It makes the preprocessing before submitting the job harder. We aim to adjust iProtGly-SS in its future versions to get the raw protein sequence as the input and generate SPD3 by itself. It can definitely make it easier for users to work. However, it also makes it much slower. As it was mentioned earlier, generating PSSM is a time taking process. In addition, it makes the version control harder since SPIDE 3.0 and PSI-BLAST consistency also needs to be checked.
3. iProtGly-SS also requires regular version control to make sure it is consistent with new versions of Python and the libraries that are used to implement it.
4. As mentioned earlier, iProtGly-SS can work with the output of different tools that predicts the local structure of the proteins as long as they are formatted similar to SPD3. However, its output will no longer be similar and will depend on the prediction accuracy of that tool for the protein secondary structure prediction task.

References

1. Khoury GA, Baliban RC, Floudas CA (2011) Proteome-wide post-translational modification statistics: frequency analysis and curation of the swiss-prot database. Sci Rep 1(1):1–5
2. Wang D, Liu D, Yuchi J, He F, Jiang Y, Cai S, Li J, Xu D (2020) MusiteDeep: a deep-learning based webserver for protein post-translational modification site prediction and visualization. Nucleic Acids Res 48(W1): W140–W146
3. Castellani RJ, Harris PL, Sayre LM, Fujii J, Taniguchi N, Vitek MP, Founds H, Atwood CS, Perry G, Smith MA (2001) Active glycation in neurofibrillary pathology of Alzheimer disease: Nε-(carboxymethyl) lysine and hexitol-lysine. Free Radic Biol Med 31(2): 175–180
4. Choudhary C, Kumar C, Gnad F, Nielsen ML, Rehman M, Walther TC, Olsen JV, Mann M (2009) Lysine acetylation targets protein complexes and co-regulates major cellular functions. Science 325(5942):834–840
5. Ulrich P, Cerami A (2001) Protein glycation, diabetes, and aging. Recent Prog Horm Res 56(1):1–22
6. Tatjewski M, Kierczak M, Plewczynski D (2017) Predicting post-translational modifications from local sequence fragments using machine learning algorithms: overview and best practices. Methods Mol Biol 1484: 275–300
7. Thornalley PJ, Battah S, Ahmed N, Karachalias N, Agalou S, Babaei-Jadidi R, Dawnay A (2003) Quantitative screening of advanced glycation endproducts in cellular and extracellular proteins by tandem mass spectrometry. Biochem J 375(3):581–592
8. Zhang Q, Ames JM, Smith RD, Baynes JW, Metz TO (2009) A perspective on the Maillard reaction and the analysis of protein glycation by mass spectrometry: probing the pathogenesis of chronic disease. J Proteome Res 8(2): 754–769
9. Eisenhaber B, Eisenhaber F (2010) Prediction of posttranslational modification of proteins from their amino acid sequence. Methods Mol Biol 609:365–384
10. Liu Z, Wang Y, Gao T, Pan Z, Cheng H, Yang Q, Cheng Z, Guo A, Ren J, Xue Y (2014) CPLM: a database of protein lysine modifications. Nucleic Acids Res 42(D1): D531–D536
11. Xue Y, Liu Z, Cao J, Ren J (2011) Computational prediction of post-translational modification sites in proteins. Syst Comput Biol Mol Cell Exp Syst 5772(6):18559
12. Liu Y, Wang M, Xi J, Luo F, Li A (2018) PTM-ssMP: a web server for predicting different types of post-translational modification sites using novel site-specific modification profile. Int J Biol Sci 14(8):946

13. Johansen MB, Kiemer L, Brunak S (2006) Analysis and prediction of mammalian protein glycation. Glycobiology 16(9):844–853
14. Liu Y, Gu W, Zhang W, Wang J (2015) Predict and analyze protein glycation sites with the mRMR and IFS methods. BioMed Res Int 2015:561547
15. Xu Y, Li L, Ding J, Wu LY, Mai G, Zhou F (2017) Gly-PseAAC: identifying protein lysine glycation through sequences. Gene 602:1–7
16. Yu J, Shi S, Zhang F, Chen G, Cao M (2019) PredGly: predicting lysine glycation sites for Homo sapiens based on XGboost feature optimization. Bioinformatics 35(16):2749–2756
17. Chen J, Zhang C, Yang R, Zhang L (2019) A new machine learning based framework to identify protein glycation sites using comprehensive features and the mRMR method. In: 2019 Chinese control and decision conference (CCDC). IEEE, pp 3605–3609
18. Chen J, Yang R, Zhang C, Zhang L, Zhang Q (2019) DeepGly: a deep learning framework with recurrent and convolutional neural networks to identify protein glycation sites from imbalanced data. IEEE Access 7: 142368–142378
19. Reddy HM, Sharma A, Dehzangi A, Shigemizu D, Chandra AA, Tsunoda T (2019) GlyStruct: glycation prediction using structural properties of amino acid residues. BMC Bioinformatics 19(13):55–64
20. Khanum S, Ashraf MA, Karim A, Shoaib B, Khan MA, Naqvi RA, Siddique K, Alswaitti M (2021) Gly-LysPred: identification of lysine glycation sites in protein using position relative features and statistical moments via Chou's 5 step rule. Comput Mater Continua 66(2): 2165–2181
21. Abrahams JL, Taherzadeh G, Jarvas G, Guttman A, Zhou Y, Campbell MP (2020) Recent advances in glycoinformatic platforms for glycomics and glycoproteomics. Curr Opin Struct Biol 62:56–69
22. UniProt Consortium (2015) UniProt: a hub for protein information. Nucleic Acids Res 43 (D1):D204–D212
23. Lee TY, Huang HD, Hung JH, Huang HY, Yang YS, Wang TH (2006) dbPTM: an information repository of protein post-translational modification. Nucleic Acids Res 34(suppl_1): D622–D627
24. Tang YR, Chen YZ, Canchaya CA, Zhang Z (2007) GANNPhos: a new phosphorylation site predictor based on a genetic algorithm integrated neural network. Protein Eng Des Sel 20(8):405–412
25. Heffernan R, Dehzangi A, Lyons J, Paliwal K, Sharma A, Wang J, Sattar A, Zhou Y, Yang Y (2016) Highly accurate sequence-based prediction of half-sphere exposures of amino acid residues in proteins. Bioinformatics 32(6): 843–849
26. Heffernan R, Yang Y, Paliwal K, Zhou Y (2017) Capturing non-local interactions by long short-term memory bidirectional recurrent neural networks for improving prediction of protein secondary structure, backbone angles, contact numbers and solvent accessibility. Bioinformatics 33(18):2842–2849
27. Heffernan R, Paliwal K, Lyons J, Dehzangi A, Sharma A, Wang J, Sattar A, Yang Y, Zhou Y (2015) Improving prediction of secondary structure, local backbone angles and solvent accessible surface area of proteins by iterative deep learning. Sci Rep 5(1):1–11
28. Yang Y, Heffernan R, Paliwal K, Lyons J, Dehzangi A, Sharma A, Wang J, Sattar A, Zhou Y (2017) Spider2: a package to predict secondary structure, accessible surface area, and main-chain torsional angles by deep neural networks. In: Prediction of protein secondary structure. Humana Press, New York, NY, pp 55–63
29. Paliwal KK, Sharma A, Lyons J, Dehzangi A (2014) Improving protein fold recognition using the amalgamation of evolutionary-based and structural based information. BMC Bioinformatics 15(16):1–9
30. Islam MM, Saha S, Rahman MM, Shatabda S, Farid DM, Dehzangi A (2018) iProtGly-SS: identifying protein glycation sites using sequence and structure based features. Proteins 86(7):777–789
31. López Y, Dehzangi A, Reddy HM, Sharma A (2020) C-iSUMO: a sumoylation site predictor that incorporates intrinsic characteristics of amino acid sequences. Comput Biol Chem 87:107235
32. Chandra A, Sharma A, Dehzangi A, Shigemizu D, Tsunoda T (2019) Bigram-PGK: phosphoglycerylation prediction using the technique of bigram probabilities of position specific scoring matrix. BMC Mol Cell Biol 20(2):1–9
33. Dipta SR, Taherzadeh G, Ahmad MW, Arafat ME, Shatabda S, Dehzangi A (2020) SEMal: accurate protein malonylation site predictor using structural and evolutionary information. Comput Biol Med 125:104022
34. Singh V, Sharma A, Dehzangi A, Tsunoda T (2020) PupStruct: prediction of pupylated lysine residues using structural properties of amino acids. Genes 11(12):1431

35. López Y, Sharma A, Dehzangi A, Lal SP, Taherzadeh G, Sattar A, Tsunoda T (2018) Success: evolutionary and structural properties of amino acids prove effective for succinylation site prediction. BMC Genomics 19(1): 105–114
36. Arafat M, Ahmad M, Shovan SM, Dehzangi A, Dipta SR, Hasan M, Mehedi A, Taherzadeh G, Shatabda S, Sharma A (2020) Accurately predicting glutarylation sites using sequential bi-peptide-based evolutionary features. Genes 11(9):1023
37. Sharma A, Lysenko A, López Y, Dehzangi A, Sharma R, Reddy H, Sattar A, Tsunoda T (2019) HseSUMO: SUMOylating site prediction using half-sphere exposures of amino acids residues. BMC Genomics 19(9):1–7
38. Taherzadeh G, Dehzangi A, Golchin M, Zhou Y, Campbell MP (2019) SPRINT-Gly: predicting N-and O-linked glycosylation sites of human and mouse proteins by using sequence and predicted structural properties. Bioinformatics 35(20):4140–4146
39. Chandra A, Sharma A, Dehzangi A, Ranganathan S, Jokhan A, Chou KC, Tsunoda T (2018) PhoglyStruct: prediction of phosphoglycerylated lysine residues using structural properties of amino acids. Sci Rep 8(1):1–11
40. Dehzangi A, Karamizadeh S (2011) Solving protein fold prediction problem using fusion of heterogeneous classifiers. Information 14(11):3611–3622
41. Uddin MR, Sharma A, Farid DM, Rahman MM, Dehzangi A, Shatabda S (2018) EvoStruct-Sub: an accurate Gram-positive protein subcellular localization predictor using evolutionary and structural features. J Theor Biol 443:138–146
42. Shatabda S, Saha S, Sharma A, Dehzangi A (2017) iPHLoc-ES: identification of bacteriophage protein locations using evolutionary and structural features. J Theor Biol 435:229–237
43. Dubchak I, Muchnik I, Mayor C, Dralyuk I, Kim SH (1999) Recognition of a protein fold in the context of the SCOP classification. Proteins 35(4):401–407
44. Cortes C, Vapnik V (1995) Support-vector networks. Machine Learning 20(3):273–297
45. Liu H, Setiono R (1998) Incremental feature selection. Appl Intell 9(3):217–230

Chapter 6

Functions of Glycosylation and Related Web Resources for Its Prediction

Kiyoko F. Aoki-Kinoshita

Abstract

Glycosylation involves the attachment of carbohydrate sugar chains, or glycans, onto an amino acid residue of a protein. These glycans are often branched structures and serve to modulate the function of proteins. Glycans are synthesized through a complex process of enzymatic reactions that occur in the Golgi apparatus in mammalian systems. Because there is currently no sequencer for glycans, technologies such as mass spectrometry is used to characterize glycans in a biological sample to ascertain its glycome. This is a tedious process that requires high levels of expertise and equipment. Thus, the enzymes that work on glycans, called glycogenes or glycoenzymes, have been studied to better understand glycan function. With the development of glycan-related databases and a glycan repository, bioinformatics approaches have attempted to predict the glycosylation pathway and the glycosylation sites on proteins. This chapter introduces these methods and related Web resources for understanding glycan function.

Key words Glycans, Glycobiology, Glycome, Glycosylation, Biosynthesis

1 Introduction

As one of the major, and certainly largest, post-translational modifications, glycosylation has been the focus of many prediction tools. Protein databases now store large amounts of glycosylation site information, enabling bioinformaticians to develop predictive tools for glycosylation sites based on amino acid sequence alone. Many such tools now exist and exhibit high levels of predictive accuracy. On the other hand, many other challenges remain in glycobiology. Firstly, no high-throughput sequencer currently exists for glycans, and thus the community depends on technologies such as mass spectrometry to characterize the glycans in a glycome. Such technologies require high levels of expertise as well as time and money. Therefore, predictive tools for assigning the glycans in a mass spectrum have been developed [1–4]. While many

Dukka B. KC (ed.), *Computational Methods for Predicting Post-Translational Modification Sites*,
Methods in Molecular Biology, vol. 2499, https://doi.org/10.1007/978-1-0716-2317-6_6,

of these tools were based on a glycan library in the past, de novo methods have recently been developed, albeit with lower predictive accuracy.

Due to the lack of a glycan sequencer, many researchers turned to the glycoenzymes involved in synthesizing glycans in an attempt to predict the glycans in the glycome based on the glycogene expression profile [5, 6]. Here, the bottleneck was the lack of information regarding glycan substrate specificity of each enzyme. Nevertheless, based on the knowledge of such specificity at the time, mathematical models have been developed, and software tools created to predict the glycan biosynthetic pathway [7, 8]. Furthermore, kinetic reaction parameter information could be incorporated to perform simulations of the fluctuations in glycan concentration over time. Again, the lack of such parameter data has limited the application of such simulations to real data, and predictive tools for supplementing the missing information have been developed.

This chapter presents a brief introduction glycobiology and its importance in understanding disease. This is followed by a description of the currently available data in glycobiology and the various attempts at developing predictive tools for glycosylation sites on amino acid sequences and glycan biosynthesis. We hope to show that despite the challenges, much progress has been made in the field.

2 Materials

2.1 Brief Introduction to Glycobiology

Glycobiology is the study of complex carbohydrates, or glycans, which are chains of monosaccharides (five- or six-carbon rings) that often form branched structures [9]. Glycans are formed in a complex process of glycosylation, which proceeds by the reactions of glycosyltransferase and glycosylhydrolase enzymes. These enzymes act on specific glycan substrates, forming the complex structures of glycans. Glycosyltransferases require a sugar-nucleotide donor to attach a monosaccharide onto a glycan. These glycosyltransferases exist in mammalian systems mainly in the Golgi apparatus, through which the glycan substrates are processed to eventually proceed to the cell surface.

Once at the cell surface, glycans serve as the "face" of cells, whereby various viruses, bacteria, hormones, proteins such as lectins, and other cells interact with the host cell. Many studies in glycobiology attempt to understand the mechanisms of these interactions, which have shown to be important in many biological processes. Changes in glycosylation are known to occur in various disease conditions, including cancer [10]. Thus, glycans are potential biomarkers for different types of cancer.

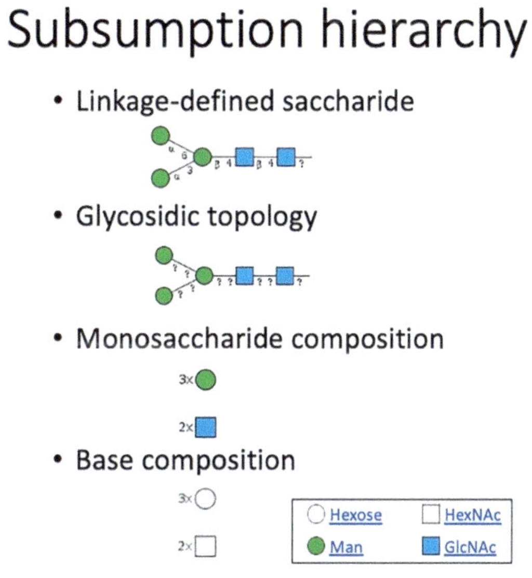

Fig. 1 The subsumption hierarchy encapsulating the varying level of ambiguity in glycan structures published in the literature and in GlyTouCan

Currently, bioinformatics research in glycobiology consist of database development, which involves the development of ontologies and database resources that include storage of the ambiguous nature of glycans. Current glycomics technologies can basically only ascertain the monosaccharide composition of a glycan, and so when the glycome of a sample is characterized, the glycans are represented in an ambiguous form, limited by the sensitivity of the techniques used (usually mass spectrometry and liquid chromatography). Figure 1 illustrates the varying level of ambiguity that can be found in the literature regarding glycan structure. The fully defined glycan with all glycosidic linkages characterized is ideal, but even some linkage-defined glycans may have some missing linkage information. Using orthogonal technologies such as glycosidase digestions, some of this information could be obtained. The second level is the glycosidic topology, whereby linkage conformations are unknown, but it is known which monosaccharide is linked to which. The more common level of glycan structure is the monosaccharide composition, obtained from mass spectrometry, and at the basic level, even the conformation of the monosaccharides may not be known, resulting in base composition data.

Regarding other types of ambiguity in glycobiology, the enzymes known to react with glycans are also often ambiguous in nature, in that certain environmental conditions must be satisfied for an enzyme to catalyze a glycosidic reaction. For example, B3GALNT2 is a glycosyltransferase that basically transfers a GalNAc residue onto GlcNAc residues on *O*-glycans, but there are certain GlcNAc substrates onto which it does not transfer

[11]. This is evident from the GGDB page listing the variety of substrates onto which this enzyme does and does not transfer. The precise localization of most of these transferase enzymes in the Golgi are also unknown, making it difficult to predict the glycome of a sample based on gene expression data alone. Moreover, many well-known lectins such as concanavalin A or wheat germ agglutinin (WGA) are known to bind to well-defined glycan patterns, but many lectins show varying levels of affinity for various types of glycans, as can be seen from the Lectin frontier DataBase (LfDB) [12], provided by ACGG-DB and by GlyCosmos, described later.

However, attempts at the prediction of glycosylation have been made, and this chapter will attempt to introduce some of these efforts. Two types of prediction will be described: glycosylation prediction involving the simulation of the glycosylation process within the Golgi, and glycosylation prediction involving the prediction of the glycosylation position on a protein sequence.

2.2 Glycan-Related Databases

2.2.1 Glycan Structures

GlyTouCan (https://glytoucan.org) is the international glycan repository, which users can access to register glycans as well as retrieve the glycans that have been reported in the field [13]. GlyTouCan uses Semantic Web technologies [14] to store glycan structures and their related metadata, especially links to GlyTouCan Partners which are glycan databases that link to and back from GlyTouCan. Additional information such as monoisotopic mass, glycan motifs, and a concept called subsumption have been applied to glycans in GlyTouCan, and this information is available under the Glycans dataset in the GlyCosmos Portal (https://glycosmos.org) [15]. GlyCosmos pulls in the glycans registered in GlyTouCan on a weekly basis, adding such metadata to make it easier to search for glycans. GlyCosmos also provides a Download page for various datasets including glycans and glycoproteins.

2.2.2 Glycoproteins

There are a number of data resources of glycoprotein information, including GlyConnect [16], GlyGen [17] and GlycoProtDB [18]. GlyConnect and GlyGen have accumulated glycoprotein information from the literature, including glycosylation sites and the glycans at those sites. GlycoProtDB is a database of glycoprotein sites and lectin microarray results on each site, characterized by the IGOT-MS [19] and lectin microarray [20] technologies at AIST in Japan. Lectins are known to recognize specific patterns of glycans, and since they are easier to handle than tandem MS technologies, lectin microarray technologies are a semi–high-throughput method for characterizing glycans. GlycoProtDB is available from ACGG-DB as well as GlyCosmos, which provides both download links as well as a SPARQL endpoint for access to its data. GlyGen provides tabular data of all the data integrated in its portal, and GlyConnect provides a SPARQL endpoint by which data can be retrieved. GlyConnect is a part of the Glycomics@ExPASy portal, which, together with GlyGen and GlyCosmos, forms the GlySpace Alliance [21]. This alliance aims to share the data from each of the member portals so that the glycoscience community can obtain the consistent data from any of these resources.

2.2.3 Glycoenzymes and Diseases

One of the earliest developed resources of carbohydrate active enzymes is CAZy (http://cazy.org) [22, 23]. Based on structural and sequence data of known glycoenzymes, five classes of enzymes (glycoside hydrolases (GH), glycosyltransferases (GT), polysaccharide lyases (PL), carbohydrate esterases (CE), and other auxiliary activities (AA)) are covered. Semiautomated tools predict the CAZymes in genomic databases, the predictions of which are available from their website.

Probably the second oldest glycoenzyme database is GlycoGene DataBase, or GGDB (https://acgg.asia/db/ggdb/), originally provided by JCGGDB, and now by ACGG-DB. GGDB information has been integrated into GlyCosmos as well. GGDB consists of glycosyltransferases, sugar-nucleotide synthases, sugar-nucleotide transporters, and sulfotransferases, among others. The activities of all human glycogenes have been characterized by the developers of GGDB, which stores all the obtained information from these experiments. Thus, expression information, substrate specificity, and orthologous gene information are provided. In particular, substrate specificity is represented using GlyTouCan accession numbers, providing a link between enzymes and glycans.

ACGG-DB also provides the GDGDB database of genetic diseases known to be related to glycan metabolism. Also integrated in GlyCosmos, this database forms an ontology of 120 genetic diseases, with links to OMIM, GGDB, NCBI, and UniProtKB.

GlycoEnzDB (http://www.virtualglycome.org/GlycoEnzDB) is a manually curated database of glycoenzymes in human, providing pathway diagrams of these enzymes, along with known substrate specificities, reaction parameters and localization information. Transcription factor ChiP-Seq data and microRNA data have also been integrated.

KEGG GLYCAN (https://www.genome.jp/kegg/glycan/) provides pathway-based glycan structure and enzyme information. The basic types of glycans (*N*-glycans, *O*-glycans, etc.) can be viewed with the enzymes known to synthesize specific glycosidic linkages mapped onto them. KEGG BRITE also has a table for cancer-associated carbohydrates as well as Lectins.

3 Methods

3.1 Glycan Biosynthesis Prediction

Because of the difficulty in characterizing glycans and the glycome at the glycosidic linkage level, tools to predict the glycome from a set of glycogenes have been developed. One of the earliest mathematical models of *N*-glycan biosynthesis is now used as the basis of many systems glycobiology research today. This model [24] used the LinearCode [25] representation of glycans as sequences to define glycan substrates, and it was used as the basis of the Glycan Pathway Predictor (GPP) tool [8] in RINGS, a web resource for data mining and computational tools for glycans [26]. The GPP tool was further applied to the entire dataset of human glycogenes to predict the human glycome [27]. As a result, over one million glycans of up to 15 monosaccharides each were computed. However, glycan databases only contain a couple of thousand glycans characterized in human. This discrepancy is most likely due to the ambiguity of glycoenzymes and thus the lack of sufficient constraints on the rules for predicting the biosynthesis pathway. Thus, improved prediction tools and more data are needed to explain this gap.

In order to provide a Web-based version of GPP which could further used to perform simulations, the GlycoSim Web tool was developed (https://rings.glycoinfo.org/glycosims/index). Similar to the pathway prediction tool in GlycoEnzDB, users can select a set of enzymes, an initial glycan structure, and the biosynthetic pathway can be predicted based on known substrate specificity information. GlycoSim used the LinearCode method for specifying substrate specificity, whereas GlycoEnzDB uses the IUPAC nomenclature [28] which could be considered more flexible due the freedom of being able to specify a wider variety of monosaccharides and substituents compared to LinearCode. Because members of the Systems Glycobiology Consortium (https://sysglyco.org) tended to use various extended versions of LinearCode as well, a standardized format called LiCoRR was proposed, with the IUPAC-version of LiCoRR called LiCoRRICE [29]. Table 1 is a listing of select enzymes, their substrates and products in LinearCode format, and their constraint rules in both LiCoRR and LiCoRRICE formats. Details can be found in the original publication, but here we note that this table only specifies the rules for these enzymes in human, whereas other organisms will have different enzymes with different specificities. These specificities may also change based on other conditions such as temperature and pH. However, this was a first step to enable the community to use a consistent format for glycan substrate specificity rules so that such rules could be exchanged more easily.

Table 1
A selected list of human glycoenzymes and their substrate specificity, described in LiCoRR and LiCoRRICE formats. The substrate and products are represented in LinearCode format, which is a single-letter code for compactly representing glycans. Constraint rules are described in the original publication [29]

Enzyme	Substrate	Product	Constraints (LiCoRR)	Constraints (LiCoRRICE)
ManI	(Ma2Ma	(Ma	!@2Ma3 (…Ma6) Ma6 & !Ga3	nMan(a1-?)>4 & nMan(a1-?)<8 & !Man (a1-2)Man(a1-3)...Man(a1-6) & !Glc (a1-3)
ManI	(Ma3 (Ma2Ma3 (Ma6) Ma6)	(Ma3(Ma3 (Ma6)Ma6)	!Ga3	!Glc(a1-3)
ManII	(Ma3(Ma6) Ma6	(Ma6Ma6	(GNb2+Ma3 & !Gnbis	!Gal(b1-?) & !GlcNAc(b1-4)...Man(b1-4) & GlcNAc(b1-2)Man(a1-3)
ManII	(Ma6Ma6	(Ma6	(GNb2+Ma3 & !Gnbis	
a6FucT	GNb4GN	GNb4(Fa6)GN	GNb2+Ma3 & #A=0 & ! Gnbis	GlcNAc(b1-2)Man(a1-3)...Man(b1-4) & !GlcNAc(b1-4)...Man(b1-4) & !Fuc (a1-3)
GnTI	(Ma3(Ma3 (Ma6) Ma6)Mb4	(GNb2Ma3 (Ma3(Ma6) Ma6)Mb4		nMan(a1-?)=4
GnTII	(GNb2+Ma3 (Ma6)Mb4	(GNb2+Ma3 (GNb2Ma6) Mb4		nMan(a1-?)=2 & !GlcNAc(b1-4)...Man (b1-4) & !Fuc(a1-3) & !Gal(b1-?)
GnTIII	GNb2+Ma3	GNb2+Ma3 (GNb4)	!Ab & !Gnbis	GlcNAc(b1-2)Man(a1-3)...Man(b1-4) & !Gal(b1-?)
GnTIV	(GNb2Ma3	(GNb2(GNb4) Ma3	!Gnbis	!Gal(b1-?) & !GlcNAc(b1-4)...Man(b1-4)
GnTV	(GNb2Ma6	(GNb2(GNb6) Ma6	!Gnbis	!Gal(b1-?) & !GlcNAc(b1-4)...Man(b1-4)
iGnT	(Ab4GN	(GNb3Ab4GN	!@_Ma3+Mb4	
b4GalT	(GN	(Ab4GN	!@GNb4)(… Ma6)Mb4	!Gal(b1-3)GlcNAc(b1-?) & !@GlcNAc (b1-4)...Man(b1-4)
b3GalT	(GN	(Ab3GN	!@GNb4)(… Ma6)Mb4	!Gal(b1-4)GlcNAc(b1-?) & !@GlcNAc (b1-4)...Man(b1-4)

3.2 Glycosylation Site Prediction

The problem of glycosylation site prediction has existed for many years, dating back to the start of the century. The "marker sequence" for N-glycosylation was determined in the 1980s as the triplet Asn-Xaa-Thr(Ser) [30, 31]. In the late 1990s, it was found that protein databases showed that the majority of proteins are glycosylated, based on an analysis of this marker sequence and the frequency at which glycans have been characterized at these sequons [32]. Then, since a large amount of carbohydrate data were found in the Protein Data Bank (PDB) [33], statistical analyses of the carbohydrates and surrounding amino acids were performed, and these tools and data made available from the GLYCOSCIENCES.de portal [34, 35].

Due to the accumulation of glycosylation site data in protein databases, a plethora of machine learning methods have been developed to predict glycosylation sites from amino acid sequences. Support vector machine classifiers [36, 37], random forests [38–40], and combinations of these with neural networks [41] and with structural data [42] have been developed. Since glycobiologists may not necessarily be computer-savvy, Web tools for glycosylation prediction have also been developed. In addition to some of those mentioned above, NetNGlyc [43], NetOGlyc [44], and NetCGlyc [45] have all been made available since the early 2000s and continue to make improvements. NetNGlyc predicts *N*-glycosylation sites in human proteins using artificial neural networks surrounding the marker sequence. NetOGlyc predicts *O*-glycosylation, which is a class of small glycans ranging from one to five monosaccharides, often starting with a GalNAc residue found on Ser and Thr residues. NetCGlyc predicts C-mannosylation, which is the attachment of a Man residue to a tryptophan via a C-C linkage. The WXXW sequon is known to be C-mannosylated at the first Trp. However, due to the lack of true positive data regarding glycosylation, many of these methods are dependent on the dataset at hand.

In conclusion, the lack of data is the greatest bottleneck in developing tools for biosynthesis and glycosylation site prediction. In particular, reaction parameters for glycoenzymes are lacking; while experimental techniques exist to establish these parameters, the efforts required to obtain them far outweigh the results. Thus, tool developers are left with developing parameter estimation and sensitivity analyses to establish these parameters. GlycoSim has started implementing such functionality to make it available from the Web. On the other hand, nonparametric models have also been developed to avoid such details altogether by using flux models and Markov models [46, 47].

Regarding glycosylation site prediction, while there is a lot of data on *N*-glycosylation, fewer datasets of *O*-glycosylation, and even fewer of C-mannosylation and even other types of glycosylation exist. Glycan and protein databases have gradually started incorporating such information. However, until high-throughput technologies for sequencing such glycosylation patterns mature, prediction tools to better understand the basic mechanisms of how glycoenzymes function may be needed. Indeed, while there have been attempts to predict glycoenzymes from sequence [48], the results have been limited. This may require further analysis of the 3D structural conformations of these enzymes, many of which are available from the PDB as well as CAZy. Moreover, PDB has recently undergone a Carbohydrate Remediation, enhancing the validation of carbohydrates in their database [33]. This should provide a stable resource of data to better understand glycosylation.

References

1. Kumozaki S, Sato K, Sakakibara Y (2015) A machine learning based approach to de novo sequencing of glycans from tandem mass spectrometry spectrum. Comput Biol Bioinformatics 12:1267–1274
2. Apte A, Meitei NS (2010) Bioinformatics in glycomics: glycan characterization with mass spectrometric data using SimGlycan. Methods Mol Biol 600:269–281
3. Wei J, Tang Y, Bai Y et al (2020) Toward automatic and comprehensive glycan characterization by online PGC-LC-EED MS/MS. Anal Chem 92:782–791
4. Hong P, Sun H, Sha L et al (2017) GlycoDeNovo—an efficient algorithm for accurate de novo glycan topology reconstruction from tandem mass spectra. J Am Soc Mass Spectrom 28:2288–2301
5. Krambeck FJ, Bennun SV, Narang S et al (2009) A mathematical model to derive N-glycan structures and cellular enzyme activities from mass spectrometric data. Glycobiology 19:1163–1175
6. Umaña P, Bailey JE (1997) A mathematical model of N-linked glycoform biosynthesis. Biotechnol Bioeng 55:890–908
7. Liu G, Neelamegham S (2014) A computational framework for the automated construction of glycosylation reaction networks. PLoS One 9:e100939
8. Aoki-Kinoshita KF (2015) Analyzing glycan structure synthesis with the glycan pathway predictor (GPP) tool. Methods Mol Biol 1273:139–147
9. Varki A, Cummings RD, Esko JD et al (2017) Essentials of glycobiology. Cold Spring Harbor Laboratory Press
10. Azevedo R, Silva AMN, Reis CA et al (2018) In silico approaches for unveiling novel glycobiomarkers in cancer. J Proteomics 171:95–106
11. Hiruma T, Togayachi A, Okamura K et al (2004) A novel human β1,3-N-acetylgalactosaminyltransferase that synthesizes a unique carbohydrate structure, GalNAcβ1-3GlcNAc. J Biol Chem 279:14087–14095
12. Hirabayashi J, Tateno H, Shikanai T et al (2015) The lectin frontier database (LfDB), and data generation based on frontal affinity chromatography. Molecules 20:951–973
13. Fujita A, Aoki NP, Shinmachi D et al (2021) The international glycan repository GlyTouCan version 3.0. Nucleic Acids Res 49:D1529–D1533
14. Aoki-Kinoshita KF, Bolleman J, Campbell MP et al (2013) Introducing glycomics data into the semantic web. J Biomed Semantics 4:39
15. Yamada I, Shiota M, Shinmachi D et al (2020) The GlyCosmos portal: a unified and comprehensive web resource for the glycosciences. Nat Methods 17:649–650
16. Alocci D, Mariethoz J, Gastaldello A et al (2019) GlyConnect: glycoproteomics goes visual, interactive, and analytical. J Proteome Res 18:664–677
17. York WS, Mazumder R, Ranzinger R et al (2020) GlyGen: computational and informatics resources for glycoscience. Glycobiology 30:72–73
18. Kaji H, Shikanai T, Suzuki Y et al (2017) GlycoProtDB: a database of glycoproteins mapped with actual glycosylation sites identified by mass spectrometry. In: A practical guide to using glycomics databases. Springer, Tokyo, pp 215–224
19. Kaji H, Shikanai T, Sasaki-Sawa A et al (2012) Large-scale identification of N-glycosylated proteins of mouse tissues and construction of a glycoprotein database, GlycoProtDB. J Proteome Res 11:4553–4566
20. Zou X, Yoshida M, Nagai-Okatani C et al (2017) A standardized method for lectin microarray-based tissue glycome mapping. Sci Rep 7:43560
21. Aoki-Kinoshita KF, Lisacek F, Mazumder R et al (2020) The GlySpace alliance: toward a collaborative global glycoinformatics community. Glycobiology 30(2):70–71
22. Lombard V, Golaconda Ramulu H, Drula E et al (2014) The carbohydrate-active enzymes

database (CAZy) in 2013. Nucleic Acids Res 42:D490–D495

23. Terrapon N, Lombard V, Drula E et al (2017) The CAZy database/the carbohydrate-active enzyme (CAZy) database: principles and usage guidelines. In: A practical guide to using glycomics databases. Springer, Tokyo, pp 117–131
24. Krambeck FJ, Betenbaugh MJ (2005) A mathematical model of {N}-linked glycosylation. Biotechnol Bioeng 92:711–728
25. Banin E, Neuberger Y, Altshuler Y et al (2002) A novel Linear Code(R) nomenclature for complex carbohydrates. Trends Glycosci Glycotechnol 14:127–137
26. Akune Y, Hosoda M, Kaiya S et al (2010) The RINGS resource for glycome informatics analysis and data mining on the Web. OMICS 14: 475–486
27. Akune Y, Lin C-H, Abrahams JL et al (2016) Comprehensive analysis of the N-glycan biosynthetic pathway using bioinformatics to generate UniCorn: atheoretical N-glycan structure database. Carbohydr Res 431:56–63
28. McNaught AD (1996) Nomenclature of carbohydrates (IUPAC Recommendations 1996). Pure Appl Chem 68:1919–2008
29. Kellman BP, Zhang Y, Logomasini E et al (2020) A consensus-based and readable extension of Linear Code for Reaction Rules (LiCoRR). Beilstein J Org Chem 16: 2645–2662
30. Bause E, Legler G (1981) The role of the hydroxy amino acid in the triplet sequence Asn-Xaa-Thr(Ser) for the N-glycosylation step during glycoprotein biosynthesis. Biochem J 195:639–644
31. Bause E (1983) Structural requirements of N-glycosylation of proteins. Studies with proline peptides as conformational probes. Biochem J 209:331–336
32. Apweiler R, Hermjakob H, Sharon N (1999) On the frequency of protein glycosylation, as deduced from analysis of the SWISS-PROT database. Biochim Biophys Acta 1473:4–8
33. Feng Z, Westbrook JD, Sala R et al (2021) Enhanced validation of small-molecule ligands and carbohydrates in the Protein Data Bank. Structure 29:393–400.e1
34. Werz DB, Ranzinger R, Herget S et al (2007) Exploring the structural diversity of mammalian carbohydrates ("glycospace") by statistical databank analysis. ACS Chem Biol 2:685–691
35. Lutteke T, Bohne-Lang A, Loss A et al (2005) {GLYCOSCIENCES.de}: an {I}nternet portal to support glycomics and glycobiology research. Glycobiology 16:71R–81R
36. Caragea C, Sinapov J, Silvescu A et al (2007) Glycosylation site prediction using ensembles of support vector machine classifiers. BMC Bioinformatics 8:438
37. Chauhan JS, Rao A, Raghava GPS (2013) In silico platform for prediction of N-, O- and C-glycosites in eukaryotic protein sequences. PLoS One 8:e67008
38. Hamby SE, Hirst JD (2008) Prediction of glycosylation sites using random forests. BMC Bioinformatics 9:500
39. Li F, Li C, Wang M et al (2015) GlycoMine: a machine learning-based approach for predicting N-, C- and O-linked glycosylation in the human proteome. Bioinformatics 31: 1411–1419
40. Chuang G-Y, Boyington JC, Joyce MG et al (2012) Computational prediction of N-linked glycosylation incorporating structural properties and patterns. Bioinformatics 28: 2249–2255
41. Taherzadeh G, Dehzangi A, Golchin M et al (2019) SPRINT-Gly: predicting N- and O-linked glycosylation sites of human and mouse proteins by using sequence and predicted structural properties. Bioinformatics 35:4140–4146
42. Li F, Li C, Revote J et al (2016) GlycoMinestruct: a new bioinformatics tool for highly accurate mapping of the human N-linked and O-linked glycoproteomes by incorporating structural features. Sci Rep 6:34595
43. Gupta R, Brunak S (2002) Prediction of glycosylation across the human proteome and the correlation to protein function. Pacific Symposium on Biocomputing. Pacific Symposium on Biocomputing, pp 310–322
44. Steentoft C, Vakhrushev SY, Joshi HJ et al (2013) Precision mapping of the human O-GalNAc glycoproteome through SimpleCell technology. EMBO J 32:1478–1488
45. Julenius K (2007) NetCGlyc 1.0: prediction of mammalian C-mannosylation sites. Glycobiology 17:868–876
46. Kellman BP, Lewis NE (2020) Big-data glycomics: tools to connect glycan biosynthesis to extracellular communication. Trends Biochem Sci 46(4):284–300
47. Spahn PN, Hansen AH, Hansen HG et al (2016) A Markov chain model for N-linked protein glycosylation—towards a low-parameter tool for model-driven glycoengineering. Metab Eng 33:52–66
48. Kikuchi N, Narimatsu H (2003) Comparison of glycosyltransferase families using the profile hidden {M}arkov model. Biochem Biophys Res Comm 310:574–579

Chapter 7

Analysis of Posttranslational Modifications in Arabidopsis Proteins and Metabolic Pathways Using the FAT-PTM Database

Madison N. Blea and Ian S. Wallace

Abstract

Posttranslational modifications (PTMs) are critical regulators of protein behavior, and over 200 different types of PTMs have been identified. Recent developments in mass spectrometry technology and sample enrichment approaches have led to a massive expansion in the number of identified PTM types and sites within eukaryotic proteins. As these types of data become increasingly available, it is important to develop additional analysis tools and data repositories to investigate PTM cross talk and larger networks of PTMs. Recently, we developed the Functional Analysis Tools for Post-translational Modifications (FAT-PTM) database, which supports data from publicly available proteomic analyses encompassing eight different types of PTMs and over 49,000 PTM sites. In this chapter, we describe the utility of FAT-PTM for analysis of posttranslationally modified proteins in three different contexts. First, a simple protein search tool is available that allows users to investigate proteins in the Arabidopsis proteome to identify types of PTMs that are associated with the query protein as well as quantitative phosphorylation site changes associated with ten different experimental conditions. Second, FAT-PTM contains a metabolic pathway analysis tool to investigate PTMs in the broader context of over 600 different metabolic pathways compiled from the Plant Metabolic Network. Finally, FAT-PTM contains a comodification tool that can be used to identify groups of proteins that are subject to two or more user-defined PTMs. Overall, FAT-PTM provides a user-friendly platform to visualize posttranslationally modified proteins at the individual, metabolic pathway, and PTM cross-talk levels.

Key words Posttranslational modifications, Protein phosphorylation, PTM cross talk, Modification database, Metabolic regulation, *Arabidopsis thaliana*

1 Introduction

Posttranslational modifications (PTMs) regulate nearly every aspect of protein behavior. Over 200 different types of posttranslational modifications have been described [1] and these regulatory modifications can control enzymatic activity, protein-protein interaction networks, protein stability, and protein localization. Protein phosphorylation is one of the most common PTMs in eukaryotic

Dukka B. KC (ed.), *Computational Methods for Predicting Post-Translational Modification Sites*,
Methods in Molecular Biology, vol. 2499, https://doi.org/10.1007/978-1-0716-2317-6_7,

systems, and it estimated that over one third of eukaryotic proteins are subject to at least one posttranslational phosphorylation event during their lifetime. Additionally, large-scale proteomic surveys to examine lysine acetylation, small ubiquitin-like modifier attachment, cysteine oxidations, and various glycosylation or lipid modification types have demonstrated that these modifications are similarly widespread [2–6] and further indicate that many proteins are subject to multiple PTMs [7].

Advances in mass spectrometry instrumentation and sample preparation technologies have revealed the breadth of posttranslational modifications in eukaryotic species. These advances include the widespread implementation of high mass accuracy mass spectrometers, the routine utilization of quantitative mass spectrometry methods, more precise fragmentation methods for the localization of posttranslational modification sites [8], and the development of enrichment technologies to isolate peptides containing posttranslational modifications [9–14]. Despite these advances, the aggregation and meta-analysis of posttranslational modification data from large scale proteomics experiments is still in its infancy.

Various databases have been constructed to collect and analyze posttranslational modification data. For example, phosphorylation is among the most common posttranslational modifications, and various databases, including PhosphoGRID [15], iPTMNet [16], and PHOSIDA [17], are specifically focused on this modification. PhosPhAt 4.0 [18] and P^3DB [19] offer exceptional repositories of plant-specific phosphorylation and additional PTM analysis tools, but currently, these databases are specifically focused on protein phosphorylation and do not consider other PTMs. Many currently existing posttranslational modification databases suffer from certain limitations, including (1) considering only one type of PTM, (2) not providing quantitative mass spectrometry data associated with PTMs, (3) not presenting PTMs in broader context outside of individual proteins, and (4) not providing advanced informatics tools for the large-scale analysis of PTM data.

Previously, we constructed the Functional Analysis Tools for Posttranslational Modifications (FAT-PTM) database [20], which circumvents some of these limitations. FAT-PTM contains publicly available data from large scale PTM proteomic studies encompassing eight different types of PTMs and over 49,000 defined PTM sites in Arabidopsis. FAT-PTM is divided into three modules: a protein search tool to examine qualitative and quantitative posttranslational modification data associated with a user defined protein, a metabolic pathway module to examine PTMs in the context of metabolic pathways, and a comodification tool to identify proteins that are subject to more than one PTM. In this chapter, we will describe the functional modules of FAT-PTM and guide users through the process of using these modules as well as harvesting data from FAT-PTM for further analysis.

2 Materials

The user must have access to a computer with internet access. FAT-PTM can be accessed at the following web address: bioinformatics.cse.unr.edu/fat-ptm.

3 Methods

3.1 Protein Search Tool

The FAT-PTM protein search tool allows users to investigate whether any protein in the Arabidopsis proteome is subject to one of eight different types of PTMs (currently limited to phosphorylation, lysine acetylation, SUMOylation, ubiquitination, N-glycosylation, S-nitrosylation, S-acylation, and O-GlcNAc modification). Additionally, this module allows users to visualize quantitative phosphoproteomic data associated with ten different experimental conditions [20]. Here, we describe how to use the protein search tool function and export the resulting graphical and numerical data.

1. Users should navigate to the FAT-PTM database, which can be accessed directly through the site or through individual gene pages in The Arabidopsis Information Resource (TAIR; www.arabidopsis.org) (*see* **Note 1**).
2. Once the database has been accessed, users should select the protein tool in the upper left section of the page.
3. Users are then able to input an Arabidopsis Gene Identified (AGI) number or gene names into the protein search bar. For the purposes of this demonstration, we will use the Arabidopsis Cellulose Synthase A1 (CESA1) protein as a test case. The FAT-PTM record for AtCESA1 can be accessed by searching "CESA1" or the corresponding AGI number, At4g32410.
4. Users should select posttranslational modifications from the drop-down menu to the right of the search bar and then click search.
5. The search result will then display a network diagram where the central green node represents the queried protein and the peripheral nodes represent PTM sites (*see* Fig. 1). The shape of the peripheral nodes corresponds to the type of PTM. The edges connecting these nodes reflect the number of spectral observations associated with each modified site. Gray peripheral nodes (*see* Fig. 1) indicate that a PTM is functionally associated with a selected protein, but due to experimental design or other technical limitations, the associated modification site was not mapped [20].

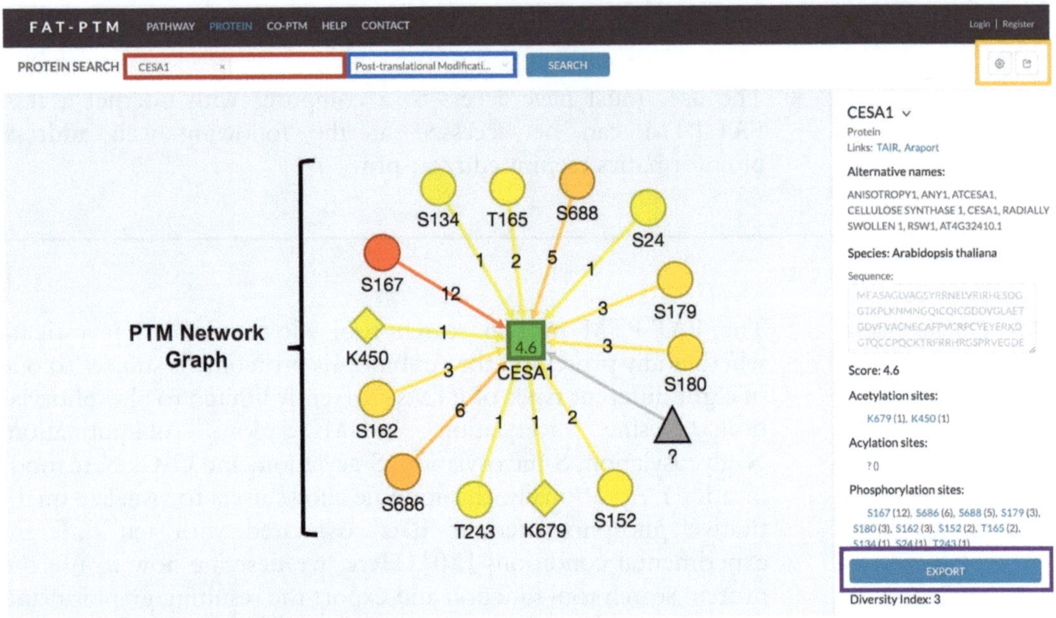

Fig. 1 Features and usage of the FAT-PTM protein search module: the FAT-PTM protein search module can be used to illustrate PTMs associated with any Arabidopsis query protein. Users can enter gene names or AGI numbers into the search box (highlighted in red) and select the "Post-translational modifications" option in the box highlighted in blue to generate a network graph that illustrates the query protein as a central green node, and different types of PTMs as peripheral nodes. The edges connecting the central node to the peripheral nodes represent the number of spectral observations supporting a PTM. Users can also select the "Quantitative phosphorylation" option in the blue box to investigate quantitative phosphorylation site changes under a variety of experimental conditions. Users can change the appearance of the network graph by selecting the gear icon in the gold-highlighted box, or export the image using the image download tab. Additionally, users can left-click on the central green node to access and export (purple box) PTM data associated with a target protein. Finally, users can right-click on the central green node to toggle between network and lollipop plot views of protein PTM data

6. Users may right-click on the central green node to select a lollipop plot to visualize the same modified protein as a primary sequence cartoon containing Pfam protein domains. In these lollipop plots, different posttranslational modifications correspond to different shapes as seen previously, and the variation in heights correlates to the number of spectral observations (*see* **Note 2**).
7. In the network view, users can select the quantitative phosphorylation option from the drop-down menu next to the search bar and conduct a search. This option displays quantitative phosphorylation data from up to ten quantitative phosphoproteomic studies [20].
8. In the quantitative phosphorylation view, a similar network diagram will be shown where the central green node is representative of the modified protein and the peripheral nodes

represent posttranslational phosphorylation sites. The edges that connect peripheral nodes to the central node represent the fold-change abundance of the site under the selected condition.

9. Users are able to once again left-click the central node to access a drop-down menu in the upper left corner below the protein search bar that allows them to select a treatment to represent. The different treatments will give the user a different quantitative phosphorylation network (*see* **Note 3**).
10. When using the protein search tool, the user may export network graphs, lollipop plots, or the underlying proteomic data by simply clicking the image download icon at the top right side of the task bar (*see* Fig. 1). Additionally, users can alter the color scheme of the network diagrams using the "Graph Settings" function in the task bar.
11. Users may also left-click on any PTM site to further investigate additional attributes of a particular modification, including the protein sequence surrounding the modification site, links to the original proteomic study that identified the modified site, and experimental conditions under which the experiment was performed.
12. Finally, users may left-click on the green central node to investigate additional attributes of the modified protein, including the protein name(s), protein sequence, AGI number, etc. Users may also click on the "Export" tab in this view to download a tab-delimited file containing additional information about modifications in the associated protein, including modification type, modification site, protein sequence surrounding the modification site, and number of spectral observations supporting each modification site (*see* Fig. 1).

3.2 Metabolic Pathway Tool

Posttranslational modifications often control protein behavior in the larger context of metabolic or signaling pathways. The Pathway Tool in FAT-PTM allows users to investigate and analyze PTMs in the broader context of 615 unique metabolic pathways collected from the Plant Metabolic Network database [21].

1. Users should navigate to the FAT-PTM database.
2. After reaching the database, users should select the Pathways tool under the search bar.
3. After selecting the pathways tool, users should be able to select one of over 600 metabolic pathway selections from the drop-down menu on the upper left section of the page. For demonstration purposes, we will select "Superpathway of cytosolic glycolysis, pyruvate dehydrogenase, and TCA cycle" as an example.

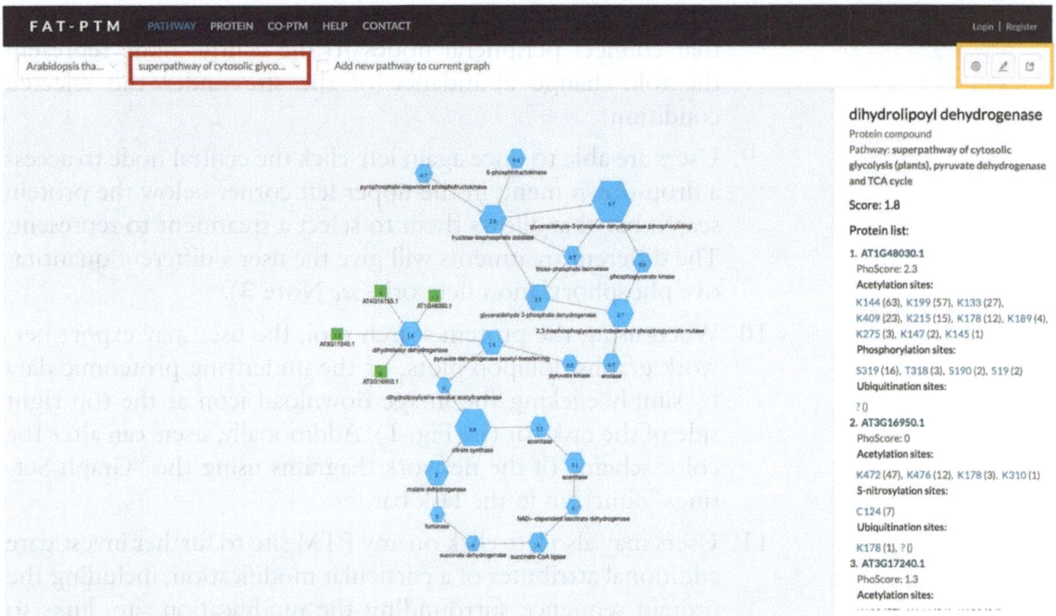

Fig. 2 Features and usage of the FAT-PTM pathway module: the pathway module of FAT-PTM allows users to investigate PTMs in the context of larger metabolic pathways. Users can select from over 600 predefined metabolic pathways using the "Pathway" box highlighted in red. This selection will result in a reaction centric metabolic pathway graph in which the hexagonal nodes represent enzyme classes and lines connecting the nodes represent hypothetical metabolic routes. Nodes can be sized based on PhoScore or users can select the gear icon (highlighted in gold) to scale nodes based on PTM diversity index. Enzyme class nodes can be expanded to view individual enzymes or individual PTMs associated with a specific enzyme by right-clicking and expanding each node. Left clicking on any hexagonal node will also open the panel to the right, which provides additional information for each enzyme class. Image data or tab-delimited data formats can be exported by using the icons in the upper right-hand corner of the screen (highlighted in gold box)

4. After a pathway is selected, a diagram of the pathway will be generated. The hexagonal shapes represent specific enzyme classes that catalyze each reaction in the metabolic pathway (*see* Fig. 2). The size of these hexagons correlates to the PhoScore value for each enzyme class, which is calculated as previously described [20] (*see* **Note 4**).
5. Users can also click on the enzyme class to expand the group of proteins that comprise each enzyme class represented by the original hexagonal node by right-clicking on the enzyme class node.
6. Left clicking on any enzyme class will also bring up information on the right side of the screen detailing the proteins that comprise each enzyme class, their AGI numbers, and associated PTM site identifications (*see* Fig. 2).
7. Right-clicking on any protein node of an enzyme class will allow the user to expand the PTM network diagram for a

selected protein, and the resulting PTM networks are displayed in the same manner as described above for the protein search tool (*see* **Note 5**).

8. Users can alter the appearance of the network graph using the "Graph Settings" tab in the top right corner of the screen. Additionally, this tab allows users to size enzyme class nodes based on PTM Diversity index, which is simply the number of unique types of PTMs associated with a selected enzyme class.
9. Users may also select the Data Download tab in the top right portion of the screen to export images of selected metabolic pathways or to export a tab-delimited file containing all PTM data (protein ID, modification site, modification type, number of spectral observations, sequence surrounding modification site) (*see* **Note 6**).

3.3 Co-modification Tool

Many proteins are subject to multiple types of PTMs, and therefore, rapid identification of comodified proteins could facilitate further analysis of PTM types that typically co-occur or even compete with one another in target protein systems. Therefore, we developed the Co-PTM comodification tool in FAT-PTM to quickly identify groups of proteins that are subject to multiple user defined PTMs.

1. Users should navigate to the FAT-PTM database.
2. After accessing the FAT-PTM database, users should select the Co-PTM tool from the top left section of the page.
3. Users are then able to select any number of posttranslational modifications that they would like to investigate from the selection of modifications present in the selection boxes at the top of the screen (up to eight). For demonstration purposes, we will generate a Co-PTM output for all proteins in the FAT-PTM database that are both ubiquitinylated and SUMOylated (*see* Fig. 3).
4. Once the PTM selections have been made, users should click the generate button to the right of the selection boxes.
5. A Venn diagram will be generated in which the number of proteins subject to each modification is represented by different colored circles with a circumference that is proportional to the total number of proteins subject to a selected modification. The shaded areas of overlap represent proteins that are subject to multiple user-defined PTMs (*see* **Note 7**).
6. Users may left-click on these shaded overlap areas to see the total number of proteins with the selected PTMs as well as the associated AGI codes for these proteins. The export tab at the top of this panel can be used to export all data associated with these proteins, including AGI numbers, all PTM data associated with these proteins, and protein sequences surrounding the modification site into a tab-delimited file.

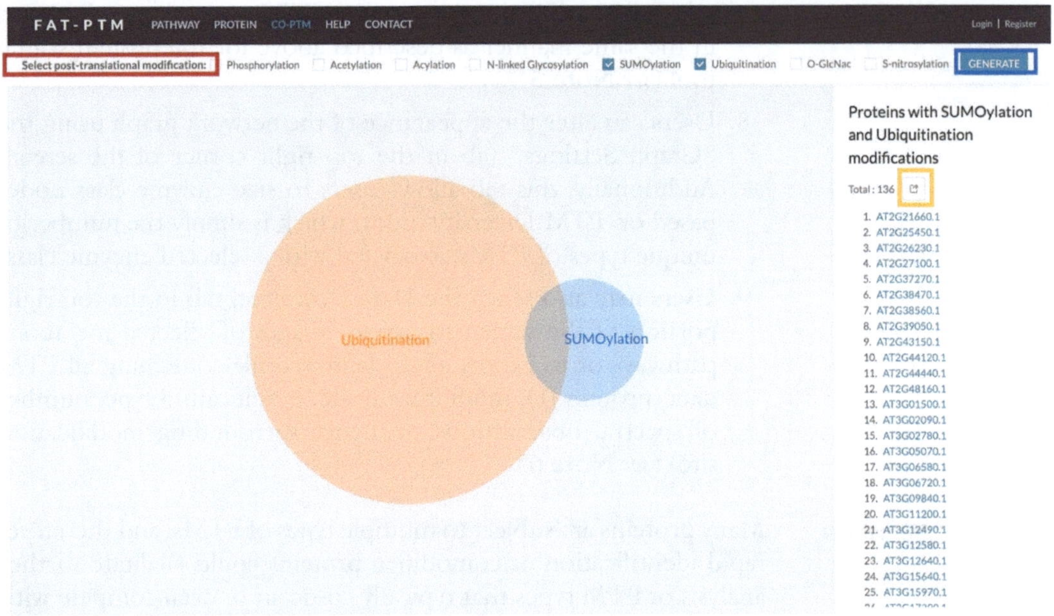

Fig. 3 Features and usage of the FAT-PTM Co-PTM module: the Co-PTM module of FAT-PTM can be used to identify groups of Arabidopsis proteins that are modified by two or more user-defined PTMs. Users can select multiple PTMs from the options at the top of the panel, which will result in the construction of a Venn diagram as shown above. The size of each circle correlates with the total number of proteins that are subject to a selected PTM, and the union of the Venn diagram (shaded region) indicates the number of proteins that are subject to both types of modifications. A user may left-click on the shaded region to bring up the panel on the right, which contains links to each comodified protein as well as a data download option (gold box) to export all PTM data associated with the comodified proteins to a tab-delimited file

7. Users are also able to select these AGI codes, and the database will navigate them to the protein page for that specific AGI code, where it will display all known PTMs associated with that protein including the modifications selected by the user for analysis using the protein tool.

4 Notes

1. FAT-PTM records for individual proteins can be accessed from any gene page in TAIR using the External Links section at the bottom of the page. Additionally, users can quickly access other excellent PTM databases in this section, including P^{3}DB [19] and Plant PTM Viewer [22] from the same location.
2. Users can mouse over any of the sites on the lollipop plot to collect information about the residue index and number of observations associated with that modification. They can also left-click the site and get a new window which indicates the modified sequence and links to the experiments that show

where the modification was identified as well as links to PubMed which indicates the papers where these posttranslational modifications are described.

3. Users are able to visualize multiple treatments at once by clicking the "add new treatment to current graph" button located next to the drop-down menu for the treatment selection after they have left-clicked on the central node of this display.
4. The PhoScore is associated with the number of phosphorylation sites in a selected protein. Higher numbers of phosphorylation sites and larger numbers of spectral observations associated with a selected protein produces a higher PhoScore.
5. Expanded nodes can be collapsed by simply right-clicking on the node again and selecting the collapse option.
6. FAT-PTM users can create their own user defined pathways by registering for a free FAT-PTM account and using the MyPathways function that is described in the Help section of the FAT-PTM web server. This option is particularly useful for creating pathways or protein complex architectures that are not represented in the predefined metabolic pathways available in the Pathway module. User-created pathways are not shared on the FAT-PTM website, and data can be exported as described above for the Pathway analysis tool.
7. When three or greater PTMs are selected in the Co-PTM module, it is possible that multiple areas of overlap in the Venn diagram will be generated. These areas correspond to subgroups of proteins that are modified with two or more PTMs and the resulting data can be exported independently. Mousing over the shaded areas reveals a pop-up box that indicates which groups of modified proteins are being selected.

Acknowledgments

This work was supported by a National Science Foundation Award (MCB #1750359) to I.S.W.

References

1. Garavelli JS (2004) The RESID database of protein modifications as a resource and annotation tool. Proteomics 4(6):1527–1533. https://pubmed.ncbi.nlm.nih.gov/15174122/
2. Kühn C (2016) Review: Post-translational cross-talk between brassinosteroid and sucrose signaling. Plant Sci 248:75–78. https://pubmed.ncbi.nlm.nih.gov/27181949/
3. Strasser R (2016) Plant protein glycosylation. Glycobiology 26(9):926–939. https://pubmed.ncbi.nlm.nih.gov/26911286/
4. Turnbull D, Hemsley PA (2017) Fats and function: protein lipid modifications in plant cell signalling. Curr Opin Plant Biol 40:63–70. https://pubmed.ncbi.nlm.nih.gov/28772175/

5. Withers J, Dong X (2017) Post-translational regulation of plant immunity. Curr Opin Plant Biol 38:124–132. https://pubmed.ncbi.nlm.nih.gov/28538164/
6. Augustine RC, Vierstra RD (2018) SUMOylation: re-wiring the plant nucleus during stress and development. Curr Opin Plant Biol 45 (Pt A):143–154. https://pubmed.ncbi.nlm.nih.gov/30014889/
7. Beltrao P, Bork P, Krogan NJ et al (2013) Evolution and functional cross-talk of protein post-translational modifications. Mol Syst Biol 9:714. https://pubmed.ncbi.nlm.nih.gov/24366814/
8. Coon JJ, Shabanowitz J, Hunt DF et al (2005) Electron transfer dissociation of peptide anions. J Am Soc Mass Spectrom 16:880–882
9. Larsen MR, Thingholm TE, Jensen ON et al (2005) Highly selective enrichment of phosphorylated peptides from peptide mixtures using titanium dioxide microcolumns. Mol Cell Proteomics 4:873–886
10. Xu SL, Chalkley RJ, Maynard JC et al (2017) Proteomic analysis reveals O-GlcNAc modification on proteins with key regulatory functions in Arabidopsis. Proc Natl Acad Sci U S A 114:E1536–E1543
11. Xu G, Paige JS, Jaffrey SR (2010) Global analysis of lysine ubiquitination by ubiquitin remnant immunoaffinity profiling. Nat Biotechnol 28:868–873
12. Kim DY, Scalf M, Smith LM et al (2013) Advanced proteomic analyses yield a deep catalog of ubiquitylation targets in Arabidopsis. Plant Cell 25:1523–1540
13. Majeran W, Le CJP, Ponnala L et al (2018) Targeted profiling of Arabidopsis thaliana subproteomes illuminates co- and posttranslationally N-terminal myristoylated proteins. Plant Cell 30:543–562
14. Zeng W, Ford KL, Bacic A et al (2018) N-linked glycan micro-heterogeneity in glycoproteins of arabidopsis. Mol Cell Proteomics 17:413–421
15. Sadowski I, Breitkreutz BJ, Stark C et al (2013) The PhosphoGRID Saccharomyces cerevisiae protein phosphorylation site database: version 2.0 update. Database 2013:bat026
16. Huang H, Arighi CN, Ross KE et al (2018) IPTMnet: an integrated resource for protein post-translational modification network discovery. Nucleic Acids Res 46:D542–D550
17. Gnad F, Gunawardena J, Mann M (2011) PHOSIDA 2011: the posttranslational modification database. Nucleic Acids Res 39: D253–D260
18. Zulawski M, Braginets R, Schulze WX (2013) PhosPhAt goes kinases-searchable protein kinase target information in the plant phosphorylation site database PhosPhAt. Nucleic Acids Res 41(Database issue):D1176–D1184
19. Yao Q, Ge H, Wu S et al (2014) P3DB 3.0: from plant phosphorylation sites to protein networks. Nucleic Acids Res 42(Database issue):D1206–D1213
20. Cruz ER, Nguyen H, Nguyen T et al (2019) Functional analysis tools for post-translational modification: a post-translational modification database for analysis of proteins and metabolic pathways. Plant J 99:1003–1013
21. Schläpfer P, Zhang P, Wang C et al (2017) Genome-wide prediction of metabolic enzymes, pathways, and gene clusters in plants. Plant Physiol 173:2041–2059
22. Willems P, Horne A, Van Parys T et al (2019) The plant PTM viewer, a central resource for exploring plant protein modifications. Plant J 99:752–762

Chapter 8

Bioinformatic Analyses of Peroxiredoxins and RF-Prx: A Random Forest-Based Predictor and Classifier for Prxs

Hussam AL-Barakati, Robert H. Newman, Dukka B. KC, and Leslie B. Poole

Abstract

Peroxiredoxins (Prxs) are a protein superfamily, present in all organisms, that play a critical role in protecting cellular macromolecules from oxidative damage but also regulate intracellular and intercellular signaling processes involving redox-regulated proteins and pathways. Bioinformatic approaches using computational tools that focus on active site-proximal sequence fragments (known as active site signatures) and iterative clustering and searching methods (referred to as TuLIP and MISST) have recently enabled the recognition of over 38,000 peroxiredoxins, as well as their classification into six functionally relevant groups. With these data providing so many examples of Prxs in each class, machine learning approaches offer an opportunity to extract additional information about features characteristic of these protein groups.

In this study, we developed a novel computational method named "RF-Prx" based on a random forest (RF) approach integrated with K-space amino acid pairs (KSAAP) to identify peroxiredoxins and classify them into one of six subgroups. Our process performed in a superior manner compared to other machine learning classifiers. Thus the RF approach integrated with K-space amino acid pairs enabled the detection of class-specific conserved sequences outside the known functional centers and with potential importance. For example, drugs designed to target Prx proteins would likely suffer from cross-reactivity among distinct Prxs if targeted to conserved active sites, but this may be avoidable if remote, class-specific regions could be targeted instead.

Key words Peroxiredoxin classification, Prx bioinformatics, Feature selection, Random forest, Machine learning

1 Introduction

Peroxide reductases, also known as peroxidases, which rely on cysteine (Cys) redox reactions rather than heme-bound iron

---

Supplementary Information The online version contains supplementary material available at [https://doi.org/10.1007/978-1-0716-2317-6_8].

Dukka B. KC (ed.), *Computational Methods for Predicting Post-Translational Modification Sites*,
Methods in Molecular Biology, vol. 2499, https://doi.org/10.1007/978-1-0716-2317-6_8,

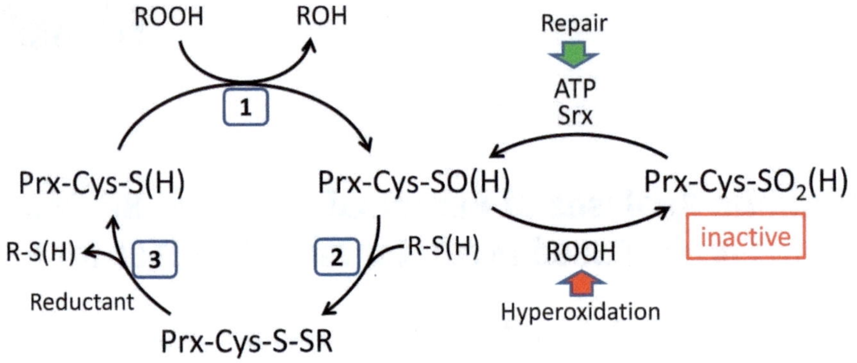

Fig. 1 Peroxidatic (left) and regulatory cycles (right) of peroxiredoxins (Prxs). Shown are (1) the initial reaction with peroxide to generate the alcohol (or water) and sulfenic acid (SOH) on the enzyme, (2) condensation of the sulfenic acid with a thiol to form a disulfide, and (3) reduction by thioredoxin (Trx) or a Trx-like protein to regenerate the reduced, active protein. In conditions of high peroxide (ROOH) concentration, hyperoxidation to form sulfinic acid (SO_2H) can occur, which inactivates the Prx, although this moiety can be repaired by the enzyme sulfiredoxin (Srx). Protonation state is not explicitly designated as (H) represents either the protonated (neutral) form (R-SH) or deprotonated (anionic) form ($R\text{-}S^-$). R'-S(H) represents a resolving thiol group which can come from the Prx or from another molecule

(as do catalase and most plant peroxidases), were first recognized as an enzyme class in lactic acid bacteria possessing both the catalytic Cys and a tightly bound flavin [1, 2] prior to the discovery of the phylogenetically widespread superfamily of Cys-containing (but nonflavin) peroxiredoxins (Prxs) [3, 4]. Through biochemical and biophysical studies of these cofactor-free Prxs, features influencing the stability and reactivity of thiol (also known as sulfhydryl, R-SH), sulfenic acid (R-SOH), disulfide (R-SS-R'), and sulfinic acid ($R\text{-}SO_2H$) moieties, in particular, have revealed that all these species are involved in the Prx catalytic and/or regulatory cycle (Fig. 1) [5–7]. In other research findings relevant to redox regulation which developed in parallel, more and more cellular proteins undergoing redox regulation through Cys posttranslational modifications (PTMs), including reversible SOH and SS formation, have been identified with roles in regulating metabolism and signaling [8, 9]. While Prxs have served as protein models to analyze the chemistry, they have also been shown to be actively involved in redox regulatory processes occurring in other proteins, as they can mediate or modulate the transfer of oxidizing equivalents from signal-generated H_2O_2 to target proteins, directly or indirectly [10, 11]. The presence of Prxs in all organisms, almost always with more than one Prx, and typically three to six distinct Prxs (e.g., three in *E. coli*, six in mammals, and 10 in *Arabidopsis thaliana*, the model angiosperm [7, 12]), argues that these enzymes are vital, and not likely simply as peroxide removal systems given their redundancy [13]. During the 1980s and 1990s, Prxs were turning up in diverse settings and linked to multiple functions or processes

[3, 14, 15]; their common molecular function was shown to be the Cys-dependent reduction of H_2O_2, lipid or protein hydroperoxides, and small-molecule peroxides and/or peroxynitrite [7, 16–21]. As the number of known Prx structures grew along with the rapidly expanding number of known sequences in databases, it became more and more evident that Prxs could be subdivided into natural subclasses or groupings [12, 22–24]. With an eye toward understanding evolved and distinct biochemical and structural features of these proteins, bioinformatics approaches were applied, as discussed further below.

It should first be noted that the specialized Prx active site, with conserved Pro, Thr (or sometimes Ser), Cys and Arg residues and additional features that activate and reduce the incoming peroxide substrates, have significant "signatures" and active site architectures that can be used to definitively identify Prxs. Protein interfaces relevant to oligomerization and interactions with protein reductants and other interacting proteins also contribute to specialized features in Prx structures [7]. On the other hand, for redox-sensitive Cys residues in other proteins for which peroxide reduction is not the primary function, but rather which serve regulatory roles, those Cys residues are found in a wide variety of structural and chemical contexts within the directly reactive H_2O_2-target proteins. Information regarding sensitive sites is still rather limited, and primarily relies on their identification by use of chemical probes to detect Cys-SOH, or crystallographic evidence for the presence of this moiety [9, 25–27].

Reactivity at Cys sites is quite variable and modulated by the protein microenvironment. In fact, it has been shown that the reactivity toward H_2O_2 of the Cys thiolate anion (R-S^-), which is more prevalent at low pH and in protein Cys residues with depressed pK_as, is limited to ~20 M^{-1} s^{-1} when there are no other nearby activating or stabilizing chemical groups [28]. In a simple mixture of proteins, most Prxs, with their high rate constants of ~10^6 to 10^8 M^{-1} s^{-1} in reactions with peroxides, would outcompete other cellular proteins for reactivity with H_2O_2 in a simple chemical competition [29]. However, Prxs require recycling by reductase proteins like thioredoxins (Trxs) for continuing turnover and may not always be in their active state [30]; in addition, there is considerable compartmentalization in cells that could localize peroxide-generating enzymes like NADPH oxidases near their downstream targets to promote reaction [31]. Moreover, Prxs are sensitive to PTMs, including phosphorylation and hyperoxidation, which can decrease or otherwise modulate their reactivity. Likewise, scaffolding or other interacting proteins can have considerable influence on the "channels" down which oxidizing equivalents can flow. Although we as yet have an incomplete understanding, reactions of various protein Cys residues with H_2O_2 to form SOH (as shown in Fig. 1, reaction 1) are increasingly shown to be relevant in biological systems as they have been observed by chemical trapping experiments and in X-ray crystal structures (which notably could be generated due to radiation exposure during data acquisition and must be independently proven to be physiologically relevant) [25, 27, 32]. While sorting out the contexts in which each oxidation event/target would have relevance in cell signaling and regulation will require considerable additional research, it has been clearly demonstrated that the reactivity of even relatively slowly oxidized Cys sites can be modulated by the surrounding amino acids; in GAPDH, an important metabolic enzyme and regulatory hub, the presence or absence of nearby Cys156 modulates the H_2O_2 sensitivity of Cys 152, and further experiments in yeast showed that this redox regulation is biologically important [33]. The likelihood that a rather wide range of activating contexts exists in proteins from diverse fold families that are redox regulated highlights the challenge in comprehensively predicting such sites by bioinformatics methods alone.

Focusing on the Prxs, the locally conserved motif just upstream of the peroxide-reactive (peroxidatic) Cys, PXXX(T/S)XXC, is definitive for the Prx superfamily [23]. This peroxidatic Cys directly attacks the –OOH terminal end of the hydroperoxide substrate to break the -O-O- bond and release the ROH product; in the process, this Cys becomes oxidized to sulfenic acid (Fig. 1). It is the precise arrangement of these conserved residues, as well as a conserved Arg contributed by a sequence fragment close in structure but distant in sequence, that provides the potent active site architecture and hydrogen-bonding atoms and angles that serve to activate the Cys sulfur and the substrate hydroperoxide for reaction and selectively stabilize the transition state, accelerating the peroxidatic reaction [6]. Once the sulfenic acid is formed (Fig. 1, reaction 1), most, but not all, Prxs also have a second Cys, termed a resolving Cys, that forms a disulfide bond with the Cys-SOH through condensation once conformational changes allow the two to come together (Fig. 1, reaction 2). Surprisingly, with information from many Prxs now available, it is clear that the resolving Cys can be found in at least five distinct locations within the core Prx structure [7]. The final step (Fig. 1, reaction 3) is the reduction of the disulfide by thioredoxin (Trx) or other Trx-like reductase proteins to reset the enzyme for another catalytic cycle. Prxs lacking a resolving Cys must instead form a disulfide with a thiol group from another protein or small molecule on the return to their reduced (active) state. Biochemical evidence also reveals that Prxs are variably sensitive to hyperoxidation (sulfinic acid formation), a reaction occurring under conditions of high peroxide concentrations which inactivates the protein (Fig. 1) [34, 35]. Some Prxs can in fact be repaired and recover activity as many organisms express an enzyme known as sulfiredoxin (Srx), which reverses the hyperoxidation (i.e., sulfinic acid product), an otherwise biologically irreversible modification.

Structurally, Prxs are one "superfamily" (a designation based on bioinformatics analyses), among other Trx-related superfamilies, that possess a common fold. This fold therefore defines the Trx "suprafamily" whose members include not only thioredoxins and Prxs but also cytochrome maturation proteins, glutaredoxins, glutathione peroxidases, and glutathione-S-transferases [23, 36, 37]. The common fold is built on a β sheet flanked by α helices, and, for Prxs, an active site TXXC (or sometimes SXXC) at the N-terminal end of the α2 helix that aligns with and arguably evolved from the CXXC active site motif of ancestral Trxs [24, 38–40]. Detailed sequence comparisons using hidden Markov models (HMM) originally identified four classes of Prx, designated Prx1, Prx2, Prx3, and Prx4, in the first bioinformatics study of these proteins reported in 2004 [23]. Analyses of structures as described several years later [41] provided evidence for six subgroups, named for canonical or founding members (Prx1, Prx5, Prx6, Tpx, PrxQ and AhpE), the last of which had not been known at the time of the HMM study. In that 2004 study, two of these groups with significant similarities, Prx1 and Prx6, were combined as a single group (the "Prx4" group by their designation), accounting for the differences in subgroup numbers. Subsequently, more definitive bioinformatics analyses based on "active site profiling" were conducted and able to distinguish the six subgroups, initially assigning over 3500 Prx sequences unambiguously into one of the six groups (from GenBank in January 2008) [42], and later with more refined and automated methods, classifying more than 38,000 Prx sequences into one of these six subgroups (March 2016 GenBank) as reported by Harper et al. [43].

To elaborate on these bioinformatics analyses, active site profiling (also known as functional site profiling, as it is not restricted to enzyme active sites) begins with known structures of the protein family of interest, extracting fragments of sequence surrounding key conserved residues within a radius of 10 Å (Fig. 2). With 29 distinct Prx structures available in the Protein Data Bank in January, 2008, the Deacon Active Site Profiler (DASP) tool was used to create an active site profile (ASP) for each of the six groups previously identified based on structural information, followed by sequence database searches to identify sequences containing active site features similar to those in the ASP. In a single search approach, more than 3500 Prx proteins were identified and categorized [42]; these results are available for web-based searches in the PREX database developed by members of our bioinformatics group (www.csb.wfu.edu/PREX) [44]. These results were subsequently expanded in the Structure Function Linkage Database, SFLD [45], by adding sequences to these groups using the HMM approach.

A

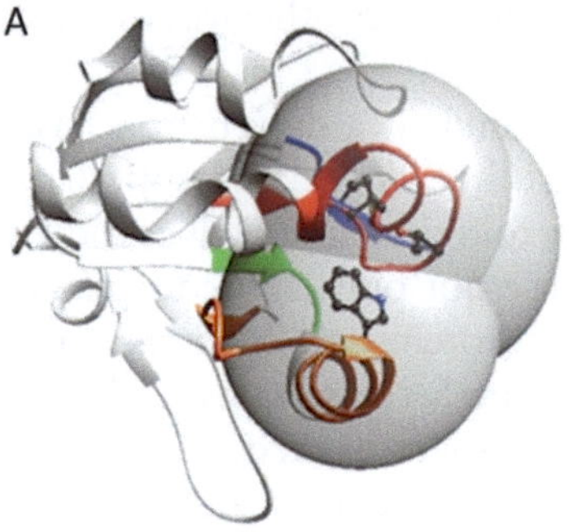

B

```
Active Site Signature:

2PWJ FGLPGAYTGVCSSKHVP-CVAIN-TVNAWAEKIQAKFTLG-LGIRSER

Active Site Profile:

2PWJ FGLPGAYTGVCSSKHVP-CVAIN-TVNAWAEKIQAKFTLG-LGIRSER
4F82 FGLPGAFTPTCSAQHVP-CVSVN-VMGAWGRDL--TMQLR-MGIRSLR
1TP9 FGVPGAFTPTCSLKHVP-CISVN-VMKAWAK----SYFLL-LGTRSRR
1NM3 FSLPGAFTPTCSSSHLP-VVSVN-VMNAWKEDE--SFVGK-LFGRSWR
-------------------------------------------------
3DRN YFYPKDDTPGCTREASP--VSSD-SHKRFKEKYKLFLRAL--PARQPA
3GKN YFYPKDSTPGCTTEGLP--VSRD-SHDNFCAKQG-FLVGI--PARRSV
```

Fig. 2 Active Site Profiling identifies molecular features around a protein's functional site. (**a**) In an enzyme structure, key functional residues (black side chains) are identified from sequence and structural analysis (conserved Pro, Thr, Cys, and Trp/Phe for Prxs). All residues within 10 Å of any key residue (light gray spheres) are identified, (**b**) then extracted and concatenated to form an active site signature (top right). Signatures from a protein family are then aligned to create an active site profile (ASP) (bottom right). Within the profile, molecular features that are common across the superfamily (blue arrows), as well as features that serve to divide the profile into two distinct groups (red arrows), can be identified. The black line separates two functional families of Prxs, with Prx5 proteins on top of the line and PrxQ proteins below the line. From Harper et al., PLoS Computational Biology, 2017 [43]

In 2017, new tools and approaches to define Prx groups were reported built around the DASP tool, working toward more automated approaches that could ultimately be applied to other protein families and superfamilies. As a first step, rather than predefining the six groups based on expert analysis, the starting set of isofunctional clusters is identified by iterative clustering of the active site signatures (Fig. 2b) of proteins of known structure that share common active site features, in this case for all known Prxs at the time (47 nonredundant Prx structures). This process, called TuLIP (Two-Level Iterative clustering Procedure), starts with all structures from a protein superfamily and iteratively subdivides those into smaller and smaller groups based on active site features, continuing until the DASP search for a given cluster self identifies (i.e., retrieves all previously clustered proteins, and no others, from the PDB). If the clusters do not pass this criterion, they are further subdivided and searched again. In the case of the Prx superfamily, this approach identified four starting clusters (Fig. 3a) based solely on the iterative clustering approach rather than on expert knowledge [43], as had been done in the previous DASP study [42]. These clusters were then the starting point for iterative searches of sequence databases using a method called MISST (Multilevel Iterative Sequence Searching Technique); this process was developed to be both agglomerative (adding sequences containing similar active site features to the cluster) and divisive (subdividing TuLIP groups when active site features suggest distinct clusters). In the case of Prxs, five iterations of MISST identified the six known

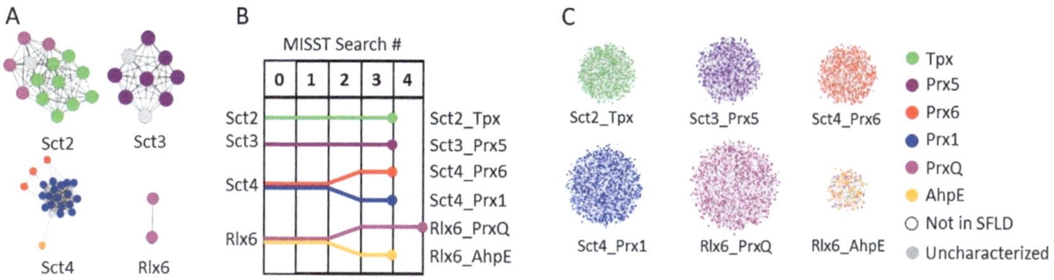

Fig. 3 Four TuLIP groups split into six functionally relevant groups after five MISST iterations. (**a**) The four TuLIP groups are represented by networks in which each node represents a Prx protein of known structure. Edges are pairwise profile scores (as defined in Knutson et al. [46]) and node colors represent expert functional annotations (*see* legend). (**b**) A dendrogram of the iterative MISST process illustrates how the initial TuLIP groups evolved into the final MISST groups. Vertical lines represent GenBank searches and dendrogram lines are colored based on the majority subgroup in each MISST cluster. Dendrogram branches represent the cluster subdivision. The circle at each line terminus represents the iteration at which the group met self-identification criteria. (**c**) The final six Prx groups are represented as networks in which nodes represent the proteins and edges represent the DASP2 search scores from the final search; the nodes are colored by expert subgroup annotation. From Harper et al., PLoS Computational Biology, 2017 [43]

functionally relevant Prx groups from the four starting groups provided by TuLIP (Fig. 3).

As the number of Prx representatives keeps growing and this large, well defined dataset of members of each class with over 38,000 Prxs total is now available, the opportunity exists to use machine learning to identify new Prxs and their groups, and to recognize sequence/amino acid features that help define these classes. In this work, we developed a random forest (RF)-based classifier termed "RF-Prx" for identification and class prediction of Prx enzymes using K-space amino acid pairs (KSAAP). This approach was implemented in two phases, identification of Prx superfamily members, and prediction of their classes. Comparing our method with other state-of-the-art machine learning algorithms, using ten-fold cross-validation and an independent test dataset, RF-Prx provided favorable accuracy and achieved better results than these other algorithms.

2 Methods

This method is split into two phases. The first phase identified Prxs and their active site from protein sequence databases. In the second phase, if the protein was a Prx, it was classified into one of six classes of Prx: class 1 (Prx1), class 2 (AhpE), class 3 (PrxQ), class 4 (Tpx), class 5 (Prx5), and class 6 (Prx6). The flowchart of our method is shown in Fig. 4. To our knowledge, this is the first machine learning method designed to identify and classify Prxs.

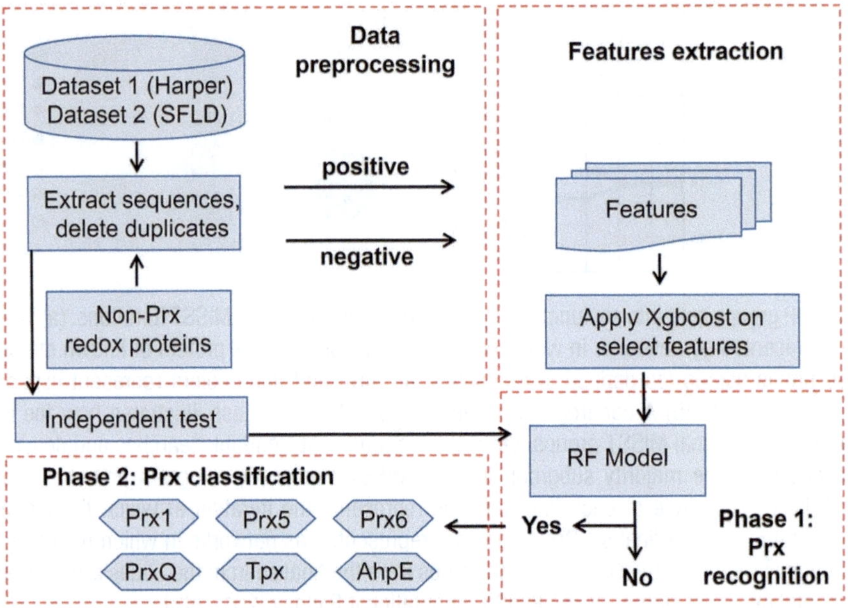

Fig. 4 Flowchart of the RF-Prx method which is split into two phases

Table 1
Sequence numbers for Datasets 1 and 2 and their combination between datasets and the final number of sequences used for each class in our method

Class	Dataset 1 (Harper)	Dataset 2 (SFLD)	Dataset 1 ∩ Dataset 2	Dataset 1 ∪ Dataset 2
Class 1 (Prx1)	9660	2260	2154	9766
Class 2 (AhpE)	1489	112	100	1447
Class 3 (PrxQ)	12,014	1954	1827	12,087
Class 4 (Tpx)	4930	956	867	5019
Class 5 (Prx5)	5434	1076	1051	5459
Class 6 (Prx6)	5212	987	963	5236
Total	38,739	7345	6962	39,014

2.1 Dataset and Preprocessing

Dataset 1, named "Harper," was retrieved from the published study of Harper et al. [43]. This dataset contained 38,739 total protein sequences, representing all six classes. Dataset 2, named "SFLD," was retrieved from The Structure-Function Linkage Database [45]. The total protein sequences retrieved from the database last updated in October 2018 included 7345 sequences for all classes. In this study, we combined Datasets 1 and 2 and removed duplicate sequences. The totals for each dataset and each class before and after combination are shown in Table 1. We also removed

Table 2
Final number of benchmark dataset entries after removal of duplicate sequences and those with dummy residues

Class	Dataset 1 ∪ Dataset 2
Class 1 (Prx1)	9683
Class 2 (AhpE)	1443
Class 3 (PrxQ)	12,071
Class 4 (Tpx)	5005
Class 5 (Prx5)	5428
Class 6 (Prx6)	5201
Total Prxs (Phase 1)	38,831
Negative sequences (Non-Prx)	25,665

sequences that appeared in more than one class. Additionally, we discarded sequences with "dummy" residues for each class. The final number of sequences used for each class is shown in Table 2.

After retrieving sequences for each class, we combined all unique Prx sequences and constructed the final positive sequences for the general Prx classifier. Supervised algorithms work best when negative sequences that correspond well with the positive set are used [47] and overall performance of the predictor depends on the quality of both the positive and negative samples [48]. Therefore, we gathered our negative sequences from cytochrome maturation proteins, glutaredoxins, and glutathione-S-transferases because they are also redox proteins in the Trx fold family, and are generally ~75–200 amino acid residues long. Sequences matching or similar to known cytochrome maturation proteins, glutaredoxins, and glutathione-S-transferases were retrieved from the UniProt database [49] using PSI-Blast, yielding a total of 30,963 sequences. We then removed duplicate sequences and any sequences that had a Prx site, and discarded sequences that contained dummy residues. The final number of negative sequences for the Non-Prx (negative) set was 25,665.

Before proceeding, we split our dataset; we used 80% for training and 20% for testing as shown in Tables 3 and 4, respectively. With different numbers of positive and negative samples for each class and the general classifier (Phase 1) we balanced our classes (including the combined total) using a random under-sampling technique [50] that randomly removes samples from the larger class until the number of samples equals the smaller class. The benefit of this technique is that it guards against the overfitting problem.

Table 3
The number of positive and negative sequences in the training set[a]

Class	Positive sequences	Negative sequences
Class 1 (Prx1)	7746	23,318
Class 2 (AhpE)	1154	29,910
Class 3 (PrxQ)	9656	21,408
Class 4 (Tpx)	4004	27,060
Class 5 (Prx5)	4342	26,722
Class 6 (Prx6)	4160	26,904

[a] For each class, negative sequences used are those for all five other classes

Table 4
The number of positive and negative sequences in the independent set[a]

Class	Positive sequences	Negative sequences
Class1 (Prx1)	1937	5830
Class2 (AhpE)	289	7478
Class3 (PrxQ)	2415	5352
Class4 (Tpx)	1001	6766
Class5 (Prx5)	1086	6681
Class6 (Prx6)	1041	6726

[a] For each class, negative sequences used are those for all five other classes

2.2 Features Construction

Supervised algorithms such as RF [51] can only be applied with numerical weights. In the initial steps before using RF, we converted all protein sequence residues into numerical data by using our Feature Extraction for Protein Sequences (FEPS) tool [52]. The features applied in this work were amino acid composition (AAC), conjoint triad (CT), and K-space amino acid pairs (KSAAP). The dimension of the entire feature was 2932. All these features were described in an earlier study [53] (Table 5).

2.3 Feature Selection

Finding ideal features for comparisons can assist in informative analysis of target sequences. If features are expanded, that will degrade the quality of the method since the probability of having correlations between features will be increased. Furthermore, growing the number of elements might cause computational issues with the training model. Hence, it is valuable to decrease the number of irrelevant attributes during supervised techniques [54, 55]. Consequently, we deployed the Gradient Boosted Trees method, xgboost [56], to discover nonlinear links between features

Table 5
Features used for method development [53]

	Feature class	Acronym	Number of features
1	Amino Acid Composition	AAC	20
2	Conjoint triad	CT	512
3	K-space amino acid pairs	KSAAP	2400
	Total features	ALL	2932

and outcomes from the cumulative features set. Xgboost is used in many comparative studies to extract the most informative attributes [53, 57, 58]. We used xgboost [56] in Python with the Scikit-learn (v0.19.0) package [59] to identify the weighty attributes from our cumulative features set.

Broadly, we estimated Gini impurity for each node of our trees to observe weights of features and to reduce the uncertainty of finding an accurate label. Subsequently, we calculated the information gain for each attribute. We chose the top value to be selected for the initial tree. We then repeated the same process for the remaining trees. Finally, we determined the average prominent features from the collected trees and verified these features by calculating the Gini impurity. Gini impurity can be defined by:

$$\Upsilon = \sum_{i=1}^{L_n} p_i(1 - p_i) \tag{1}$$

where L_n is the number of labels and p_i is the probability value of i. Meanwhile, information gain can be described by:

$$F = \Upsilon_{\text{parent}} - \Upsilon_{\text{left child}} - \Upsilon_{\text{right child}} \tag{2}$$

Any feature (F) with a weight less than 0.0001 was considered an irrelevant feature and discarded from our feature set. We chose a small number of features that defined the most significant properties for each class.

2.4 Model Construction and Assessment

Random forest (RF) [51] is a supervised method which has been widely used in many bioinformatics problems [52, 53, 60]. In this work, we used RF to identify Prx family members and classify their types. RF is constructed from decision trees. A decision tree contains leaves and nodes. Every node applies a rule to choose between several routes. The final rule produces the final class of data points. Many decision trees are made of random samples and random features. At each node, random selection is employed and expands the trees. Finally, all trees are tallied, and the most votes will select for the specific sample. We set optimal values for parameters by using the grid search technique from Scikit-learn [59]. Additionally, we compared the performance of other machine learning algorithms such as support vector machine (SVM) [61], naïve Bayes (NB) [62], and k-nearest neighbor (KNN) [63]. In this study, we used tenfold cross-validation to assess the performance of our method [64]. To measure the performance of our predictor, we calculated accuracy (ACC), sensitivity (SN), specificity (SP), and Matthews's correlation coefficient (MCC), respectively. TP indicates true positive, TN indicates true negative, FP indicates false positive, and FN indicates false negative [65]. All metrics are shown below.

$$\mathrm{ACC} = \frac{\mathrm{TP}+\mathrm{TN}}{\mathrm{TP}+\mathrm{TN}+\mathrm{FP}+\mathrm{FN}} \times 100 \tag{3}$$

$$\mathrm{SN} = \frac{\mathrm{TP}}{\mathrm{TP}+\mathrm{FN}} \times 100 \tag{4}$$

$$\mathrm{SP} = \frac{\mathrm{TN}}{\mathrm{TN}+\mathrm{FP}} \times 100 \tag{5}$$

$$\mathrm{MCC} = \frac{(\mathrm{TP})(\mathrm{TN}) - (\mathrm{FP})(\mathrm{FN})}{\sqrt{(\mathrm{TP}+\mathrm{FP})(\mathrm{TP}+\mathrm{FN})(\mathrm{TN}+\mathrm{FP})(\mathrm{TN}+\mathrm{FN})}} \tag{6}$$

3 Results and Discussion

We implemented two phases to create an RF model to (1) define Prx proteins (phase 1) and (2) classify them (phase 2) (Fig. 4).

3.1 Phase 1. Defining Prx Proteins from Non-Prx

To define Prx proteins, we first used tenfold cross-validation to examine a variable number of feature sets (i.e., AAC, KSAAP, CT), both individually and cumulatively (the latter are denoted as "ALL" in Table 6). During these analyses, K-space amino acid pairs (KSAAP) alone outperformed the other individual features, particularly with respect to ACC, SP, and MCC. Interestingly, no additional increase in performance was observed between KSAAP and the cumulative features based on tenfold cross-validation. Consistently, KSAAP exhibited higher ACC, SN, SP, and MCC scores than the other individual features alone and comparable scores compared to the cumulative features when using an independent test set (Table 7). Together, these data suggested that KSAAP alone is able to capture the salient features needed for Prx identification. Therefore, we used only KSAAP as input for RF-Prx. However, to enhance our method's performance, we applied xgboost on this feature. To this end, we assessed different thresholds with the goal of identifying the setting that produced optimal performance (Tables S1 and S2). The best results using xgboost over KSAAP to define Prx (Phase 1) for tenfold cross-validation were for ACC, SN, SP, and MCC (Table 8). We also received higher values for all metrics using an independent test set (Table 9). Reducing the KSAAP features from 2400 to 469 (after xgboost) yields slightly

Table 6
Results of tenfold cross-validation using individual and cumulative features

	Performance (%)			
Features	**ACC**	**SN**	**SP**	**MCC**
AAC	98.49	98.53	98.45	0.96
CT	99.63	99.77	99.50	0.992
KSAAP	*99.98*	*99.98*	*99.99*	*0.999*
ALL	99.93	99.95	99.99	0.99

Table 7
Independent test result using individual and cumulative features

	Performance (%)			
Features	**ACC**	**SN**	**SP**	**MCC**
AAC	98.58	98.39	98.88	0.97
CT	99.68	99.72	99.61	0.993
KSAAP	*99.99*	*99.98*	*100*	*0.999*
ALL	99.94	99.95	100	0.99

Table 8
Results of tenfold cross-validation using selected features for Level 1

Class	Length	ACC	SN	SP	MCC	AUC
Phase 1	469	99.98	99.98	99.99	0.999	0.999

Table 9
Independent test result using selected features for Level 1

Class	Length	ACC	SN	SP	MCC	AUC
Phase 1	469	100	100	100	1	1

better results but also improves the method by reducing the computational time and speeding up the classifier process to identify Prxs and their classes.

3.2 Phase 2. Class Assignment

In the second phase, RF models were created to predict which of the six classes each Prx belonged to. To this end, we first constructed six RF models designed to classify the Prx proteins (i.e., one model per class). Each RF model was then trained using Prx members from each class. For instance, for the class 1 RF model,

Table 10
Results of tenfold cross-validation using selected features from Class 1 to Class 6

Class	ACC	SN	SP	MCC	AUC
Class 1 (Prxl)	99.96	99.96	99.98	0.99	0.99
Class 2 (AhpE)	99.87	99.74	100	0.99	0.99
Class 3 (PrxQ)	99.94	99.90	99.97	0.99	0.99
Class 4 (Tpx)	100	100	100	1	1
Class 5 (Prx5)	99.98	99.97	100	0.99	0.99
Class 6 (Prx6)	99.92	99.85	100	0.99	0.99

Table 11
Independent test result using selected features Class 1 to Class 6

Class	ACC	SN	SP	MCC	AUC
Class 1 (Prxl)	99.98	100	99.98	0.99	0.99
Class 2 (AhpE)	99.92	99.65	99.93	0.99	0.99
Class 3 (PrxQ)	99.98	99.95	100	0.99	0.99
Class 4 (Tpx)	100	100	100	1	1
Class 5 (Prx5)	99.98	99.91	100	0.99	0.99
Class 6 (Prx6)	99.97	99.90	99.98	0.99	0.99

known class 1 Prx family members were reflected as positive instances, while the remaining examples of the other five classes were reflected as negative instances (Table 3). We replicated the same process for each class in order to define positive and negative classes. Each Prx protein that had been identified in the first phase was then analyzed using each of the RF-based classification models. We also examined a different number of thresholds using xgboost to select optimal features. The range of threshold started from 0.001 to 0.004 (Tables S3–S14). The best results using xgboost over KSAAP to classify Prxs for tenfold cross-validation and for the independent set are shown in Tables 10 and 11, respectively.

3.3 Feature Analysis and Importance

Unlike other popular machine learning methods, such as SVM and NB, RF provides insights into the contribution of each feature to the overall method performance. This metric, which is assessed by measuring the relative importance of each feature, was used to identify the top features for each class. These analyses revealed interesting relationships among members of the same class and provided a substantial enhancement in the prediction of Prx family members and their specific classes extracted from the KSAAP space

Table 12
Top ten features for each class

	Phase 1	Class 1 (Prx1)	Class 2 (AhpE)	Class 3 (PrxQ)	Class 4 (Tpx)	Class 5 (Prx5)	Class 6 (Prx6)
1	PXXXT	VC	PXXXT	YXY	FXP	CXXE	WXXXI
2	TXXC	CXXXW	FXP	GC	FXXXXXC	TC	HXXXXXP
3	FXXXXD	CP	FXXXXXG	YXXXXXP	CXXXG	HXP	TP
4	YP	PXXW	LXXXXW	CXXXA	FC	CXXXXXP	CP
5	SXD	EXC	PXD	SXXXXF	RXC	GXXXXXC	HXXW
6	KXXXXF	DXXF	KXXXXF	HXXW	RF	NXXXXM	PXXT
7	FXXK	PXC	WXXG	FXXXC	FXXXXD	PXXXXXH	PXXXXE
8	DF	YXXXF	VF	PXXF	DXXXC	CXXXN	TXXXW
9	TXV	FXF	FXL	FXXXXW	CXXS	VXXXW	PXXXXXL
10	FXXXC	DXXXXXP	AXXXXC	PXXW	FXXXR	HXXW	PXXXD

(Table 12 and Fig. 5). Interestingly, the top ten attributes in each class are largely distinct, with a few notable exceptions, as illustrated in Figs. S1–S7. This is consistent with the notion that each Prx class exhibits unique sequence features that could contribute to functional differences between classes.

Looking more specifically into the top features and their locations, we note that some similarities were also observed across classes. For example, among the top ten features, cysteine and proline were most prominent across all classes. Not surprisingly, the feature PXXXT, representing proline and threonine residues separated by any three amino acids, was enriched in all Prxs, as this is the sequence just upstream of the essential catalytic Cys residue (Fig. 6). Regarding class-specific features, we noted that for class 1 (the Prx1 class), the first five most important features, VC, CXXXW, CP, PXXW and EXC, were all critical for recognition of this class (Fig. 5) and surround the resolving Cys (R'-S(H) in Fig. 1) that lies near the C-terminus in this class (Fig. 7). Three additional top ten features that are important for defining this group, DXXF, FXF and DXXXXXP, are within the sequence containing or proximal to the critical active site Cys that is present in all Prxs, and were therefore also prominent in the Prx1 WebLogo of the Harper et al. study [42]. Interestingly, some patterns picked up here were found to coincide well with 3-mers and regions surrounding them that were shown to be enriched in the SVM-based classifier for Prxs that focused on the Harper database used here and a k-mer (k = 3) sequence representation, reported by Xiao and Turkett in 2018 [66]. This includes CPA that was enriched for Prx1

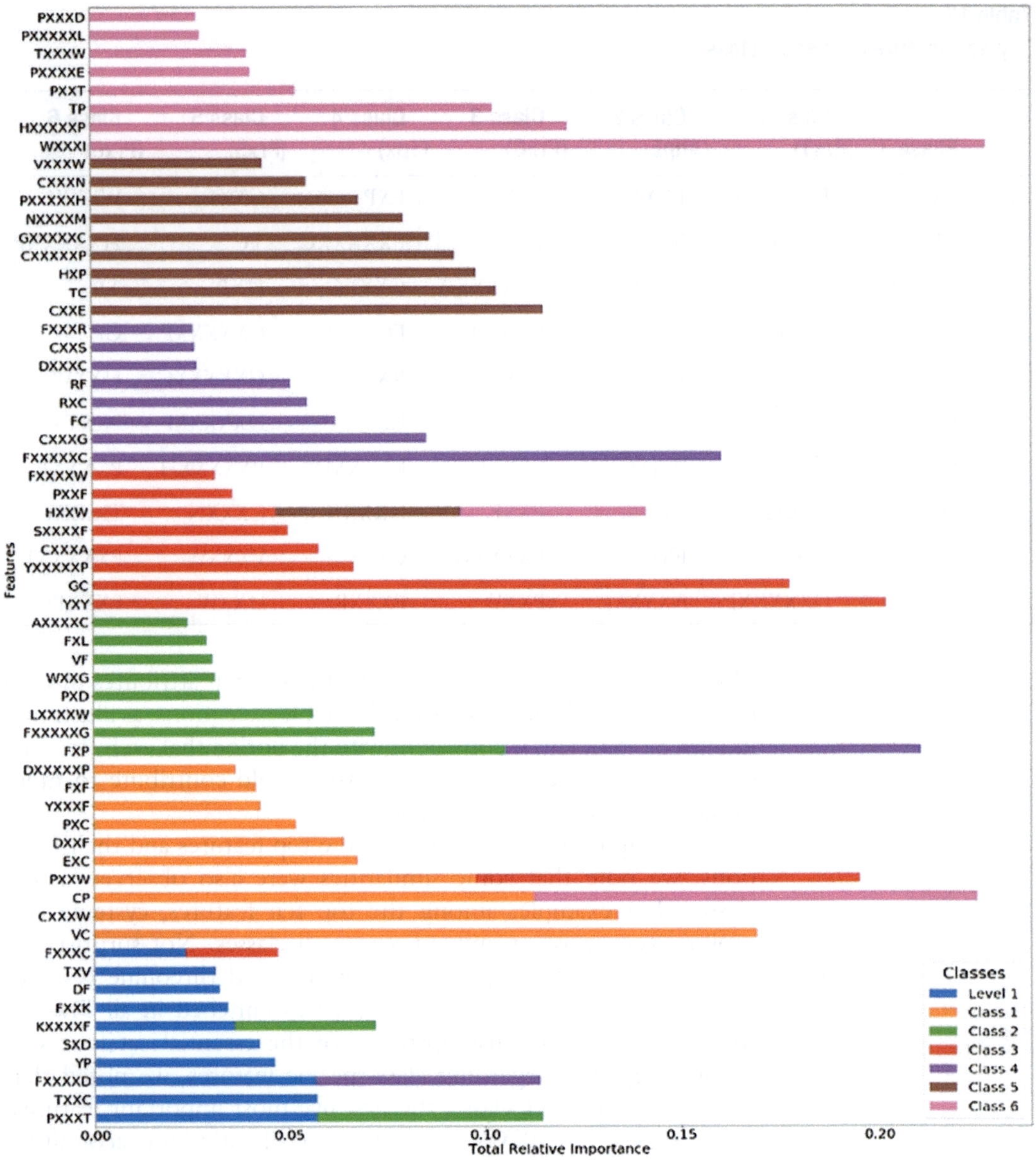

Fig. 5 Top ten features for each class from KSAAP. All KSAAP features within the top ten of at least one of the classes are shown on the left. Bars with additional colors show motifs for more than one class. Classes 1 through 6 represent classes designated Prx1, AhpE, PrxQ, Tpx, Prx5 and Prx6, respectively

group members and contains the resolving Cys, as described above (Fig. 7). For AhpE proteins, FWP was enriched in that study, and corresponds to the region containing four of the ten most important motifs identified in this study, FXP, LXXXXW, WXXG and FXL (Fig. 8). This region is located between helices α3 and α4 and has been suggested to play a role in the oligomerization interface [67]. In addition, two motifs identified by Xiao and Turkett as

Fig. 6 A WebLogo alignment of eight residues around the active site region of all Prx proteins identified in Phase 1 (with the peroxidatic Cys residue at position 16). This region is enriched in the sequence pairs PxxxT and TxxC. The logos in this and subsequent figures were created using WebLogo3 (http:/weblogo.threeplusone.com/create.cgi)

Fig. 7 A WebLogo alignment of eight residues around pattern CPA mined from Prx1 sequences. This region surrounds the resolving Cys residue (position 9) of Prx1 proteins. Five of the important features identified by our approach are enriched within this region, VC, CXXXW, CP, PXXW and EXC

Fig. 8 A WebLogo alignment of eight residues around pattern FWP mined from AhpE sequences. This region is enriched in four of the most important features identified in this study, FXP, LXXXXW, WXXG and FXL

enriched in Prx5 proteins, VND and FVM, are proximal to one another in sequence and encompass a region containing three of the ten most important motifs identified here, NXXXXM, CXXXN, and VXXXW (Fig. 9). This region includes the single noncatalytic Cys present in some members (within the β-strand–turn–helix α3 region), and is of as yet unknown significance. Importantly, these regions identified as unique to certain classes of Prxs can provide new information since some of these are relatively distant from the active site and would therefore not be captured by active site profiling approaches.

Fig. 9 A WebLogo alignment of eight residues around pattern VND and FVM mined from Prx5 sequences. This region is enriched in three of the features identified by this study, NXXXXM, CXXXN, and VXXXW

Table 13
Results of tenfold cross-validation and the independent set in terms of ACC using our method compared with other machine learning algorithms

		Training				Independent test			
	Class	**RF-Prx**	**SVM**	**NB**	**KNN**	**RF-Prx**	**SVM**	**NB**	**KNN**
1	Phase 1 (All)	*99.85*	99.65	99.43	99.79	*99.84*	99.78	99.57	99.80
2	Class 1 (Prx1)	*99.96*	99.76	99.53	99.92	*99.98*	99.89	99.67	99.96
3	Class 2 (AhpE)	*99.87*	99.08	95.14	99.65	*99.92*	99.58	99.13	99.52
4	Class 3 (PrxQ)	*99.94*	99.13	95.27	99.78	*99.98*	99.71	99.27	99.63
5	Class 4 (Tpx)	*100*	99.93	99.86	99.97	*100*	99.98	99.96	99.97
6	Class 5 (Prx5)	*99.98*	99.88	99.87	99.97	*99.98*	99.92	99.91	99.94
7	Class 6 (Prx6)	*99.92*	99.87	99.57	99.90	*99.97*	99.90	99.84	99.93

3.4 Comparison with Different Machine Learning Algorithms

Next, we compared our method with other machine learning algorithms, such as SVM, NB, and KNN. We used the same selected features in these algorithms, with optimized parameters. The results for ACC were assessed using tenfold cross-validation and the independent set (Table 13). Based on these analyses, our RF-based model exhibited the highest ACC scores for all classes, using both tenfold cross validation and an independent set.

We applied our method to the UniProtKB/SwissProt database including 559,077 sequences and predicted, in addition to those already present in our starting Prx databases, another 8199 Prx sequences, all of which could be classified into one of the six classes (Table 14).

Table 14
Prediction results (new representatives) for each class in the UniProtKB/SwissProt database of September 2021

	Number of predictions (UniProtKB/SwissProt database)
Class 1 (Prx1)	879
Class 2 (AhpE)	103
Class 3 (PrxQ)	3888
Class 4 (Tpx)	1731
Class 5 (Prx5)	1402
Class 6 (Prx6)	196

4 Conclusions

In this study, a novel computational method termed RF-Prx was developed to recognize Prx enzymes and classify them into one of six classes. We discovered from individual features that KSAAP performed well compared to AAC and CT features. The best performance was observed when implementing xgboost to choose the best variables of KSAAP to use for each of the six classes. This indicated that increasing the number of features does not necessarily help increase the quality of the method. RF-Prx outperformed other machine learning methods that were tested. It should be noted that, while RF-Prx achieved the best results, there is still room to enhance this method. For instance, inclusion of structural features could add substantial information for Prxs, which may improve prediction measurements for this method. Future iterations may include consideration of some physiochemical features that will help capture and better define Prx and its members for the future. Finally, this study provided conserved features and regions for each class, which will help researchers gain insight into each category.

Use of RF-Prx allowed identification of class-specific conserved motifs of two residues with a variable distance between them, which then allowed identification of conserved regions containing them. Like the previous method that identified classifying 3-mers [66], this approach allowed the detection of class-specific conserved sequences outside the known functional centers from the earlier bioinformatics work, and with potential biological significance. For example, drugs designed to target Prx proteins would likely suffer from cross-reactivity among distinct Prxs if targeted to conserved active sites, but this may be avoidable if remote, class-specific regions could be targeted instead. The use of multiple complementary approaches can also help highlight conserved regions not identified by other methods.

References

1. Crane EJ 3rd, Parsonage D, Poole LB, Claiborne A (1995) Analysis of the kinetic mechanism of enterococcal NADH peroxidase reveals catalytic roles for NADH complexes with both oxidized and two-electron-reduced enzyme forms. Biochemistry 34(43):14114–14124
2. Poole LB, Claiborne A (1988) Evidence for a single active-site cysteinyl residue in the streptococcal NADH peroxidase. Biochem Biophys Res Commun 153(1):261–266
3. Chae HZ, Robison K, Poole LB, Church G, Storz G, Rhee SG (1994) Cloning and sequencing of thiol-specific antioxidant from mammalian brain: alkyl hydroperoxide reductase and thiol-specific antioxidant define a large family of antioxidant enzymes. Proc Natl Acad Sci U S A 91(15):7017–7021
4. Jacobson FS, Morgan RW, Christman MF, Ames BN (1989) An alkyl hydroperoxide reductase from *Salmonella typhimurium* involved in the defense of DNA against oxidative damage. Purification and properties. J Biol Chem 264(3):1488–1496
5. Claiborne A, Yeh JI, Mallett TC, Luba J, Crane EJ 3rd, Charrier V, Parsonage D (1999) Protein-sulfenic acids: diverse roles for an unlikely player in enzyme catalysis and redox regulation. Biochemistry 38:15407–15416
6. Hall A, Parsonage D, Poole LB, Karplus PA (2010) Structural evidence that peroxiredoxin catalytic power is based on transition-state stabilization. J Mol Biol 402(1):194–209
7. Perkins A, Nelson KJ, Parsonage D, Poole LB, Karplus PA (2015) Peroxiredoxins: guardians against oxidative stress and modulators of peroxide signaling. Trends Biochem Sci 40(8): 435–445
8. Paulsen CE, Carroll KS (2013) Cysteine-mediated redox signaling: chemistry, biology, and tools for discovery. Chem Rev 113(7): 4633–4679
9. Poole LB, Nelson KJ (2008) Discovering mechanisms of signaling-mediated cysteine oxidation. Curr Opin Chem Biol 12(1):18–24
10. Stöcker S, Van Laer K, Mijuskovic A, Dick TP (2018) The conundrum of hydrogen peroxide signaling and the emerging role of peroxiredoxins as redox relay hubs. Antioxid Redox Signal 28(7):558–573. https://doi.org/10.1089/ars.2017.7162
11. Wood ZA, Poole LB, Karplus PA (2003) Peroxiredoxin evolution and the regulation of hydrogen peroxide signaling. Science 300(5619):650–653
12. Dietz KJ (2011) Peroxiredoxins in plants and cyanobacteria. Antioxid Redox Signal 15(4): 1129–1159
13. Randall LM, Ferrer-Sueta G, Denicola A (2013) Peroxiredoxins as preferential targets in H2O2-induced signaling. Methods Enzymol 527:41–63
14. Kim K, Kim IH, Lee KY, Rhee SG, Stadtman ER (1988) The isolation and purification of a specific "protector" protein which inhibits enzyme inactivation by a thiol/Fe(III)/O2 mixed-function oxidation system. J Biol Chem 263(10):4704–4711
15. Wood ZA, Schröder E, Harris JR, Poole LB (2003) Structure, mechanism and regulation of peroxiredoxins. Trends Biochem Sci 28(1): 32–40
16. Carvalho LAC, Truzzi DR, Fallani TS, Alves SV, Toledo JC Jr, Augusto O, Netto LES, Meotti FC (2017) Urate hydroperoxide oxidizes human peroxiredoxin 1 and peroxiredoxin 2. J Biol Chem 292(21):8705–8715. https://doi.org/10.1074/jbc.M116.767657
17. Netto LES, Chae HZ, Kang SW, Rhee SG, Stadtman ER (1996) Removal of hydrogen peroxide by thiol-specific antioxidant enzyme (TSA) is involved with its antioxidant properties. TSA possesses thiol peroxidase activity. J Biol Chem 271(26):15315–15321
18. Poole LB (2007) The catalytic mechanism of peroxiredoxins. Subcell Biochem 44:61–81
19. Trujillo M, Ferrer-Sueta G, Thomson L, Flohe L, Radi R (2007) Kinetics of peroxiredoxins and their role in the decomposition of peroxynitrite. Subcell Biochem 44:83–113
20. Peskin AV, Cox AG, Nagy P, Morgan PE, Hampton MB, Davies MJ, Winterbourn CC (2010) Removal of amino acid, peptide and protein hydroperoxides by reaction with peroxiredoxins 2 and 3. Biochem J 432(2):313–321
21. Hofmann B, Hecht H-J, Flohé L (2002) Peroxiredoxins. Biol Chem 383:347–364
22. Knoops B, Loumaye E, Van der Eecken V (2007) Evolution of the peroxiredoxins: taxonomy, homology and characterization. In: Flohé L, Harris JR (eds) Peroxiredoxin systems. Springer, New York, pp 27–40
23. Copley SD, Novak WR, Babbitt PC (2004) Divergence of function in the thioredoxin fold suprafamily: evidence for evolution of peroxiredoxins from a thioredoxin-like ancestor. Biochemistry 43(44):13981–13995

24. Hall A, Nelson K, Poole LB, Karplus PA (2011) Structure-based insights into the catalytic power and conformational dexterity of peroxiredoxins. Antioxid Redox Signal 15(3): 795–815
25. Furdui CM, Poole LB (2014) Chemical approaches to detect and analyze protein sulfenic acids. Mass Spectrom Rev 33(2):126–146
26. Poole LB, Furdui CM, King SB (2020) Introduction to approaches and tools for the evaluation of protein cysteine oxidation. Essays Biochem 64(1):1–17. https://doi.org/10.1042/EBC20190050
27. Yang J, Gupta V, Carroll KS, Liebler DC (2014) Site-specific mapping and quantification of protein S-sulphenylation in cells. Nat Commun 5:4776
28. Winterbourn CC, Metodiewa D (1999) Reactivity of biologically important thiol compounds with superoxide and hydrogen peroxide. Free Radic Biol Med 27(3-4): 322–328
29. Winterbourn CC (2008) Reconciling the chemistry and biology of reactive oxygen species. Nat Chem Biol 4(5):278–286
30. Portillo-Ledesma S, Randall LM, Parsonage D, Dalla Rizza J, Karplus PA, Poole LB, Denicola A, Ferrer-Sueta G (2018) Differential kinetics of two-cysteine peroxiredoxin disulfide formation reveal a novel model for peroxide sensing. Biochemistry 57(24):3416–3424. https://doi.org/10.1021/acs.biochem.8b00188
31. Heppner DE, Janssen-Heininger YM, van der Vliet A (2017) The role of sulfenic acids in cellular redox signaling: reconciling chemical kinetics and molecular detection strategies. Arch Biochem Biophys 616:40–46
32. Salsbury FR Jr, Knutson ST, Poole LB, Fetrow JS (2008) Functional site profiling and electrostatic analysis of cysteines modifiable to cysteine sulfenic acid. Protein Sci 17(2):299–312
33. Peralta D, Bronowska AK, Morgan B, Doka E, Van Laer K, Nagy P, Grater F, Dick TP (2015) A proton relay enhances H2O2 sensitivity of GAPDH to facilitate metabolic adaptation. Nat Chem Biol 11(2):156–163
34. Nelson KJ, Parsonage D, Karplus PA, Poole LB (2013) Evaluating peroxiredoxin sensitivity toward inactivation by peroxide substrates. Methods Enzymol 527:21–40
35. Poynton RA, Peskin AV, Haynes AC, Lowther WT, Hampton MB, Winterbourn CC (2016) Kinetic analysis of structural influences on the susceptibility of peroxiredoxins 2 and 3 to hyperoxidation. Biochem J 473(4):411–421
36. Atkinson HJ, Babbitt PC (2009) An atlas of the thioredoxin fold class reveals the complexity of function-enabling adaptations. PLoS Comput Biol 5(10):e1000541
37. Atkinson HJ, Babbitt PC (2009) Glutathione transferases are structural and functional outliers in the thioredoxin fold. Biochemistry 48(46):11108–11116. https://doi.org/10.1021/bi901180v
38. Choi HJ, Kang SW, Yang CH, Rhee SG, Ryu SE (1998) Crystal structure of a novel human peroxidase enzyme at 2.0 Å resolution. Nat Struct Biol 5(5):400–406
39. Fomenko DE, Gladyshev VN (2003) Identity and functions of CxxC-derived motifs. Biochemistry 42(38):11214–11225
40. Schröder E, Ponting CP (1998) Evidence that peroxiredoxins are novel members of the thioredoxin fold superfamily. Protein Sci 7(11): 2465–2468
41. Karplus PA, Hall A (2007) Structural survey of the peroxiredoxins. In: Flohé L, Harris JR (eds) Peroxiredoxin systems. Springer, New York, pp 41–60
42. Nelson KJ, Knutson ST, Soito L, Klomsiri C, Poole LB, Fetrow JS (2011) Analysis of the peroxiredoxin family: using active-site structure and sequence information for global classification and residue analysis. Proteins 79(3): 947–964
43. Harper AF, Leuthaeuser JB, Babbitt PC, Morris JH, Ferrin TE, Poole LB, Fetrow JS (2017) An atlas of peroxiredoxins created using an active site profile-based approach to functionally relevant clustering of proteins. PLoS Comput Biol 13(2):e1005284
44. Soito L, Williamson C, Knutson ST, Fetrow JS, Poole LB, Nelson KJ (2011) PREX: PeroxiRedoxin classification indEX, a database of subfamily assignments across the diverse peroxiredoxin family. Nucleic Acids Res 39 (Database issue):D332–D337
45. Akiva E, Brown S, Almonacid DE, Barber AE 2nd, Custer AF, Hicks MA, Huang CC, Lauck F, Mashiyama ST, Meng EC, Mischel D, Morris JH, Ojha S, Schnoes AM, Stryke D, Yunes JM, Ferrin TE, Holliday GL, Babbitt PC (2014) The structure-function linkage database. Nucleic Acids Res 42(Database issue):D521–D530. https://doi.org/10.1093/nar/gkt1130
46. Knutson ST, Westwood BM, Leuthaeuser JB, Turner BE, Nguyendac D, Shea G, Kumar K, Hayden JD, Harper AF, Brown SD, Morris JH, Ferrin TE, Babbitt PC, Fetrow JS (2017) An approach to functionally relevant clustering of

the protein universe: active site profile-based clustering of protein structures and sequences. Protein Sci 26(4):677–699. https://doi.org/10.1002/pro.3112

47. Youngs N, Penfold-Brown D, Bonneau R, Shasha D (2014) Negative example selection for protein function prediction: the NoGO database. PLoS Comput Biol 10(6):e1003644. https://doi.org/10.1371/journal.pcbi.1003644
48. Li F, Zhang Y, Purcell AW, Webb GI, Chou KC, Lithgow T, Li C, Song J (2019) Positive-unlabelled learning of glycosylation sites in the human proteome. BMC Bioinformatics 20(1):112. https://doi.org/10.1186/s12859-019-2700-1
49. UniProt C (2014) Activities at the Universal Protein Resource (UniProt). Nucleic Acids Res 42(Database issue):D191–D198. https://doi.org/10.1093/nar/gkt1140
50. KrishnaVeni CV, Sobha Rani T (2011) On the classification of imbalanced datasets. IJCST 2 (SP1):145–148
51. Breiman L (2001) Random forests. Mach Learn 45(1):5–32
52. Ismail HD, Jones A, Kim JH, Newman RH, Kc DB (2016) RF-Phos: a novel general phosphorylation site prediction tool based on random forest. Biomed Res Int 2016:3281590. https://doi.org/10.1155/2016/3281590
53. Al-Barakati HJ, Saigo H, Newman RH, Kc DB (2019) RF-GlutarySite: a random forest based predictor for glutarylation sites. Mol Omics 15(3):189–204. https://doi.org/10.1039/c9mo00028c
54. Wang R, Perez-Riverol Y, Hermjakob H, Vizcaino JA (2015) Open source libraries and frameworks for biological data visualisation: a guide for developers. Proteomics 15(8):1356–1374. https://doi.org/10.1002/pmic.201400377
55. Barbu A, She Y, Ding L, Gramajo G (2017) Feature selection with annealing for computer vision and big data learning. IEEE Trans Pattern Anal Mach Intell 39(2):272–286. https://doi.org/10.1109/TPAMI.2016.2544315
56. Chen T, Guestrin C (2016) XGBoost: a scalable tree boosting system: KDD '16. In: Proceedings of the 22nd ACM SIGKDD international conference on knowledge discovery and data mining. ACM pp 785–794. https://doi.org/10.1145/2939672.2939785
57. Stahl K, Schneider M, Brock O (2017) EPSILON-CP: using deep learning to combine information from multiple sources for protein contact prediction. BMC Bioinformatics 18(1):303. https://doi.org/10.1186/s12859-017-1713-x
58. White C, Ismail HD, Saigo H, Kc DB (2017) CNN-BLPred: a convolutional neural network based predictor for beta-lactamases (BL) and their classes. BMC Bioinformatics 18(Suppl 16):577. https://doi.org/10.1186/s12859-017-1972-6
59. Pedregosa F, Varoquaux G, Gramfort A, Michel V, Thirion B (2011) Scikit-learn: machine learning in Python. J Mach Learn Res 12:2825–2830
60. Al-Barakati H, Thapa N, Hiroto S, Roy K, Newman RH, Kc D (2020) RF-MaloSite and DL-malosite: methods based on random forest and deep learning to identify malonylation sites. Comput Struct Biotechnol J 18:852–860. https://doi.org/10.1016/j.csbj.2020.02.012
61. Cortes C, Vapnik V (1995) Support-vector networks. Mach Learn 20:273–297
62. Geng H, Lu T, Lin X, Liu Y, Yan F (2015) Prediction of protein-protein interaction sites based on naive bayes classifier. Biochem Res Int 2015:978193. https://doi.org/10.1155/2015/978193
63. Venables WN, Ripley BD (2013) Modern applied statistics with S-PLUS, 3rd edn. Springer-Verlag
64. Kohavi R (1995) A study of cross-validation and bootstrap for accuracy estimation and model selection. In: IJCAI'95: Proceedings of the 14th international joint conference on artificial intelligence—volume 2. ACM, pp 1137–1143
65. Baldi P, Brunak S, Chauvin Y, Andersen CA, Nielsen H (2000) Assessing the accuracy of prediction algorithms for classification: an overview. Bioinformatics 16(5):412–424. https://doi.org/10.1093/bioinformatics/16.5.412
66. Xiao J, Turkett WH (2018) K-mer based classifiers extract functionally relevant features to support accurate Peroxiredoxin subgroup distinction. bioRXiv https://doi.org/10.1101/387787
67. Li S, Peterson NA, Kim MY, Kim CY, Hung LW, Yu M, Lekin T, Segelke BW, Lott JS, Baker EN (2005) Crystal structure of AhpE from *Mycobacterium tuberculosis*, a 1-Cys peroxiredoxin. J Mol Biol 346(4):1035–1046

Chapter 9

Computational Prediction of N- and O-Linked Glycosylation Sites for Human and Mouse Proteins

Ghazaleh Taherzadeh, Matthew Campbell, and Yaoqi Zhou

Abstract

Protein glycosylation is one of the most complex posttranslational modifications (PTM) that play a fundamental role in protein function. Identification and annotation of these sites using experimental approaches are challenging and time consuming. Hence, there is a demand to build fast and efficient computational methods to address this problem. Here, we present the SPRINT-Gly framework containing the largest dataset and a prediction model of glycosylation sites for a given protein sequence. In this framework, we construct a large dataset containing N- and O-linked glycosylation sites of human and mouse proteins, collected from different sources. We then introduce the SPRINT-Gly method to predict putative N- and O-linked sites. SPRINT-Gly is a machine learning-based approach consisting of a number of trained predictive models for glycosylation sites in both human and mouse proteins, separately. The method is built by incorporating sequence-based, predicted structural, and physicochemical information of the neighboring residues of each N- and O-linked glycosylation site and by training deep learning neural network and support vector machine as classifiers. SPRINT-Gly outperformed other existing methods by achieving 18% and 50% higher Matthew's correlation coefficient for N- and O-linked glycosylation site prediction, respectively. SPRINT-Gly is publicly available as an online and stand-alone predictor at https://sparks-lab.org/server/sprint-gly/.

Key words Glycosylation, Posttranslational modifications, N- and O-linked glycosylation sites, Glycosylation sites prediction, Machine learning, Deep learning

1 Introduction

Protein glycosylation is the most diverse form of posttranslational modification and essential for biological processes such as immune responses, intercellular signaling, and host–pathogen interaction. It is also observed as an important factor in several diseases including many cancers [1]. Although glycosylation is an abundant posttranslational modification, studies show that only a limited number of amino acids undergo such modifications [2, 3]. So far, several classes of glycosylation have been identified. These classes vary in

Dukka B. KC (ed.), *Computational Methods for Predicting Post-Translational Modification Sites*,
Methods in Molecular Biology, vol. 2499, https://doi.org/10.1007/978-1-0716-2317-6_9,

terms of their structures, biosynthesis, locations of sites, and the types of amino acid residues [4]. In addition to less abundant classes like C-glycosylation and O-glycosylation, there are two main classes of protein glycosylation known as N-linked (N-glycosylation), and mucin-type O-linked (O-glycosylation) [2, 3, 5]. N-linked glycosylation involves the covalent attachment of glycan structures to the *Asparagine* residues in the Asn-X-Ser or Asn-X-Thr sequon where X is the occurrence of any residue except proline. This glyco-attachment occurs before protein folding and impacts the tertiary structure and stability of the protein. In contrast, O-linked glycosylation does not involve any specific motif and attaches to *Serine* or *Threonine* residues after a protein fold [6]. Yet not all *Asparagine, Serine*, or *Threonine* residues in a protein undergo glycosylation due to their limited accessibility, substrate availability, and intricate regulation of glycosylation-related enzymes [7]. An important characteristic of protein glycosylation is the distribution of glycans on a protein glycosylation site, commonly known as its microheterogeneity, where an individual glycoprotein can exist in many glycoforms. These glycoforms differ from each other based on the exact glycan structures found at each glycosylation site.

Due to the biological importance of glycosylation, accurate identification of glycosylation sites is essential to understanding the role of glycosylation in biological processes and to assist the development of disease treatments. There are a number of techniques available to profile glycoproteins and to identify sites of glycosylation, including high-performance liquid chromatography (HPLC), capillary electrophoresis (CE), mass spectrometry (MS), isoelectric focusing (IEF), and lectin-based microarray [2, 4, 8, 9]. However, these methods are expensive and labor intensive. Meanwhile, the exponential growth in the number of discovered proteins makes the identification of glycosylated sites by experimental methods practically infeasible. Thus, computational approaches for predicting glycosylated sites become increasingly necessary as a screening tool.

Several machine learning-based approaches have been introduced to predict the N- and O-linked glycosylation sites namely, NetNglyc and NetOglyc [10], EnsembleGly [11], GPP [12], GlycoEP [13], GlycoPP [14], GlycoMine [15], N-GlyDE [16], and N-GlycoGo [17]. These models were solely trained using sequence-based features such as a binary representation of amino acid sequence, Position Specific Scoring Matrix (PSSM), and physicochemical properties as well as predicted structure-based features such as accessible surface area (ASA), Secondary Structure (SS), and pattern-based encoding. In our study, we have developed SPRINT-Gly [18], an alternative machine learning–based framework trained using deep neural network (DNN) and support vector machine (SVM) classifiers to predict the N- and O-linked glycosylation sites in human and mouse proteins, respectively. The framework contains four separate models trained with different combinations of selected sequence-based features, predicted structural features generated by SPIDER 3.0 [19], and intrinsic disorder information determined by SPOT-disorder [20]. Some of these features have been used in previous studies with promising results [21–28]. By employing these features, SPRINT-Gly framework outperformed other existing tools in predicting N- and O-linked glycosylation sites in human and mouse proteins.

2 Material and Algorithm

2.1 Dataset

A comprehensive human and mouse glycoprotein dataset was curated from several available sources including UniProt [29], N-linked glycoproteins from dbPTM [30], GlycoProtDB [31], human N-glycosylation site information from UniPep [32], and UniCarbKB [33] as well as the data collected in previous studies [10–15]. To the best of our knowledge, this dataset is the largest publicly available. It consists of 7023 glycoproteins with 2369 N-linked and 211 O-linked human proteins, and 2096 N-linked and 398 O-linked mouse proteins [34, 35]. For a given protein, each potential N-linked (asparagine amino acid) and O-linked (serine and threonine amino acids) glycosylation site and their fixed number of upstream and downstream amino acids are represented by a sequence segment named glyco-segment. Glyco-segments are then used to extract features and train different predictive models. In addition, in order to gain an insight into glycosylation sites in different species, we built cross-species independent test sets for N- and O-linked human and mouse proteins. These independent datasets are nonredundant (with 30% cutoff) with cross-species training sets.

2.2 Method

For each glyco-segment in the dataset, five groups of sequence-based features were extracted. The first is *amino acid occurrence*, which is a binary vector to represent the glycol-segment where the value of the corresponding amino acid residue position is set to "1" and the remaining positions are set to "0." The second is *evolutionary information* including Position Specific Scoring Matrix (PSSM) generated by PSI-Blast [34], information entropy [36] and close neighboring correlation coefficient from PSSM profile of the glycol-segment. The third is *25 physicochemical properties* representing [21, 37, 38] neighboring N- and O-linked amino acid residues. The fourth is *predicted structural* features generated by SPIDER 3.0 including Half-Sphere Exposure (HSE) representing contact numbers in the upper and lower sphere [39], Secondary Structure (SS) probabilities and the fraction of each SS type in the glycol-segment, relative Accessible Surface Area (rASA) and average rASA of each glycol-segment) [19]. The fifth is SPOT-Disorder (Intrinsically disordered region) [20] to represent the probability of each glycol-segment amino acids being in the intrinsically disordered proteins.

Two different classifiers (four models), deep neural network (DNN) [40] and support vector machine (SVM) [41], were trained to predict N-linked and O-linked putative sites for human and mouse proteins, separately. These four models were trained on 11,175 N-linked and 687 O-linked sites for human and 13,428 N-linked and 839 O-linked sites for mouse proteins. To deal with the vacancies for amino acid residues in the N- or C-terminals we assigned a feature vector of zero value to amino acid residues before the N- or after the C-terminals. Note that all feature values are normalized using the Min-Max normalization technique.

2.2.1 N-Linked Glycosylation Site Prediction

Two DNN models were trained on the N-linked glycosylation sites in human and mouse proteins. The optimized DNN model is formed by three hidden layers containing 150 nodes in each layer using the sigmoid activation function and 0.01 as the learning rate. The input layer for the DNN in the first iterative learning layer consists of 424 features including features of target amino acids as well as upstream and downstream neighboring amino acid residues. Since the DNN adjusts the weights for each feature, we used all features in the feature set to train the model and let the network architecture choose the discriminative features.

2.2.2 O-Linked Glycosylation Site Prediction

We trained two separate SVM models to predict the O-linked glycosylation sites in human and mouse proteins. The best model achieved the highest AUC value with the regularization parameter C in the linear SVM model set at 0.0125 and incorporating features of four amino acid residues upstream and downstream. To optimize the feature set and improve the model performance, we applied sequence forward feature selection [42] to eliminate redundant features. We started with training the models with each feature group individually and added the feature groups with the highest AUC values. The process continues sequentially until there is no improvement in the AUC value when adding a new feature group. The best feature groups selected for the human dataset includes physicochemical properties, evolutionary information, HSE, SS, rASA feature groups with 503 features in total. On the other hand, for the mouse dataset employing all features (726 features) were necessary to achieve the highest AUC values. This illustrates different characteristics of the O-linked glycosylation sites in human and mouse proteins.

Table 1
Performance of SPRINT-Gly on (1) tenfold cross-validation, (2) independent test set, and (3) cross-species test on N- and O-linked glycosylation site for human and mouse proteins

Dataset			MCC	AUC	Acc	Sn	Sp	Pr
Human	N-linked	Tenfold CV	0.814	0.966	0.938	0.974	0.930	0.737
		Variance	0.009	0.002	0.007	0.010	0.016	0.021
		Ind. set	0.939	0.984	0.978	0.978	0.979	0.927
	Mouse trained	Human N-link test set	0.941	0.981	0.979	0.975	0.981	0.934
	O-linked	Tenfold CV	0.240	0.812	0.924	0.363	0.947	0.218
		Variance	0.007	0.011	0.008	0.016	0.018	0.004
		Ind. set	0.191	0.824	0.94	0.329	0.95	0.144
	Mouse trained	Human O-link test set	0.067	0.602	0.973	0.038	0.995	0.158
Mouse	N-linked	Tenfold CV	0.988	0.997	0.996	0.98	0.99	0.985
		Variance	0.005	0.001	0.010	0.008	0.003	0.009
		Ind. set	0.986	0.993	0.995	0.99	0.997	0.989
	Human trained	Mouse N-link test set	0.988	0.995	0.996	0.99	0.998	0.993
	O-linked	Tenfold CV	0.142	0.772	0.960	0.238	0.971	0.108
		Variance	0.004	0.009	0.010	0.011	0.015	0.009
		Ind. set	0.135	0.795	0.986	0.17	0.992	0.119
	Human trained	Mouse O-link test set	0.024	0.575	0.624	0.52	0.624	0.009

Acc accuracy, *Sn* sensitivity, *Sp* specificity, *Pr* precision, *CV* cross-validation

2.2.3 Cross-Species Site Prediction

We also evaluated the performance of SPRINT-Gly on cross-species independent test sets. We trained the N- and O-linked models on human and mouse proteins and tested them on the independent cross-species test sets.

Table 1 shows the performance of the SPRINT-Gly framework on the prediction of N- and O-linked glycosylation sites in human and mouse proteins as well as cross-species independent test sets. The N-linked site prediction achieved high AUC values of 0.984, 0.993, 0.981, and 0.995 for human and mouse proteins and cross-species (human and mouse) independent test sets, respectively. Moreover, the MCC values for these sets are in the range of 0.81–0.98, which demonstrates the accurate prediction of N-linked glycosylation sites. On the other hand, the prediction of O-linked glycosylation sites is not as accurate because of the weak sequence patterns of these sites and the smaller size of the training dataset. SPRINT-Gly continues to outperform other existing methods in predicting O-linked glycosylation sites.

3 Web Server and Stand-Alone Package

The SPRINT-Gly web server is accessible at: https://sparks-lab.org/server/sprint-gly/. It takes a query sequence and the type of glycosylation site in the human or mouse proteins as input (*see* Subheading 4).

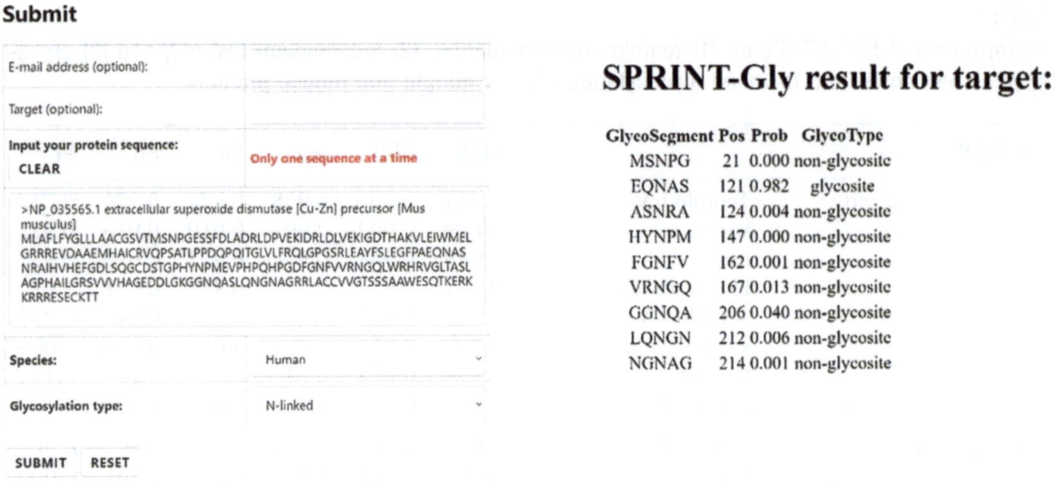

Fig. 1 The web server input (includes query sequence, type of species, and type of glycosylation, *Left*), notice Note 1 to 3 when using webserver, and the output screenshot reporting the glycol-segment, the position, the probability, and the glycol type (*Right*) for the example sequence"NP_035565.1.seq" by SPRINT-Gly

1. As shown in Fig. 1. (Left), the protein sequence can be entered or copied to the text area in a FASTA format. It is important to note that only one protein sequence can be processed at the time. The sequence must be in a FASTA format including 20 standard amino acids only and the first line starts with (">" followed by the name of protein) to identify the name of the protein. If the first line did not include the protein name, it will set to "Unknown" by default. The species is set to "Human" by default and the user is required to change it if otherwise. If the species type is not available, the species parameter should be set to "Mouse." In addition, the glycosylation type should be set to N- or O-linked depending on the prediction goal. It is optional for the user to provide "E-mail address" and "Target" name.
2. The job will be created and sent to the queue by clicking the submit button and then the page will be redirected to the to-be-available prediction results. The new page includes the job information and will be refreshed every 60 min. The results will be displayed upon the completion of the job. The "Delete Job files" button is provided for the user to terminate and delete all related files to the submitted job.
3. The total prediction time depends on the length of the given protein sequence and how busy the server is. This time can vary between 15 min to a few hours. If an E-mail address is provided in the link of job submission, the user will receive an e-mail containing the link of prediction result once the job is completed. All files and prediction results are stored in the server for a month and will be deleted after that.

4. Moreover, the "Check Job Status," "Check Queue Status," and "Check My Jobs" buttons are available for the user to get an instant update about the progress. We request users to avoid submitting their sequences more than once and save the server's computational resources.
5. Figure 1. (Right) illustrates an example of the output page. The result page includes glyco-segments containing all N- or O-linked amino acid residues (depending on the type of submitted job), position, predicted probability, and whether the amino acid is predicted as glycosite or nonglycosite. The glycotype is defined by setting a threshold on the predicted probability.

In addition to the web server, we created a GitLab project page (available at https://gitlab.com/matthew.campbell1980/sprint-gly) for user convenience. The GitLab page includes datasets (in the /dataset folder) and source code (in the /src folder) for N- and O-linked human and mouse proteins. As the input, the user should provide generated files by PSSM, SPIDER 3.0, and SPOT-Disorder programs. These files should be located in the same folder/directory and the location should be pointed out in the source code. The predicted output file will be generated after running the program. The format of the file is the same as the output page of the web server. The stand-alone version allows a user to run the SPRINT-Gly program over a list of sequences.

4 Notes

1. A given protein sequence should be provided in the FASTA format and do not contain nonstandard amino acids (e.g., X amino acid). If only DNA or RNA sequences are known, users should convert and obtain the respective protein sequence by EasyPy (http://web.expasy.org/translate) or any other translation tools.
2. In the current version of the SPRINT-Gly web server, PSI-blast, SPIDER 3.0, and SPOT-Disorder programs are executed once the job is submitted. This process takes a while depending on the length of a protein. This duration can be minimized if the output files generated by these programs are available in advance. One nice feature for the stand-alone program is to give a user the ability to upload the mentioned files so as to reduce computational time. This feature is not yet available for the web server.
3. The performance of trained models depends on the sequences and known experimentally annotated glycoproteins at the time of developing the SPRINT-Gly framework. A future update for the SPRINT-Gly models is required when more data is available.

5 Summary

Due to the importance of glycoproteins and their roles in protein function and diseases, several studies have been developed to predict putative N- and O-linked glycosylation sites. Here, we provided a brief description of a machine learning–based framework, called SPRINT-Gly. This framework includes the largest glycosylation-site repository collected from six different resources. SPRINT-Gly is built by employing sequence and structure-based features and contains four different models to predict N- and O-linked glycosylation sites in human and mouse proteins, respectively. Our results demonstrated that SPRINT-Gly is able to significantly outperform the existing prediction tools.

Acknowledgments

We acknowledge the use of the High-Performance Computing Cluster "Gowonda," resources provided by the Queensland Cyber Infrastructure Foundation, and Australian Research Data Commons. We also acknowledge the authors of previous methods for providing the web servers. This work was supported in part by Institute for Glycomics to M.P.C., and by the Australian Government Research Training Program Scholarship for G.T.

References

1. Xiao H, Sun F, Suttapitugsakul S, Wu R (2019) Global and site-specific analysis of protein glycosylation in complex biological systems with mass spectrometry. Mass Spectrom Rev 38(4–5):356–379
2. Joshi HJ, Narimatsu Y, Schjoldager KT, Tytgat HL, Aebi M, Clausen H, Halim A (2018) SnapShot: O-glycosylation pathways across kingdoms. Cell 172(3):632–632
3. Varki A (2017) Biological roles of glycans. Glycobiology 27(1):3–49
4. Thomas DR, Scott NE (2021) Glycoproteomics: growing up fast. Curr Opin Struct Biol 68:18–25
5. Woods RJ (2018) Predicting the structures of glycans, glycoproteins, and their complexes. Chem Rev 118(17):8005–8024
6. Banazadeh A, Veillon L, Wooding KM, Zabetmoghaddam M, Mechref Y (2017) Recent advances in mass spectrometric analysis of glycoproteins. Electrophoresis 38(1):162–189
7. Weerapana E, Imperiali B (2006) Asparagine-linked protein glycosylation: from eukaryotic to prokaryotic systems. Glycobiology 16(6): 91R–101R
8. Li Q, Xie Y, Wong M, Lebrilla CB (2019) Characterization of cell glycocalyx with mass spectrometry methods. Cell 8(8):882
9. Ruhaak LR, Xu G, Li Q, Goonatilleke E, Lebrilla CB (2018) Mass spectrometry approaches to glycomic and glycoproteomic analyses. Chem Rev 118(17):7886–7930
10. Blom N, Sicheritz-Pontén T, Gupta R, Gammeltoft S, Brunak S (2004) Prediction of post-translational glycosylation and phosphorylation of proteins from the amino acid sequence. Proteomics 4(6):1633–1649
11. Caragea C, Sinapov J, Silvescu A, Dobbs D, Honavar V (2007) Glycosylation site prediction using ensembles of support vector machine classifiers. BMC Bioinformatics 8(1): 1–13
12. Hamby SE, Hirst JD (2008) Prediction of glycosylation sites using random forests. BMC Bioinformatics 9(1):1–13

13. Chauhan JS, Bhat AH, Raghava GP, Rao A (2012) GlycoPP: a webserver for prediction of N-and O-glycosites in prokaryotic protein sequences. PLoS One 7(7):e40155
14. Chauhan JS, Rao A, Raghava GP (2013) In silico platform for prediction of N-, O-and C-glycosites in eukaryotic protein sequences. PLoS One 8(6):e67008
15. Li F, Li C, Wang M, Webb GI, Zhang Y, Whisstock JC, Song J (2015) GlycoMine: a machine learning-based approach for predicting N-, C-and O-linked glycosylation in the human proteome. Bioinformatics 31(9):1411–1419
16. Pitti T, Chen CT, Lin HN, Choong WK, Hsu WL, Sung TY (2019) N-GlyDE: a two-stage N-linked glycosylation site prediction incorporating gapped dipeptides and pattern-based encoding. Sci Rep 9(1):1–11
17. Chien CH, Chang CC, Lin SH, Chen CW, Chang ZH, Chu YW (2020) N-GlycoGo: predicting protein N-glycosylation sites on imbalanced data sets by using heterogeneous and comprehensive strategy. IEEE Access 8: 165944–165950
18. Taherzadeh G, Dehzangi A, Golchin M, Zhou Y, Campbell MP (2019) SPRINT-Gly: predicting N-and O-linked glycosylation sites of human and mouse proteins by using sequence and predicted structural properties. Bioinformatics 35(20):4140–4146
19. Heffernan R, Yang Y, Paliwal K, Zhou Y (2017) Capturing non-local interactions by long short-term memory bidirectional recurrent neural networks for improving prediction of protein secondary structure, backbone angles, contact numbers and solvent accessibility. Bioinformatics 33(18):2842–2849
20. Heffernan R, Paliwal K, Lyons J, Dehzangi A, Sharma A, Wang J et al (2015) Improving prediction of secondary structure, local backbone angles and solvent accessible surface area of proteins by iterative deep learning. Sci Rep 5(1):1–11
21. Taherzadeh G, Yang Y, Xu H, Xue Y, Liew AWC, Zhou Y (2018) Predicting lysine-malonylation sites of proteins using sequence and predicted structural features. J Comput Chem 39(22):1757–1763
22. Ahmad MW, Arafat ME, Taherzadeh G, Sharma A, Dipta SR, Dehzangi A, Shatabda S (2020) Mal-light: enhancing lysine malonylation sites prediction problem using evolutionary-based features. IEEE Access 8: 77888–77902
23. Arafat M, Ahmad M, Shovan SM, Dehzangi A, Dipta SR, Hasan M et al (2020) Accurately predicting glutarylation sites using sequential bi-peptide-based evolutionary features. Genes 11(9):1023
24. Dipta SR, Taherzadeh G, Ahmad MW, Arafat ME, Shatabda S, Dehzangi A (2020) SEMal: accurate protein malonylation site predictor using structural and evolutionary information. Comput Biol Med 125:104022
25. Chandra A, Sharma A, Dehzangi A, Ranganathan S, Jokhan A, Chou KC, Tsunoda T (2018) PhoglyStruct: prediction of phosphoglycerylated lysine residues using structural properties of amino acids. Sci Rep 8(1):1–11
26. Islam MM, Saha S, Rahman MM, Shatabda S, Farid DM, Dehzangi A (2018) iProtGly-SS: identifying protein glycation sites using sequence and structure based features. Proteins 86(7):777–789
27. Dehzangi A, López Y, Lal SP, Taherzadeh G, Michaelson J, Sattar A et al (2017) PSSM-Suc: accurately predicting succinylation using position specific scoring matrix into bigram for feature extraction. J Theor Biol 425:97–102
28. Chandra AA, Sharma A, Dehzangi A, Tsunoda T (2019) EvolStruct-Phogly: incorporating structural properties and evolutionary information from profile bigrams for the phosphoglycerylation prediction. BMC Genomics 19(9): 1–9
29. Apweiler R, Bairoch A, Wu CH, Barker WC, Boeckmann B, Ferro S et al (2004) UniProt: the universal protein knowledgebase. Nucleic Acids Res 32(suppl_1):D115–D119
30. Huang KY, Su MG, Kao HJ, Hsieh YC, Jhong JH, Cheng KH et al (2016) dbPTM 2016: 10-year anniversary of a resource for post-translational modification of proteins. Nucleic Acids Res 44(D1):D435–D446
31. Kaji H, Shikanai T, Suzuki Y, Narimatsu H (2017) GlycoProtDB: a database of glycoproteins mapped with actual glycosylation sites identified by mass spectrometry. In: A practical guide to using glycomics databases. Springer, Tokyo, pp 215–224
32. Zhang H, Loriaux P, Eng J, Campbell D, Keller A, Moss P et al (2006) UniPep-a database for human N-linked glycosites: a resource for biomarker discovery. Genome Biol 7(8): 1–12
33. Campbell MP, Peterson R, Mariethoz J, Gasteiger E, Akune Y, Aoki-Kinoshita KF et al (2014) UniCarbKB: building a knowledge platform for glycoproteomics. Nucleic Acids Res 42(D1):D215–D221
34. Altschul SF, Madden TL, Schäffer AA, Zhang J, Zhang Z, Miller W, Lipman DJ (1997) Gapped BLAST and PSI-BLAST: a

new generation of protein database search programs. Nucleic Acids Res 25(17):3389–3402

35. Li W, Godzik A (2006) Cd-hit: a fast program for clustering and comparing large sets of protein or nucleotide sequences. Bioinformatics 22(13):1658–1659
36. Taherzadeh G, Yang Y, Zhang T, Liew AWC, Zhou Y (2016) Sequence-based prediction of protein–peptide binding sites using support vector machine. J Comput Chem 37(13): 1223–1229
37. Meiler J, Müller M, Zeidler A, Schmäschke F (2001) Generation and evaluation of dimension-reduced amino acid parameter representations by artificial neural networks. Mol Model Ann 7(9):360–369
38. Ruiz-Blanco YB, Marrero-Ponce Y, García-Hernández E, Green J (2017) Novel "extended sequons" of human N-glycosylation sites improve the precision of qualitative predictions: an alignment-free study of pattern recognition using ProtDCal protein features. Amino Acids 49(2):317–325
39. Hamelryck T (2005) An amino acid has two sides: a new 2D measure provides a different view of solvent exposure. Proteins 59(1): 38–48
40. Liu W, Wang Z, Liu X, Zeng N, Liu Y, Alsaadi FE (2017) A survey of deep neural network architectures and their applications. Neurocomputing 234:11–26
41. Vapnik V (2013) The nature of statistical learning theory. Springer
42. Kudo M, Sklansky J (2000) Comparison of algorithms that select features for pattern classifiers. Pattern Recogn 33(1):25–41

Chapter 10

iPTMnet RESTful API for Post-translational Modification Network Analysis

Sachin Gavali, Karen E. Ross, Julie Cowart, Chuming Chen, and Cathy H. Wu

Abstract

iPTMnet is a resource that combines rich information about protein post-translational modifications (PTM) from curated databases as well as text mining tools. Researchers can use the iPTMnet website to query, analyze and download the PTM data. In this chapter we describe the iPTMnet RESTful API which provides a way to streamline the integration of iPTMnet data into an automated data analysis workflow. In the first section, we give an overview of the architecture of the API. In the second section, we describe various function defined by the API and provide detailed examples of using these functions.

Key words RESTful API, Post-translational modification, Database, Protein-protein interaction, Phosphorylation

1 Introduction

Proteins are among the fundamental building blocks of the molecular tools employed by biological systems to carry out their essential functions. Even though the genetic information (genome) is the source of all the proteins (proteome), the diversity of the proteome far outweighs the diversity of the genome. A major reason for the increased diversity of the proteome is protein PTM [1].

PTMs play a major role in a variety of biological process such as cellular signaling (Phosphorylation), immune system regulation (Glycosylation) and gene expression (Acetylation/Methylation) [2]. PTMs have been implicated in important diseases such as cancer [3, 4] and Alzheimer's [5, 6]. Their understanding is often essential to aid in development of novel therapeutics such as kinase inhibitors which are used in cancer treatments [7, 8]. Hence, it becomes essential to study and understand the role played by these PTMs in biological systems.

Dukka B. KC (ed.), *Computational Methods for Predicting Post-Translational Modification Sites*, Methods in Molecular Biology, vol. 2499, https://doi.org/10.1007/978-1-0716-2317-6_10,

In recent years, with the advent of techniques such as mass spectrometry, many new PTM events are being discovered, but the information about these PTMs is scattered across many databases and the literature. Hence, we have developed iPTMnet [9] as an integrated resource to provide access to PTM information. It provides several unique features such as (1) Comprehensive and continuously updated PTM information that is collected from curated databases, ontologies and literature (2) Comprehensive scoring of the quality of underlying data sources. (3) Visualization of interaction networks formed due to PTM events; and (4) Visualization of the sequence alignment of the proteoforms formed due to PTM events. Additionally, iPTMnet is the only resource that connects the proteoforms formed due to PTMs with their corresponding enzymes across multiple taxons.

In recent years there has been a tremendous increase in the amount of available biological data. To analyze this data, researchers have been increasingly using highly automated data analysis workflows [10, 11]. To make it easier for researchers to continue utilizing iPTMnet as a part of their workflows, we provide a RESTful API [12] for iPTMnet. The API provides a means to programmatically access all the functions available on the iPTMnet website. To reduce the barrier of entry, we also provide Python and R packages that make it easier to interact with the API and streamline its integration into existing bioinformatics pipelines.

2 Methods

2.1 Overview of the RESTful API Architecture

The RESTful API by its very definition is an application programming interface that is meant to allow external programs to interact directly with the iPTMnet database over the Hyper Text Transfer Protocol (HTTP). The API can be thought of as a contract that guarantees a smooth interaction irrespective of changes to the underlying software architecture. Hence, to uphold this contract and minimize the possibility of any changes in the underlying code breaking this contract we have developed the API using a modular architecture (Fig. 1).

The RESTful API consists of three modules — (1) Database module; (2) HTTP interface module; and (3) Client packages. The database module holds the raw data of the iPTMnet database. The HTTP interface module is responsible for mapping the HTTP endpoints to the underlying data sources. Client packages are responsible for mapping Python or R function calls to the respective HTTP endpoints. The API defines eight endpoints (Table 1) that allow users to interact with iPTMnet (*see* **Note 1**). By default, the client packages connect to the API servers located at Center for Bioinformatics and Computational Biology (CBCB) in University

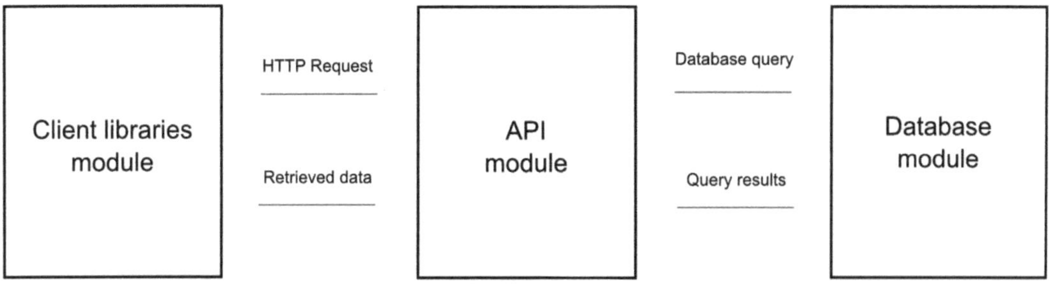

Fig. 1 Overview of iPTMnet RESTful API architecture. The client packages convert the Python or R function calls to HTTP request. These requests are then converted by the HTTP interface module to database queries. The database module processes the data queries and returns the results to HTTP interface module which transform the data into the format requested by the client packages

Table 1
HTTP endpoints defined by the iPTMnet RESTful API

HTTP endpoint	Client function	Description
\search	*search*	Search for a protein in iPTMnet given a filtering criteria
\{ID}\info	*get info*	Get information about a given protein from iPTMnet
\{ID}\substrate	*get substrates*	Get a list of PTM sites for a given substrate
\{ID}\proteoforms	*get proteforms*	Get a list of proteforms for the given protein
\{ID}\ptmppi	*get ptm dependent ppi*	Get a list of PTM dependent PPI for a given protein
\batch ptm enzymes	*get ptm enzymes from list*	Get PTM enzymes for a list of given substrates in bulk
\batch ptm ppi	*get ptm ppi from list*	Get PTM dependent PPI for a given list of protein in bulk
\variants	*get variants*	Get a list of variants for a given protein

of Delaware. If you are self-hosting the API, you can specify a custom URL (*see* **Note 2**). In the next section we provide a comprehensive tutorial demonstrating the usage of these endpoints.

2.2 Examples of Using API Endpoints

2.2.1 Setting up a Development Environment

To run the examples demonstrated in this section users are required to install a python development environment. The best way to configure a python development environment is to install the anaconda python distribution [13]. Once the anaconda distribution is installed, users must create a virtual environment, in this case named *iptmnet_env* by executing the following command in the system terminal: *conda create -n iptmnet_env python=3.8*.

Following this, users must activate the environment by executing the command: *conda activate iptmnet_env*.

Finally, the users must install the pyiptmnet python package by executing the following command: *pip install pyiptmnet*.

Table 2
Parameters for the search function

Parameter	Description
search-term	String representing the Gene/Protein name, UniprotKB ID/AC, PRO ID or PMID to search for in iPTMnet.
term-type	Enum denoting the type of search-term being supplied as the first parameter. Possible values are Termtype.All, Termtype.UniprotID, Termtype.ProteinGeneName and Termtype.PMID. [Default=Termtype.All]
role	Enum denoting the type of search-term being supplied as the first parameter. Possible values are Role.EnzymeOrSubstrate, Role.EnzymeAndSubsrate, Role.Enzyme and Role.Substrate [Default=Role.EnzymeOrSubstrate]
PTM-list	A list containing PTM types to filter upon. It can contain following values—Acetylation, C-Glycosylation, Myristoylation, Ubiquitination, N-Glycosylation, S-Glycosylation, Phosphorylation, S-Nitrosylation, O-Glycosylation, Methylation, Sumoylation. Passing an empty list or None will match against all possible values. [Default=None]
organism-list	A list containing taxon codes for organism to filter upon. Passing an empty list or None will match against all organisms. [Default=None]
Dict	If True return results as a JSON dictionary and if False returns the result as DataFrame. Default is false

By default, the pyiptmnet client package will use the latest version of the API. The user can specify a custom version if they wish to lock the version for better reproducibility (*see* **Note 3**).

2.2.2 Searching the iPTMnet Database

Users can search and explore the iPTMnet database using the *search* function from the client libraries (Listing 1). The *search* function requires one mandatory parameter which can be a string representing a Gene or Protein name, UniprotKB AC/ID [14], a Protein Ontology (PRO) ID [15] or a PubMed Reference Number (PMID) to be searched in iPTMnet database. The search function also accepts five optional parameters (Table 2). If no optional parameter is provided the function will use the default values. By default, all the functions in client libraries will return the data in DataFrame format which makes it very easy for the user to integrate the results into their existing data analysis pipelines (*see* **Note 4**). Alternatively, the user can request the data in JavaScript Object Notation (JSON) format by supplying a parameter "dict=True". JSON is a standard format for exchanging data among applications and hence is valuable if the user wishes to perform custom postprocessing on the data before integrating it into their pipelines. Listing 1 provides an example of using the *search* function to search iPTMnet database for references to SMAD2 gene. We provide the gene name to the function by specifying the text "smad2" as the first parameter. Then we specify the type of first parameter by providing Termtype.ALL as the second parameter. Finally, we specify that results can have either an enzyme or substrate role by providing Role.EnzymeOrSubstrate as the third parameter.

The above query returns five entries as shown (only top two displayed) in Table 3. The first entry is for the organism *Homo sapiens* (Human) and the second is for *Mus musculus* (Mouse). Each row of the table provides detailed information about a given iPTMnet entry. The table shows if an entry has a substrate or enzyme role, the number of interacting substrates and enzymes, presence of PTM dependent protein-protein interactions, the number of PTM dependent protein-protein interactions, the number of PTM sites and the number of PTM isoforms.

Listing 1 Search the Database for smad2 Gene

```
#import the package
import pyiptmnet.api as api from pyiptmnet.enums import *

# search the database for smad2
api.search("smad2", Termtype.ALL, Role.EnzymeOrSub-
strate)
```

2.2.3 Getting Information for a Particular Entry

Users can retrieve general information about a particular iPTMnet entry by using the *get_info* function (Listing 2). This function requires an iPTMnet entry ID as the first input parameter. Unlike the *search* function, the *get_info* function does not retrieve the data in DataFrame format, but always retrieves the data in JSON format. Hence, the user will have to perform their own post-processing if they wish to use this data for further analysis. Listings 2 and 3 demonstrate an example of using the *get_info* function to retrieve detailed information for the first entry (Q15796) from the results in the previous section (Table 3). The retrieved data contains the UniProt identifiers matching the given iPTMnet ID, the full protein name, the primary gene symbol, synonymous gene symbols, organism information, and the mapped PRO entry.

Listing 2 Get Detailed Info for the SMAD2 Gene

```
#import the package
import pyiptmnet.api as api

# retrieve information for iPTMnet ID - Q15796
api.get_info("Q15796")
```

Table 3
Results obtained from search function (Transposed for readability)

Column	Result-1	Result-2
iPTM-ID	Q15796	Q62432
Protein-name	Mothers against decapentaplegic homolog 2	Mothers against decapentaplegic homolog 2
Gene-name	SMAD2	SMAD2
Synonyms	MADH2, MADR2	MADH2
Taxon-code	9606	10090
Species	*Homo sapiens*	*Mus musculus*
Common-name	Humans	Mouse
Substrate-role	True	True
Substrate-num	0	0
Enzyme-role	True	True
Enzyme-num	17	1
PTM-dependent-ppirole	True	False
PTM-dependent-ppinum	1	0
Sites	38	15
Isoforms	2	2

Listing 3 Result Obtained by Using the Get Info Function for the SMAD2 Gene


```
{
  "uniprot_ac": "Q15796",
  "uniprot_id": "SMAD2_HUMAN",
  "protein_name": "Mothers against decapentaplegic homo-
log 2;",
  "gene_name": "SMAD2",
  "synonyms": [
    "MADH2",
    "MADR2"
  ],
  "organism": {
    "taxon_code": 9606,
```


(continued)

```
    "species": "Homo sapiens",
    "common_name": "Human"
  },
  "pro": {
    "id": "PR:Q15796",
      "name": "mothers against decapentaplegic homolog
2 (human)",
    "definition": "A smad2 that is encoded in the genome
of human.",
    "short_label": "hSMAD2",
    "category": "organism-gene"
  }
}
```


2.2.4 Getting PTM Events for a Protein Acting as a Substrate

Users can retrieve a list of PTM events for a given protein when it is acting as a substrate by using *get_substrates* function. The function requires the iPTMnet entry ID of the protein to be queried as its first parameter. Similar to *search* function, *get_substrates* can retrieve the data in either DataFrame format or JSON format. By default, the function will retrieve the data in DataFrame format, unless the user specifies "dict=True" as the second parameter.

Listing 4 shows an example of using *get_substrates* function to retrieve a list of PTM events for the protein—Q15796 when it is acting as a substrate. The retrieved data (Table 4) provides information about the residue, position and type of the PTM event. It also provides information about the exact proteoform, and enzymes involved in the event. Finally, it provides the data source, PMID and a score for the quality of the underlying information.

Listing 4 Get Substrates for Q15796

```
#import the package
import pyiptmnet.api as api

# retrieve proteoforms for iPTMnet ID - Q15796
api.get_substrates("Q15796")
```

Table 4
Substrates for iPTMnet entry—Q15796

Sub form	Residue	Sites	PTM type	Score	Sources	Enzymes	PMIDs
Q15796	S	S110	Phosphorylation	4	PSP, neXtProt, Signor	CAMK2A, Q9UQM7, WNK1, Q9H4A3	11027280, 17392271
Q15796-1	S	S245	Phosphorylation	4	PRO, HPRD	MAP2K1/iso:1/Phos:1, PR:000026196	10197981, 12193595
Q15796-2	S	S435	Phosphorylation	2	PRO	TGFBR2/iso:1/SigPep/Phos:1, PR:000025963	17074756

2.2.5 Getting Proteoforms for a Protein

A gene can give rise to many protein products, known as proteoforms due to sequence variations, alternative splicing, and/or PTMs [16]. There are often many different PTM proteoforms of a protein with one or more types of PTM on one or more sites. Users can retrieve PTM proteoforms (protein forms with different combinations of PTMs) along with the corresponding PTM events for a given iPTMnet entry by using the *get_proteoforms* functions. The first parameter of the function is the iPTMnet entry ID. Similar to the *search* function, *get_proteoforms* function can retrieve the data in DataFrame or JSON format. By default, the function will retrieve the data in DataFrame format. To retrieve the data in JSON format, users must pass "dict=True" as the second parameter. Listing 5 provides an example of using the *get_proteoforms* function to retrieve the proteoforms for iPTMnet entry—Q15796. The retrieved data (Table 5) contains the PRO ID and PRO label of the proteoform, the sites at which PTM events occur, the ID and label of the enzymes initiating these PTM events, and the data source from which the information was obtained.

Listing 5 Get Proteoforms for Q15796

```
#import the package
 import pyiptmnet.api as api from

# retrieve proteoforms for iPTMnet ID - Q15796
api.get_proteoforms("Q15796")
```

Table 5
Proteoforms of iPTMnet entry—Q15796

Pro ID	Label	Sites	PTM enzyme ID	PTM enzyme label	Source
PR:000025934	hSMAD2/ iso:1 /Phos:1	pS465, pS467	PR:000025963	Transforming growth factor beta type II receptor homodimeric complex phosphorylated form (human)	PRO
PR:000025935	hSMAD2/ iso:1 /Phos:2	pS465, pS467	PR:000025963	Transforming growth factor beta type II receptor homodimeric complex phosphorylated form (human)	PRO
PR:000025935	hSMAD2/ iso:1 /Phos:2	pT220, pS245, pS250, pS255, pT8	PR:000026189	hMAPK3/iso:1 /Phos:1	PRO

2.2.6 Getting Protein-Protein Interactions for Proteoforms of a Protein

iPTMnet integrates manually curated information about protein-protein interactions for the proteoforms of a given protein from PRO. Users can retrieve these protein-protein interactions by using the *get_ppi_for_proteoforms* function. The function requires the iPTMnet entry ID as the first parameter. Similar to the *get_proteoforms* function, it retrieves the data in DataFrame format, unless the user provides "dict=True" parameter to retrieve the data in JSON format.

Listing 6 shows an example of using the *get_ppi_for_proteoforms* function to retrieve the PPIs for proteoforms of iPTMnet entry—Q15796. The results from the above function (Table 6) show that Q15796 has four PPIs involving four different proteoforms. The retrieved data provides the PRO IDs and labels of the interacting proteins, the type of relation, the data source and the PMIDs from which the information was retrieved.

Listing 6 Get PPI for Proteoforms of Q15796

```
#import the package import pyiptmnet.api as api from

# retrieve PPIs for proteoforms of iPTMnet ID - Q15796
api.get_ppi_for_proteoforms("Q15796")
```

Table 6
PPI for proteoforms of iPTMnet entry—Q15796

Protein 1 pro ID	Protein 1 label	Relation	Protein 2 pro ID	Protein 2 pro label	Source	PMIDs
PR:000025934	hSMAD2/iso: Long /Phos:1	Interaction	PR:Q13485	hSMAD4	PRO	9311995
PR:000025935	hSMAD2/iso: Long /Phos:2	Interaction	PR:Q13485	hSMAD4	PRO	12193595
PR:000025936	hSMAD2/iso: Long /Phos:3	Interaction	PR:Q13485	hSMAD4	PRO	11027280
PR:000045371	hSMAD2/iso: Long /UnPhos:1	Interaction			PRO	8980228

2.2.7 Getting PTM Dependent Protein-Protein Interactions for a Protein

iPTMnet integrates information about phosphorylation-dependent PPIs obtained by full scale text-mining of the PubMed abstracts using eFIP [17]. Users can retrieve these phosphorylation dependent PPIs by using the *get_ptm_dependent_ppi* function. Similar to *get_ppi_for_proteoforms*, it requires the iPTMnet entry ID of the target protein and retrieves the data in the DataFrame format by default.

Listing 7 provides an example of using the *get_ptm_dependent_ppi* function to retrieve the PTM dependent PPI for Q15796. The retrieved data in Table 7 shows that Q15796 has two PTM dependent PPIs. The data provides information about the type and site of the PTM event, UniProt ID and name of the substrate and interactant, the type of association and the data source and PMID from which the information was obtained.

Listing 7 Get PTM Dependent PPI for Proteoforms of Q15796

```
#import the package import pyiptmnet.api as api

# retrieve PTM dependent PPIs for iPTMnet ID - Q15796 api.
get_ptm_dependent_ppi("Q15796")
```

Table 7
PTM dependent PPI for proteoforms of Q15796 (Transposed for readability)

Column	Result-1	Result-2
PTM type	Phosphorylation	Phosphorylation
Substrate uniprot ID	P49841	Q15796
Substrate name	GSK3B	SMAD2
Site	S9	S467
Interactant uniprot ID	Q15796	Q13485
Interactant name	SMAD2	SMAD4
Association type	Increased association	Association
Source	eFIP	eFIP
PMID	21996745	9346908

2.2.8 Getting PTM Sites Affected in Variants of a Protein

iPTMnet integrates information about single nucleotide variations and their associated diseases from the Biomuta database [18]. Users can retrieve a list of PTM sites that overlap with cancer-associated variants in a protein by using *get_variants* function. Similar to the *get_proteoforms* function, this function requires the iPTMnet entry ID as the first parameter. By default, the function will retrieve the data in DataFrame format. The user can specify "dict=True" as the second parameter to retrieve the data in JSON format. Listing 8 provides an example of using *the get_variants* function to get a list of PTM sites affected in variants of Q15796. The retrieved data (Table 8) provides information about the position and residue of the affected PTM sites, the disease and the sample source in which the variant was found and if available the PMID of the literature source from which the information was mined.

Listing 8 Get PTM Sites Affected in Variants of Q15796

```
#import the package
import pyiptmnet.api as api

# retrieve proteoforms for iPTMnet ID - Q15796
api.get_variants("Q15796")
```

Table 8
PTM sites affected in variants of Q15796

AC	Position	Residue sequence	Residue mutated	Disease	Sample source	PMIDs
Q15796-1	110	S	Y	DOID:3571/liver cancer	icgc	
Q15796-1	156	K	E	DOID:1909/melanoma	icgc	
Q15796-1	157	K	T	DOID:9256/colorectal cancer	tcga	
Q15796-1	165	Y	*	DOID:10534/stomach cancer	cosmic	

Table 9
PTM enzyme information for a list of PTM sites

PTM type	Site	Site position	Score	Source	PMID	Enz name	Enz ID	Sub name	Sub ID
Acetylation	K120	120	2	UniProt	23431171	KAT6A	Q92794	TP53	P04637
Phosphorylation	T8	8	3	nextProt	19201832, 15241418	CDK2	P24941	SMAD2	Q15796

2.2.9 Getting Enzymes Corresponding to a List of PTM Sites

Users can retrieve enzymes corresponding to a list of PTM sites by using the *get_ptm_enzymes_from_list* function. The function requires a list of python dictionary objects representing the PTM sites to be queried. The python dictionary object should contain three keys—substrate_ac, site_residue, site_position as shown in Listing 9. Alternatively, instead of creating a list of PTM site objects manually, users can load a list of PTM sites from file by using the function *get_ptm_enzymes_from_file* (*see* **Note 5**). The input to the function should be the absolute path to a file containing a list of comma separated values. Further, the first line of the file should be a header with following attributes—substrate_ac, site_residue, site_position. An example of the input file can be found in Listing 10. The function retrieves the data in the DataFrame format by default unless the user passes "dict=True" to retrieve the data in JSON format.

Listings 11 and 12 provide examples of retrieving enzymes corresponding to a list of PTM sites. The retrieved data (Table 9) provides information about the name and ID of the enzyme, the data source and PMID used to collect the information and the type, site and position of the corresponding PTM events.

Listing 9 Python Dictionary Object Representing a PTM Site

```
# create a list of sites
sites = [
  {
    "substrate_ac":"Q15796",
    "site_residue":"K",
    "site_position":"19"
  },
  {
    "substrate_ac": "Q15796",
    "site_residue": "T",
    "site_position": "8"
  },
  {
    "substrate_ac": "P04637",
    "site_residue": "K",
    "site_position": "120"
  },
]
```

Listing 10 CSV File Containing a List of PTM Sites

```
substrate_ac,site_residue,site_position
    Q15796,K,19
    Q15796,T,8
    P04637,K,120
```

Listing 11 Get PTM Enzymes for a List of PTM Sites

```
#import the package
import pyiptmnet.api as api
```

(continued)

```
# create a list of PTM site objects sites = [{
        "substrate_ac":"Q15796",
        "site_residue":"K",
        "site_position":"19"
       },
       {
        "substrate_ac": "Q15796",
        "site_residue": "T",
        "site_position": "8"
       },
       {
        "substrate_ac": "P04637",
        "site_residue": "K",
        "site_position": "120"
       },
     ]
# retrieve PTM enzymes from a list of PTM sites
api.get_ptm_enzymes_from_list(sites)
```

Listing 12 Get PTM Enzymes from a File Containing a List of PTM Sites

```
#import the package import pyiptmnet.api as api
# retrieve PTM enzymes from a file containing PTM sites
api.get_ptm_enzymes_from_file('/path/to/file.csv')
```

2.2.10 Getting PTM Dependent PPI for a List of PTM Sites

Users can retrieve PTM dependent PPIs involving a list of PTM sites by using the function *get_ptm_ppi_from_list*. Similar to *get_ptm_enzymes_from_list* function, the first parameter to this function is a list of python dictionary objects representing the PTM sites (Listing 9) or a path to the file containing a list of comma separated values (Listing 10). The function retrieves the data in DataFrame format by default unless the user provides a "dict=True" to retrieve the data in JSON format. Similar to the *get_ptm_enzymes_from_file* function, users can use the *get_ptm_ppi_from_file* function to load PTM sites from a comma-separated values file.

Table 10
PTM dependent PPIs for a list of PTM sites (Transposed for readability)

Column	Result-1	Result-2
Interactant ID	Q92540	P08047
Interactant name	SMG7	SP1
Substrate ID	P04637	P04637
Substrate name	TP53	TP53
Association type	Unknown	Unknown
Source	eFIP	eFIP
PMID	27462439	22911849
PTM type	Phosphorylation	Phosphorylation
Site	S378	S378
Position	378	378

Listings 13 and 14 show examples of using the *get_ptm_ppi_-from_list* and *get_ptm_ ppi_from_file* functions, respectively. The retrieved data (Table 10) contains ID and name of the interactant and substrate involved in the PPI. It also contains the type of the association, the data source and PMIDs used to retrieve the information and the type, site and position of the corresponding PTM events.

Listing 13 Get PTM Dependent PPI from a List of PTM Sites

```
#import the package import pyiptmnet.api as api
# create a list of PTM site objects sites = [{
       "substrate_ac":"Q15796",
       "site_residue":"K",
       "site_position":"19"
      },
      {
       "substrate_ac": "P04637",
       "site_residue": "S",
       "site_position": "378"
      },
      {
       "substrate_ac": "P04637",
       "site_residue": "S",
       "site_position": "392"
```

(continued)

```
    },
  ]
# retrieve PTM dependent PPI from a list of PTM sites
api.get_ptm_ppi_from_list(sites)
```

Listing 14 Get PTM Dependent PPI from a File Containing a List of PTM Sites

```
#import the package
import pyiptmnet.api as api

# retrieve PTM dependent PPI from a file containing PTM
sites api.get_ptm_ppi_from_file('/path/to/file.csv')
```

3 Conclusion

In this chapter, we have provided a comprehensive tutorial on using the iPTMnet RESTful API. Users can use the iPTMnet API to automate various functions available on the iPTMnet website. We show that users can use the API to search the database, get detailed information about a particular iPTMnet entry, get a list of PTM events for a given protein, get a list of proteoforms and their PTM dependent PPIs, get a list of PTMs in cancer associated variants and finally get all the enzymes as well as PTM-dependent PPIs associated with a bunch of PTM sites.

4 Notes

1. You can explore the API interactively in your web browser using the Swagger [19] documentation explorer available on the iPTMnet website [20].
2. You can specify a custom URL for the API server by using the *set_host_url* function. This is especially helpful if you wish to deploy the iPTMnet API on your own servers. The exact steps for deployment of the API server are dependent on the

required type of deployment. Since these steps change in response to the evolving technological landscape, we have documented the recommended best practices on the public GitHub project repository page available at—https://github.com/udel-cbcb/iPTMnet-API.

3. You can specify the version of API to make the HTTP calls by using the *set_api_version* function. Specifying an API version is considered a good practice to prevent problems with incompatible code and erroneous data results once the user is no longer actively maintaining the code. The downside is possibly not being able to use new features in the API.
4. The API calls are not guaranteed to succeed every time especially because of network or server availability issues. If the user wishes to catch the failure and know the reasons behind the failures, they must use the python *try..except* mechanism to catch the exceptions raised by the function calls.
5. The API server is a shared resource and is utilized by many users concurrently. Hence, the user must refrain from sending too big and too many requests to the server in a short span of time. An easy way to reduce the size of requests is to split the data into small batches of preferably less than 50 items per batch. To prevent sending too many requests in a short span, the user can introduce short pauses in the program using *time.sleep* function.

Acknowledgments

This work has been supported by the National Institute of Health grant R35GM141873.

References

1. Karve TM, Cheema AK (2011) Small changes huge impact: the role of protein posttranslational modifications in cellular homeostasis and disease. J Amino Acids 2011:1–13. https://doi.org/10.4061/2011/207691
2. Barber KW, Rinehart J (2018) The ABCs of PTMs. Nat Chem Biol 14:188–192. https://doi.org/10.1038/nchembio.2572
3. Zhang X, Belkina N, Jacob HKC et al (2015) Identifying novel targets of oncogenic EGF receptor signaling in lung cancer through global phosphoproteomics. Proteomics 15: 340–355. https://doi.org/10.1002/pmic.201400315
4. Serrano-Gomez SJ, Maziveyi M, Alahari SK (2016) Regulation of epithelial-mesenchymal transition through epigenetic and post-translational modifications. Mol Cancer 15: 18. https://doi.org/10.1186/s12943-016-0502-x
5. Marcelli S, Corbo M, Iannuzzi F et al (2018) The involvement of post-translational modifications in Alzheimer's disease. Curr Alzheimer Res 15:313–335. https://doi.org/10.2174/1567205014666170505095109
6. Martin L, Latypova X, Terro F (2011) Post-translational modifications of tau protein: implications for Alzheimer's disease. Neurochem Int 58:458–471. https://doi.org/10.1016/j.neuint.2010.12.023
7. Wu P, Nielsen TE, Clausen MH (2015) FDA-approved small-molecule kinase

inhibitors. Trends Pharmacol Sci 36:422–439. https://doi.org/10.1016/j.tips.2015.04.005
8. Rane CK, Minden A (2019) P21 activated kinase signaling in cancer. Semin Cancer Biol 54:40–49. https://doi.org/10.1016/j.semcancer.2018.01.006
9. Huang H, Arighi CN, Ross KE et al (2018) iPTMnet: an integrated resource for protein post-translational modification network discovery. Nucleic Acids Res 46:D542–D550. https://doi.org/10.1093/nar/gkx1104
10. Allali I, Arnold JW, Roach J et al (2017) A comparison of sequencing platforms and bioinformatics pipelines for compositional analysis of the gut microbiome. BMC Microbiol 17:194. https://doi.org/10.1186/s12866-017-1101-8
11. Computational Pan-Genomics Consortium (2018) Computational pan-genomics: status, promises and challenges. Brief Bioinform 19:118–135. https://doi.org/10.1093/bib/bbw089
12. Richardson L, Amundsen M (2013) RESTful web APIs, 1st edn. O'Reilly, Beijing
13. Anaconda | The World's most popular data science platform. In: Anaconda. https://www.anaconda.com/. Accessed 18 Apr 2021
14. The UniProt Consortium (2019) UniProt: a worldwide hub of protein knowledge. Nucleic Acids Res 47:D506–D515. https://doi.org/10.1093/nar/gky1049
15. Natale DA, Arighi CN, Barker WC et al (2011) The protein ontology: a structured representation of protein forms and complexes. Nucleic Acids Res 39:D539–D545. https://doi.org/10.1093/nar/gkq907
16. The Consortium for Top Down Proteomics, Smith LM, Kelleher NL (2013) Proteoform: a single term describing protein complexity. Nat Methods 10:186–187. https://doi.org/10.1038/nmeth.2369. Tudor CO, Ross KE et al (2015)
17. Tudor CO, Ross KE, Li G et al (2015) Construction of phosphorylation interaction networks by text mining of full-length articles using the eFIP system. Database 2015: bav020. https://doi.org/10.1093/database/bav020
18. Dingerdissen HM, Torcivia-Rodriguez J, Hu Y et al (2018) BioMuta and BioXpress: mutation and expression knowledgebases for cancer biomarker discovery. Nucleic Acids Res 46:D1128–D1136. https://doi.org/10.1093/nar/gkx907
19. API Documentation & Design Tools for Teams | Swagger. https://swagger.io/. Accessed 21 Apr 2021
20. Gavali S, Cowart J (2020) iPTMnet Swagger UI. https://research.bioinformatics.udel.edu/iptmnet/api/doc/#/. Accessed 19 Apr 2021

Chapter 11

Systematic Characterization of Lysine Post-translational Modification Sites Using MUscADEL

Zhen Chen, Xuhan Liu, Fuyi Li, Chen Li, Tatiana Marquez-Lago, André Leier, Geoffrey I. Webb, Dakang Xu, Tatsuya Akutsu, and Jiangning Song

Abstract

Among various types of protein post-translational modifications (PTMs), lysine PTMs play an important role in regulating a wide range of functions and biological processes. Due to the generation and accumulation of enormous amount of protein sequence data by ongoing whole-genome sequencing projects, systematic identification of different types of lysine PTM substrates and their specific PTM sites in the entire proteome is increasingly important and has therefore received much attention. Accordingly, a variety of computational methods for lysine PTM identification have been developed based on the combination of various handcrafted sequence features and machine-learning techniques. In this chapter, we first briefly review existing computational methods for lysine PTM identification and then introduce a recently developed deep learning-based method, termed MUscADEL (Multiple Scalable Accurate Deep Learner for lysine PTMs). Specifically, MUscADEL employs bidirectional long short-term memory (BiLSTM) recurrent neural networks and is capable of predicting eight major types of lysine PTMs in both the human and mouse proteomes. The web server of MUscADEL is publicly available at http://muscadel.erc.monash.edu/ for the research community to use.

Key words Post-translational modification, Lysine, Bioinformatics, Machine learning, Long short-term memory, Deep learning

1 Introduction

Protein post-translational modifications (PTMs) are associated with diverse biological processes. PTMs occurring on protein lysine (K) residues represent major subtypes of PTMs, including acetylation, glutarylation, glycation, malonylation, methylation, succinylation, sumoylation, and ubiquitination. Due to the advances in quantitative proteomic techniques, more than 20 kinds of lysine PTMs have been characterized, and accordingly a large number of

Dukka B. KC (ed.), *Computational Methods for Predicting Post-Translational Modification Sites*, Methods in Molecular Biology, vol. 2499, https://doi.org/10.1007/978-1-0716-2317-6_11,

diverse substrates and PTM sites have been experimentally identified and documented in public repositories, such as PhosphoSitePlus [1] and Protein Lysine Modifications Database (PLMD) [2].

Among the different types of lysine PTMs, acetylation, glycation, ubiquitination and methylation have been extensively studied. These PTMs play diverse roles in different biological processes. For example, it is well known that lysine acetylation functions as an important regulator and plays crucial roles in diverse biological processes [3–8]. Glycation is associated with aging and pathogenesis of diabetes [9, 10]. Ubiquitination is associated with regulation of fundamental cellular processes including protein degradation [11–13], gene transcription, DNA repair, and DNA replication [12, 14, 15], while methylation is involved in a broad spectrum of biological and physiological processes including transcriptional and epigenetic regulation, cell metabolism and the development of human diseases [16–19]. Recently, three new types of lysine PTMs, namely malonylation [20], succinylation [21, 22], and glutarylation [2], have been discovered. In addition, a growing number of studies have been focused on the elucidating the cross talks within lysine PTMs and between lysine PTMs and other PTMs [23–25], suggesting the functional importance of lysine PTMs in regulating diverse biological processes.

With the availability of rapidly accumulated protein sequences resulting from the whole-genome sequencing projects, significant research efforts have been made to develop computational methods for predicting lysine PTM substrates and sites for annotation of the candidate proteins and guidance of experimental validation. In this chapter, we will briefly review and summarize the existing approaches for prediction of lysine PTM substrates and sites. In addition, we will also introduce a deep learning-based approach, termed MUscADEL (Multiple Scalable Accurate Deep Learner for lysine PTMs) [26]. It employs bidirectional long short-term memory recurrent neural networks [27, 28] to train deep learning models, aiming to predict eight major types of lysine PTMs in both the human and mouse proteomes.

2 Overview of Existing Methods and Tools

As described in the Subheading 1, a variety of computational methods and tools have been developed for predicting lysine PTM substrates and sites. In this section, we will briefly review these approaches and tools. A summary of major methods/servers is provided in Table 1. For a comprehensive review of existing methods, refer to our previous review [26].

The existing methods can be categorized into two major classes: machine learning–based approaches and peptide similarity–

Table 1
A summary of major methods and tools for the prediction of lysine PTM proteins/sites

	Tool	PTM	Algorithm
Machine learning–based approaches	SSPKA [29]	Acetylation	RF
	NetGlycate-1.0 [30]	Glycation	ANN
	MaloPred [31]	Malonylation	SVM
	SSPFS [32]	Sumoylation	KNN
	mRMR_UbSite [33]	Ubiquitination	KNN
	RUBI [34]	Ubiquitination	SVM/BRNN
	hCKSAAP_UbSite [35]	Ubiquitination	SVM
	KA-predictor [36]	Acetylation	SVM
	PSKAcePred [37]	Acetylation	SVM
Peptide similarity–based approaches	GPS-MSP [38]	Methylation	GPS
	SUMOsp [39]	Sumoylation	GPS
	ASEB [40]	Acetylation	ASEB
	PAIL [41]	Acetylation	BDM

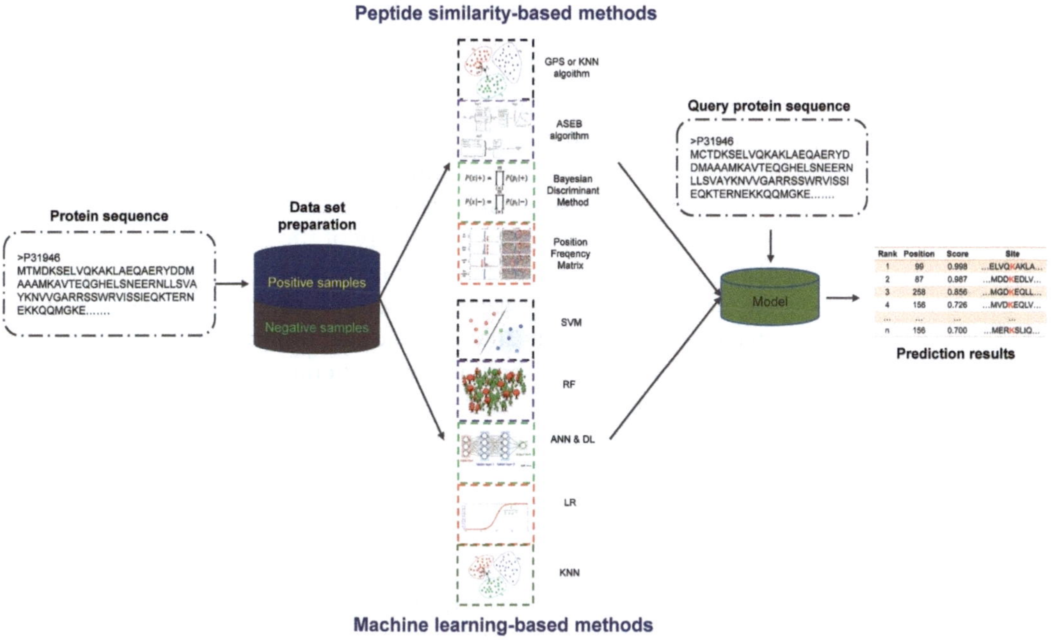

Fig. 1 A schematic illustration of peptide similarity–based and machine learning–based approaches for the prediction of lysine PTM sites

based approaches (Fig. 1). For the majority of the machine learning–based methods, each protein sequence or each fixed-length sequence fragment is transformed into a fixed size feature vector and is labeled by the corresponding PTM type(s) (including non-PTM cases). The feature vectors often consist of multifaceted features extracted from different aspects or sources, including but are not limited to (1) protein primary sequence-derived features, for example, the frequency of 20 amino acid occurrences; (2) predicted protein structural features; (3) protein physicochemical properties; (4) protein position-specific scoring matrix (PSSM); (5) peptide similarity features; and (6) protein functional annotations. Since a variety of such multifaceted features are available, feature selection becomes a crucial step during the development of many of these approaches, whose primary aim is to reduce the dimensionality of the extracted feature vectors and further remove the redundant, noisy and misleading features for a better predictive performance. Major feature selection methods include correlation-based feature subset selection (CfsSubsetEval) [42, 43], maximum relevance and minimum redundancy (mRMR) [44], information gain (IG) [45], Pearson correlation coefficient (PCC) [36], and RELIEFF [46], among others.

Finally, a machine-learning algorithm is constructed and evaluated using the training and testing datasets, respectively. Commonly used machine-learning algorithms used during this step include support vector machines (SVMs) [47], random forests (RFs) [48], artificial neural networks (ANNs) [49], k-nearest neighbor (KNN) [50], and logistic regression (LR) method [51]. The representative methods that used these algorithms for lysine PTM prediction are listed in Table 1. Among these algorithms, SVM is the most commonly deployed framework that aims to accurately classify samples by generating optimal hyperplanes based on the feature dimensionality of the training data. MaloPred [31] was developed to predict malonylation sites for three species by integrating and selecting various informative features using SVM. RF, essentially an ensemble of a number of decision trees, is another well-established and widely employed algorithm for predicting lysine PTMs. SSPKA [29] used RF algorithms to predict species-specific acetylation sites by incorporating various informative features, including sequence-derived features, predicted secondary structure and relevant function features on both residue and protein levels. An ANN usually contains multiple nodes as input and multiple layers to connect these input nodes, mimicking neurons and their functions/connectivity in human brains. Among the reviewed predictors for lysine PTM prediction, NetGlycate-1.0 [30] and RUBI [34] employed ANN as their core algorithm. KNN algorithm is another commonly employed unsupervised algorithm that clusters samples by calculating their similarities/distances. SSPFS [32] and mRMR_UbSite [33] were built using the KNN algorithm. LAceP [42] constructed an LR model for lysine acetylation prediction. In addition, some methods integrate multiple algorithms to build an ensemble model. For example, hCKSAAP_UbSite [35] used LR approach to integrate the prediction scores of four SVM classifiers for an improved prediction performance. Performance evaluation is another important step, where a variety of performance evaluation strategies can be applied to objectively assess and compare the predictive outcomes by these tools, including but not limited to *k*-fold cross-validation, jackknife validation test, and independent test (i.e., the testing datasets are not overlapped with the training data).

The second category of approaches is based on peptide similarity calculation between the testing peptide and the peptides with experimentally annotated lysine PTMs. Various similarity measures have been utilized including the BLOcks SUbstitution Matrix (BLOSUM62) matrix and position frequency matrix (PFM). GPS, ASEB and BDM are the representative methods using peptide similarity to predict lysine PTMs. Generally, GPS deploys the BLOSUM62 matrix to measure the similarity among the peptides with experimentally validated lysine PTMs, which are then divided into different groups by the Markov clustering algorithm or *k*-means clustering algorithm. Then the similarity scores are calculated between the testing peptide and different groups. The peptide will be predicted to harbor a certain PTM type based on the group which has the highest similarity score with the testing peptide [38, 39]. GPS-MSP [38] and SUMOsp [39] are the representative approaches that implemented GPS to predict methylation and sumoylation sites, respectively. ASEB [40] is another representative peptide similarity–based approach for acetylation prediction by estimating the significance of the peptide similarity with respect to known acetylation site sets. BDM [41] assumes that all flanking residues around the PTM site are mutually independent and then estimates the probability that an unlabeled input belong to the positive samples or the negative samples. It is worth noticing that some hybrid methods, PSKAcePred [37], MaloPred [31], and KA-predictor [36], integrate both peptide-based similarity and machine-learning algorithms for lysine PTM prediction. In these methods, the similarity scores between peptides are used as the input features for the following construction of machine-learning algorithms.

Despite the availability of the predictors introduced above, the PTM types covered by these approaches are limited. In addition, there is much space to further improve the prediction performance for lysine PTM prediction. To address these issues, we have recently developed a new deep-learning approach, termed MUscADEL. In the following sections, we will decipher the details regarding dataset curation, the model architecture of MUscADEL and provide step-by-step instructions of the usability of the MUscADEL web server for predicting lysine PTM sites. Users are also suggested to refer to our previous publication [26] regarding the relevant content of MUscADEL.

3 Materials

In order to build the prediction models of MUscADEL, we constructed eight lysine PTMs datasets of *Homo sapiens* and *Mus musculus* extracted from the PhosphoSitePlus database [1] and then mapped these protein sequences to human and mouse proteomes annotated by the UniProt database [52]. Although the PhosphoSitePlus database covers more than 20 lysine PTM types, we only employed eight types with sufficient training data for the deep-learning models (i.e., with more than 150 proteins in the training dataset). In addition, to reduce the sequence redundancy in the retrieved datasets and avoid potential overfitting, we applied the CD-HIT program [53] to cluster the sequences and remove the similar ones with the sequence identity threshold of 30%. Subsequently, we randomly split the datasets to training and testing datasets. In each data set, experimentally validated PTM sites were considered as the positive samples, whereas the rest sequences with lysine residues that have not been annotated as experimentally validated PTM sites were regarded as the negative samples. A statistical summary of the data sets used in this study is provided in Table 2.

Table 2
A statistical summary of benchmarking and independent test datasets used for constructing and assessing the MUscADEL models

	Fivefold cross-validation		Independent test	
PTM type	Number of proteins	Number of PTM sites	Number of proteins	Number of PTM sites
Acetylation	3289	10,845	1645	5052
Glutarylation	124	432	31	123
Glycation	1621	3209	811	1530
Malonylation	1702	4300	425	1050
Methylation	902	1352	226	337
Succinylation	772	1728	193	382
Sumoylation	1286	3863	640	1890
Ubiquitination	4768	19,949	2385	9229

4 Methods

In this section, we will describe the architecture of the MUscADEL model, the procedure for model training, the web server implementation, and step-by-step instruction for conducting prediction of lysine PTM sites using our web server.

4.1 The Model Architecture

MUscADEL is constructed based on the long short-term memory RNN (LSTM-RNN). The LSTM-RNN architecture is typically composed of a series of basic units, each of which consists of an input gate, a neuron with a self-recurrent connection, a forget gate, and an output gate. LSTM-RNN has been extensively and successfully applied in a variety of natural language processing and bioinformatic tasks [54, 55]. MUscADEL has two predictive models, including the full-sequence model and the sequence-fragment model. Both models are based on bidirectional LSTM-RNN and contain five layers: the input layer, the word embedding layer, the recurrent layer, the fully connected layer, and the output layer (Fig. 2).

The input layer (not composed of neurons) receives the residues in protein sequences directly as input features. In the full sequence model, all residues in each protein sequence are delivered into the input layer, where the maximum protein sequence length was set as $L_{\text{full}} = 1000$. In the sequence fragment-model, a fixed length sequence fragment centering the lysine residue is given to the input layer, where the length of the fragment was optimized as $L_{\text{frag}} = 27$. If the length of a protein sequence or a fragment is less than L (L_{full} or L_{frag}), the remaining positions will be filled with a gap symbol '-'.

The embedding layer is to convert each input sequence/fragment into a sequence of real valued vectors using a "dictionary" with the vocabulary size of 21 (corresponding to 20 types of

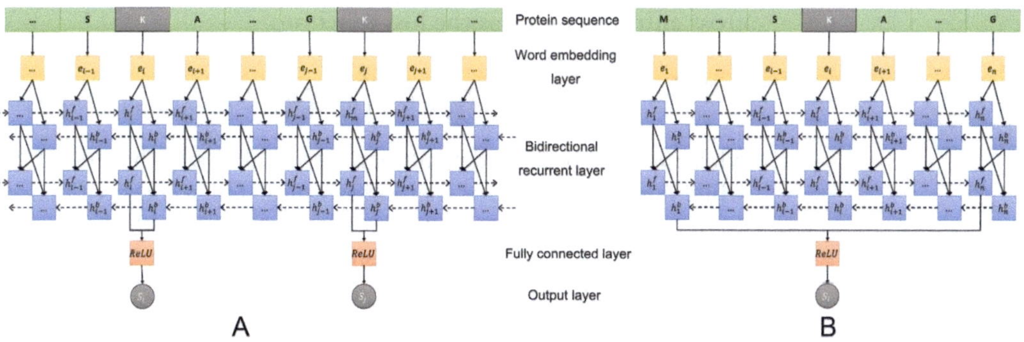

Fig. 2 The architectures of (**a**) the full-sequence model and (**b**) the fragment model embedded in the MUscADEL framework

residue and gap "-"). As such, each residue is converted into a real vector and then each sequence is therefore converted into a series of real vectors.

The recurrent layer is built using the bidirectional LSTM approach, where each recurrent unit contains 64 hidden neurons. The structure of LSTM unit is composed of a memory cell, an input gate (i), an output gate (o), and a forget gate (f) as shown in [56]. The relationships between these gates at the t-th residue in a protein sequence can be summarized as following formulas.

$$f_t = \sigma(W_f X_t + U_f h_{t-1} + b_f),$$

$$i_t = \sigma(W_i X_t + U_i h_{t-1} + b_i),$$

$$o_t = \sigma(W_o X_t + U_o h_{t-1} + b_o),$$

$$c_t = f_t \odot c_{t-1} + tanh(W_c X_t + U_c h_{t-1} + b_c),$$

$$h_t = o_t \odot tanh(c_t),$$

where $\odot$ denotes the entry-wise product, W_f, U_f, W_i, U_i, W_o, U_o, W_c, U_c denote the weighs matrices for the forget, input and output gates, and the hidden state, respectively, while b_f, b_i, b_o, b_c denote their biases, and σ is the standard sigmoid function.

The fully connected layer receives and integrates the hidden state vectors from forward and backward LSTM using the "ReLU" activate function. For the full-sequence model, only the hidden state vectors at lysine sites were used, whereas for the fragment model, the last hidden state vectors were used.

The output layer has only one neuron activated by the sigmoid function. This neuron outputs the score (ranging from 0 to 1) as the probability of the corresponding lysine being posttranslationally modified as a specified PTM type.

4.2 Model Training

In order to avoid the model overfitting, we added the dropout units with the drop-rate of 0.3 between two neighboring layers during the model training. Specifically, we employed the following binary cross entropy (BCE) as the loss function:

$$\text{BCE}(\hat{y}, y) = -\frac{1}{\sum_{i=1}^{n}\left(y_i * \log\left(\hat{y}_i\right) + \left(1 - y_i\right) * \log\left(1 - \hat{y}_i\right)\right)},$$

where $\hat{y}_i$ is the predicted result of the true label y_i of the i-th sample, while n is the number of training samples. The weights of neurons were optimized using the Adam algorithm [57] with the learning rate set to 10^{-3} and the maximum of epoch set to 1000. We terminated the process of model training if the model performance did not improve within 200 epochs. We trained two respective models (i.e., the full-sequence model and fragment model) for the eight different types of lysine PTMs respectively, using the data shown in Table 3.

Table 3
Prediction performance of the full-sequence and fragment models based on the benchmark datasets via fivefold cross-validation with a fixed specificity (*Sp*) of 90%

	Full-sequence model			Fragment model		
PTM type	*Sn* (%)	*MCC*	*AUC*	*Sn* (%)	*MCC*	*AUC*
Acetylation	36.46	0.2057	0.754	31.62	0.1915	0.742
Glutarylation	18.29	0.0913	0.602	26.45	0.1733	0.739
Glycation	31.77	0.1282	0.705	16.67	0.0622	0.610
Malonylation	45.79	0.2544	0.822	40.36	0.2616	0.807
Methylation	28.80	0.0926	0.703	20.42	0.0960	0.626
Succinylation	33.98	0.1822	0.734	32.67	0.2001	0.748
Sumoylation	42.66	0.2326	0.764	36.25	0.2291	0.741
Ubiquitination	43.77	0.2968	0.799	37.11	0.2432	0.769

4.3 Implementation of the MUscADEL Web Server

An online web server of MUscADEL has been developed and is made publicly available at http://muscadel.erc.monash.edu/. This web server was implemented using the open-source web platform LAMP (Linux-Apache-MySQL-PHP). The server resides on the extensible cloud computing facility supported by the eResearch Centre at Monash University, currently equipped with 16 cores, 64 GB memory, and a 2 TB hard disk, where the configuration of the server may be upgraded in response to the increasing demand of users in the future. This web server was tested using major web browsers, including Internet Explorer® (Versions 7 or higher), Microsoft Edge®, Mozilla Firefox®, Google Chrome®, and Safari®.

4.4 Usage of the MUscADEL Web Server

The MUscADEL web server is user friendly and easy to use in the following manner (Fig. 3).

1. At the front page of MUscADEL, click 'Server' button on the top of the page, or click 'Go to use it' button to show the main page of the page.
2. Input one or more protein sequences in the FASTA format to the textbox or upload a file with multiple sequences in the FASTA format via the file selection dialog box. The maximum number of protein sequences allowed currently is 100 and the maximum length for each protein sequence is 2000.
3. Select the target PTM type(s) (i.e., acetylation, glutarylation, glycation, malonylation, methylation, succinylation, sumoylation, and ubiquitination).
4. Select one of four cutoff values (i.e., "VERY HIGH" = 98% *Sp*, "HIGH" = 95% *Sp*, "MEDIUM" = 90% *Sp* and "LOW" = 70% *Sp*).

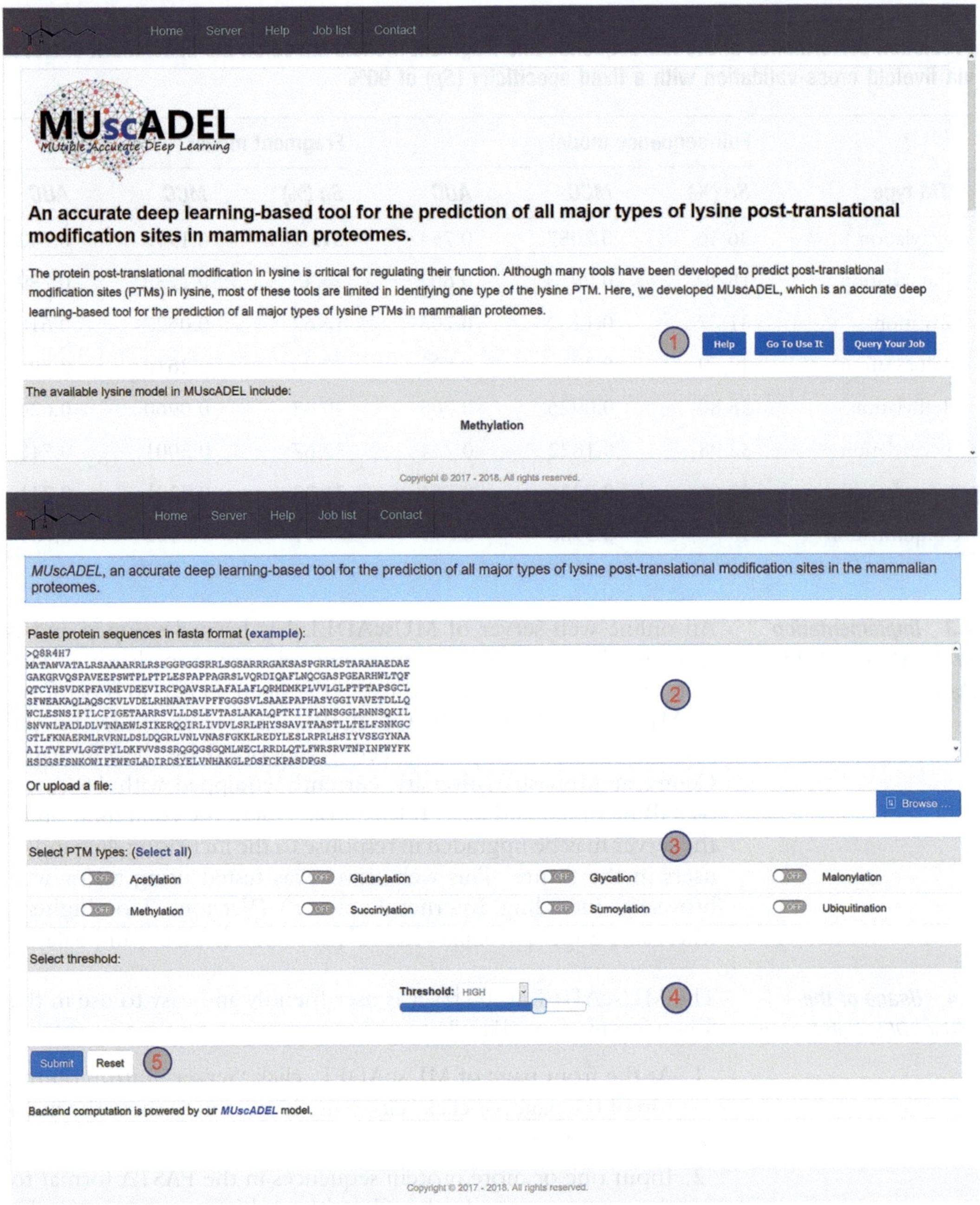

Fig. 3 A screenshot of four steps for submitting the prediction job to the MUscADEL web server using protein sequences

5. After clicking the "Submit" button, a Job ID will be generated and the current job status will be displayed. After the prediction is complete, the prediction results will be presented in a tabular form containing the information of protein IDs, positions of predicted PTM sites, predicted scores, and the prediction cutoff.

6. By clicking the "Download" button, all the results will be stored to the local computer in the plain text format. In addition, MUscADEL offers three graphical options to visualize the prediction results and the statistics. By clicking the "Statistics" button, the ratios of selected PTM sites will be displayed in pie charts. By clicking the "PTMs Sites Distribution" button, the distribution of PTMs sites will be displayed on a horizontal bar for each protein. The distribution of scores of all predicted positions will be shown as a curve for each of selected PTM types, once the "Visualize" button was clicked.
7. The prediction result can be queried in the job list page or by providing the job ID in the query form. The prediction results will be stored in our server for a month.

Figure 4 illustrates an example of prediction output generated by the MUscADEL web server. As shown in Fig. 4, the prediction result harbors four parts including the prediction result in text format (Fig. 4a), the basic statistics for the eight type PTMs (Fig. 4b), the PTM sites distribution bar plot (Fig. 4c), and prediction score visualization curves (Fig. 4d). The text-format prediction results contain four items, including protein sequence name, lysine position, prediction score for the eight types of PTM sites, and their predictive confidence (Fig. 4a). The basic statistics for the predicted PTM sites are obtained based on the calculation of the percentage of predicted PTM sites in all lysine residues for each PTM type (Fig. 4b). A bar plot for each input protein sequence is generated to illustrate the distribution of the predicted PTM sites in the protein sequences (Fig. 4c). The detailed prediction scores of potential lysine PTM sites are demonstrated in the line chart shown in Fig. 4d.

5 Performance Evaluation

Two different models were built, that is, the full-sequence model and the fragment model, for the MUscADEL framework. For the full-sequence model, the full-length protein sequences were used as input; while for the fragment model, on the other hand, motifs with the fixed window size with centered lysine residues were used. The full-sequence model considers the contributions of both full-length sequences and motifs surrounding the lysine PTM sites, while the fragment model only extracts the motif patterns surrounding the lysine PTM sites. The results are shown in Table 3. The AUC value of the full sequence model ranged from 0.602 to 0.822, while for the fragment model the AUC value range was from 0.610 to 0.807. Comparatively, the full-sequence model outperformed the fragment model for predicting lysine acetylation,

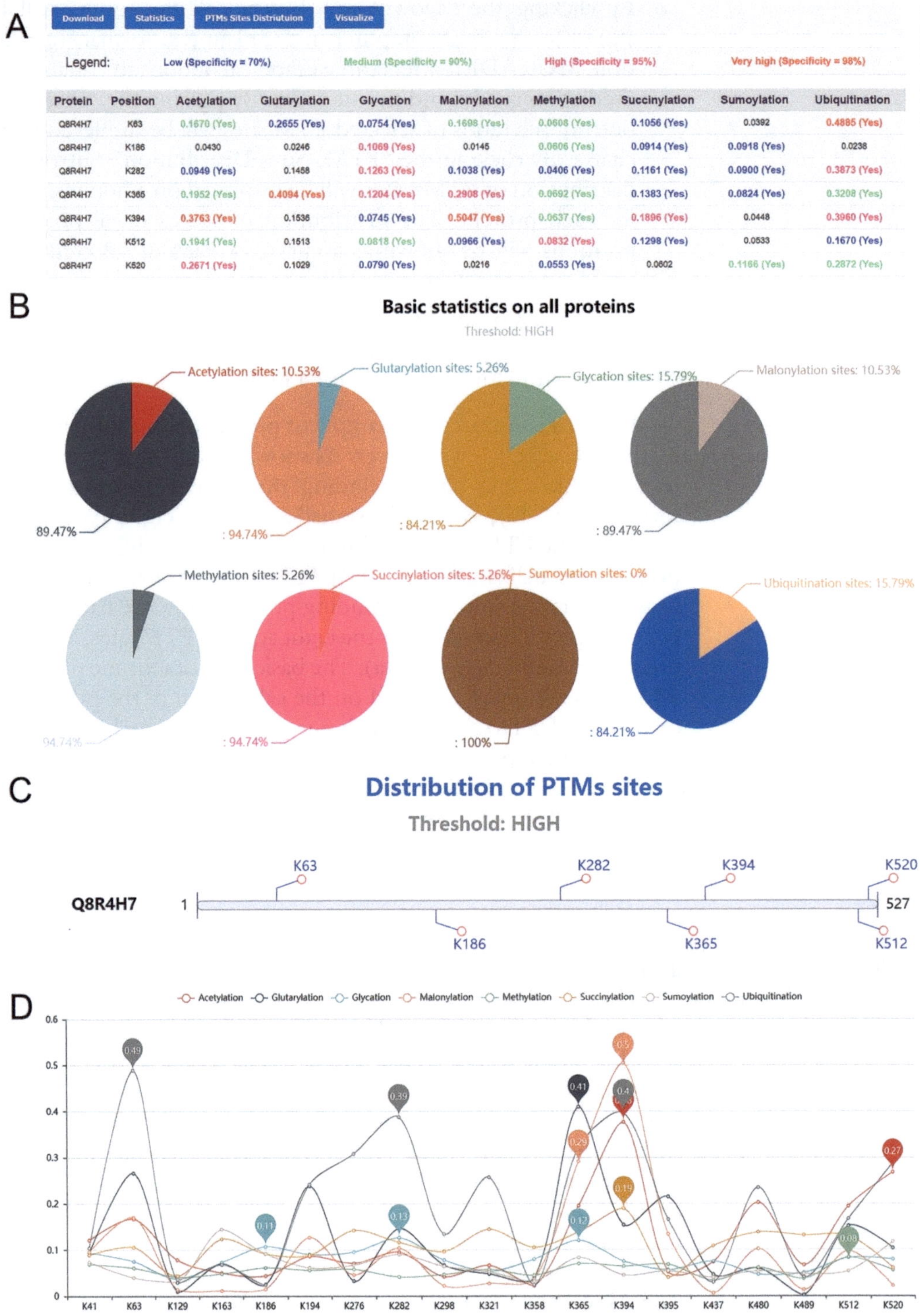

Protein	Position	Acetylation	Glutarylation	Glycation	Malonylation	Methylation	Succinylation	Sumoylation	Ubiquitination
Q8R4H7	K63	0.1670 (Yes)	0.2655 (Yes)	0.0754 (Yes)	0.1698 (Yes)	0.0608 (Yes)	0.1056 (Yes)	0.0392	0.4885 (Yes)
Q8R4H7	K186	0.0430	0.0246	0.1069 (Yes)	0.0145	0.0606 (Yes)	0.0914 (Yes)	0.0918 (Yes)	0.0238
Q8R4H7	K282	0.0949 (Yes)	0.1458	0.1263 (Yes)	0.1038 (Yes)	0.0406 (Yes)	0.1161 (Yes)	0.0900 (Yes)	0.3873 (Yes)
Q8R4H7	K365	0.1952 (Yes)	0.4094 (Yes)	0.1204 (Yes)	0.2908 (Yes)	0.0692 (Yes)	0.1383 (Yes)	0.0824 (Yes)	0.3208 (Yes)
Q8R4H7	K394	0.3763 (Yes)	0.1536	0.0745 (Yes)	0.5047 (Yes)	0.0637 (Yes)	0.1896 (Yes)	0.0448	0.3960 (Yes)
Q8R4H7	K512	0.1941 (Yes)	0.1513	0.0818 (Yes)	0.0966 (Yes)	0.0832 (Yes)	0.1298 (Yes)	0.0533	0.1670 (Yes)
Q8R4H7	K520	0.2671 (Yes)	0.1029	0.0790 (Yes)	0.0216	0.0553 (Yes)	0.0602	0.1166 (Yes)	0.2872 (Yes)

Fig. 4 A illustration of the prediction results delivered by the MUscADEL web server using the protein sequence of Q8R4H7, including (**a**) the detailed prediction results in tabular form, (**b**) the basic statistics of prediction results for selected lysine PTM types, (**c**) a graphical illustration of the distribution of predicted lysine PTMs, and (**d**) the line chart demonstrating the detailed prediction scores of each lysine PTM site

glycation, malonylation, methylation, sumoylation, and ubiquitination, by improving approximately 1–15% of sensitivity when using the same specificity (i.e., 90%). As such, for these PTMs, both full-length sequence and local motif information contributed to the prediction of PTM sites. For predicting glutarylation, on the other hand, the fragment model generally outperformed the full-sequence model, indicating that the determinative sequence patterns of these PTM types can be captured by mining the local motif environment surrounding the lysine PTM resides.

6 Notes

1. The independent test results described in Subheading 3 show that MUscADEL has outperformed 19 existing tools for seven different types of lysine PTMs.
2. In addition, MUscADEL was the only tool for predicting lysine glutarylation sites. Detailed performance comparison results between these tools can be found in our previous study [26].
3. It generally takes only a few seconds for the web server to generate and deliver the prediction result for a given protein. On average, it takes approximately 40 s or longer to generate the prediction outcomes of a hundred of input protein sequences.

Funding

This work was financially supported by grants from the Australian Research Council (LP110200333 and DP120104460), the Young Scientists Fund of the National Natural Science Foundation of China (31701142), the National Natural Science Foundation of China (31770821), the National Health and Medical Research Council of Australia (NHMRC) (490989), the National Institute of Allergy and Infectious Diseases of the National Institutes of Health (R01 AI111965), a major interdisciplinary research project awarded by Monash University and the collaborative research program of the Institute for Chemical Research, Kyoto University (grant # 2021-28). J.S. is supported by the JSPS Long-Term Invitational Fellowship (L21503). C.L. is currently supported by an NHMRC CJ Martin Early Career Research Fellowship (1143366). T.T.M.L. and A.L.'s work was supported in part by the Informatics Institute of the School of Medicine at the University of Alabama at Birmingham.

References

1. Hornbeck PV, Zhang B, Murray B et al (2015) PhosphoSitePlus, 2014: mutations, PTMs and recalibrations. Nucleic Acids Res 43: D512–D520
2. Xu H, Zhou J, Lin S et al (2017) PLMD: an updated data resource of protein lysine modifications. J Genet Genomics 44:243–250
3. Das C, Kundu TK (2005) Transcriptional regulation by the acetylation of nonhistone proteins in humans—a new target for therapeutics. IUBMB Life 57:137–149
4. Glozak MA, Sengupta N, Zhang X et al (2005) Acetylation and deacetylation of non-histone proteins. Gene 363:15–23
5. Kim SC, Sprung R, Chen Y et al (2006) Substrate and functional diversity of lysine acetylation revealed by a proteomics survey. Mol Cell 23:607–618
6. Polevoda B, Sherman F (2002) The diversity of acetylated proteins. Genome Biol 3: reviews0006
7. Spange S, Wagner T, Heinzel T et al (2009) Acetylation of non-histone proteins modulates cellular signalling at multiple levels. Int J Biochem Cell Biol 41:185–198
8. Zhao S, Xu W, Jiang W et al (2010) Regulation of cellular metabolism by protein lysine acetylation. Science 327:1000–1004
9. Munch G, Gerlach M, Sian J et al (1998) Advanced glycation end products in neurodegeneration: more than early markers of oxidative stress? Ann Neurol 44:S85–S88
10. Nicolls MR (2004) The clinical and biological relationship between Type II diabetes mellitus and Alzheimer's disease. Curr Alzheimer Res 1: 47–54
11. Chau V, Tobias JW, Bachmair A et al (1989) A multiubiquitin chain is confined to specific lysine in a targeted short-lived protein. Science 243:1576–1583
12. Pickart CM (2001) Ubiquitin enters the new millennium. Mol Cell 8:499–504
13. Welchman RL, Gordon C, Mayer RJ (2005) Ubiquitin and ubiquitin-like proteins as multifunctional signals. Nat Rev Mol Cell Biol 6: 599–609
14. Chen ZJ, Sun LJ (2009) Nonproteolytic functions of ubiquitin in cell signaling. Mol Cell 33: 275–286
15. Hicke L (2001) Protein regulation by monoubiquitin. Nat Rev Mol Cell Biol 2:195–201
16. Longo VD, Kennedy BK (2006) Sirtuins in aging and age-related disease. Cell 126: 257–268
17. Mastronardi FG, Wood DD, Mei J et al (2006) Increased citrullination of histone H3 in multiple sclerosis brain and animal models of demyelination: a role for tumor necrosis factor-induced peptidylarginine deiminase 4 translocation. J Neurosci 26:11387–11396
18. Suzuki A, Yamada R, Yamamoto K (2007) Citrullination by peptidylarginine deiminase in rheumatoid arthritis. Ann N Y Acad Sci 1108:323–339
19. Wang Z, Tang WH, Cho L et al (2009) Targeted metabolomic evaluation of arginine methylation and cardiovascular risks: potential mechanisms beyond nitric oxide synthase inhibition. Arterioscler Thromb Vasc Biol 29: 1383–1391
20. Peng C, Lu Z, Xie Z et al (2011) The first identification of lysine malonylation substrates and its regulatory enzyme. Mol Cell Proteomics 10:M111:012658
21. Xie Z, Dai J, Dai L et al (2012) Lysine succinylation and lysine malonylation in histones. Mol Cell Proteomics 11:100–107
22. Zhang Z, Tan M, Xie Z et al (2011) Identification of lysine succinylation as a new post-translational modification. Nat Chem Biol 7: 58–63
23. Caron C, Boyault C, Khochbin S (2005) Regulatory cross-talk between lysine acetylation and ubiquitination: role in the control of protein stability. BioEssays 27:408–415
24. Hunter T, Sun H (2008) Crosstalk between the SUMO and ubiquitin pathways. Ernst Schering Found Symp Proc:1–16
25. Nie Q, Gong XD, Liu M et al (2017) Effects of crosstalks between sumoylation and phosphorylation in normal cellular physiology and human diseases. Curr Mol Med 16:906–913
26. Chen Z, Liu X, Li F et al (2019) Large-scale comparative assessment of computational predictors for lysine post-translational modification sites. Brief Bioinform 20:2267–2290
27. Hochreiter S, Schmidhuber J (1997) Long short-term memory. Neural Comput 9: 1735–1780
28. Schuster M, Paliwal KK (1997) Bidirectional recurrent neural networks. IEEE Trans Signal Process 45:2673–2681
29. Li Y, Wang M, Wang H et al (2014) Accurate in silico identification of species-specific acetylation sites by integrating protein sequence-derived and functional features. Sci Rep 4:5765

30. Johansen MB, Kiemer L, Brunak S (2006) Analysis and prediction of mammalian protein glycation. Glycobiology 16:844–853
31. Wang LN, Shi SP, Xu HD et al (2017) Computational prediction of species-specific malonylation sites via enhanced characteristic strategy. Bioinformatics 33:1457–1463
32. Lu L, Shi XH, Li SJ et al (2010) Protein sumoylation sites prediction based on two-stage feature selection. Mol Divers 14: 81–86
33. Cai Y, Huang T, Hu L et al (2012) Prediction of lysine ubiquitination with mRMR feature selection and analysis. Amino Acids 42: 1387–1395
34. Walsh I, Di Domenico T, Tosatto SC (2014) RUBI: rapid proteomic-scale prediction of lysine ubiquitination and factors influencing predictor performance. Amino Acids 46:853–862
35. Chen Z, Zhou Y, Song J et al (2013) hCKSAAP_UbSite: improved prediction of human ubiquitination sites by exploiting amino acid pattern and properties. Biochim Biophys Acta 1834:1461–1467
36. Wuyun Q, Zheng W, Zhang Y et al (2016) Improved species-specific lysine acetylation site prediction based on a large variety of features set. PLoS One 11:e0155370
37. Suo SB, Qiu JD, Shi SP et al (2012) Position-specific analysis and prediction for protein lysine acetylation based on multiple features. PLoS One 7:e49108
38. Deng W, Wang Y, Ma L et al (2017) Computational prediction of methylation types of covalently modified lysine and arginine residues in proteins. Brief Bioinform 18:647–658
39. Xue Y, Zhou F, Fu C et al (2006) SUMOsp: a web server for sumoylation site prediction. Nucleic Acids Res 34:W254–W257
40. Wang L, Du Y, Lu M et al (2012) ASEB: a web server for KAT-specific acetylation site prediction. Nucleic Acids Res 40:W376–W379
41. Li A, Xue Y, Jin C et al (2006) Prediction of nepsilon-acetylation on internal lysines implemented in Bayesian discriminant method. Biochem Biophys Res Commun 350:818–824
42. Hou T, Zheng G, Zhang P et al (2014) LAceP: lysine acetylation site prediction using logistic regression classifiers. PLoS One 9:e89575
43. Li S, Li H, Li M et al (2009) Improved prediction of lysine acetylation by support vector machines. Protein Pept Lett 16:977–983
44. Liu Y, Gu W, Zhang W et al (2015) Predict and analyze protein glycation sites with the mRMR and IFS methods. Biomed Res Int 2015: 561547
45. Wen PP, Shi SP, Xu HD et al (2016) Accurate in silico prediction of species-specific methylation sites based on information gain feature optimization. Bioinformatics 32:3107–3115
46. Yavuz AS, Sezerman OU (2014) Predicting sumoylation sites using support vector machines based on various sequence features, conformational flexibility and disorder. BMC Genomics 15(Suppl 9):S18
47. Cortes C, Vapnik V (1995) Support-vector networks. Mach Learn 20:273–297
48. Breiman L (2001) Random forests. Mach Learn 45:5–32
49. McCulloch WS, Pitts W (1943) A logical calculus of the ideas immanent in nervous activity. Bull Math Biol 5:115–133
50. Altman NS (1992) An introduction to kernel and nearest-neighbor nonparametric regression. Am Stat 46:175–185
51. Freedman AD (2006) Statistical models: theory and practice. Technometrics 48:315–315
52. UniProt Consortium T (2018) UniProt: the universal protein knowledgebase. Nucleic Acids Res 46:2699
53. Li W, Godzik A (2006) Cd-hit: a fast program for clustering and comparing large sets of protein or nucleotide sequences. Bioinformatics 22:1658–1659
54. Chen Z, He N, Huang Y et al (2018) Integration of a deep learning classifier with a random forest approach for predicting malonylation sites. Genomics Proteomics Bioinformatics 16:451–459
55. Maas AL, Daly RE, Pham PT et al (2011) Learning word vectors for sentiment analysis. In: Proceedings of the 49th annual meeting of the association for computational linguistics: human language technologies - volume 1. Association for Computational Linguistics, Portland, OR, pp 142–150
56. Greff K, Srivastava RK, Koutník J et al (2016) LSTM: a search space odyssey. IEEE Trans Neural Netw Learn Syst 28:2222–2232
57. Kingma D, Ba J (2014) Adam: a method for stochastic optimization. International conference on learning representations. arXiv:1412.6980

Chapter 12

Enhancing the Discovery of Functional Post-Translational Modification Sites with Machine Learning Models – Development, Validation, and Interpretation

Nolan English and Matthew Torres

Abstract

Protein posttranslational modifications (PTMs) are a rapidly expanding feature class of significant importance in cell biology. Due to a high burden of experimental proof, the number of functionals PTMs in the eukaryotic proteome is currently underestimated. Furthermore, not all PTMs are functionally equivalent. Computational approaches that can confidently recommend PTMs of probable function can improve the heuristics of PTM investigation and alleviate these problems. To address this need, we developed SAPH-ire: a multifeature heuristic neural network model that takes community wisdom into account by recommending experimental PTMs similar to those which have previously been established as having regulatory impact. Here, we describe the principle behind the SAPH-ire model, how it is developed, how we evaluate its performance, and important caveats to consider when building and interpreting such models. Finally, we discus current limitations of functional PTM prediction models and highlight potential mechanisms for their improvement.

Key words Posttranslational modification, PTM, Machine learning, Functional prediction, Mass spectrometry, Proteins

1 Introduction

Posttranslational modifications (PTMs), chemical or proteinaceous alterations to amino acid residues in a protein, have the potential to expand the function and regulatory control of proteins beyond the limits of the genome [1]. PTMs can act on long or short timescales that allow for dynamic control and response of a cellular proteome to changing environments or cellular phases that ultimately shape cellular phenotype, often by modulating changes in protein interaction, localization, or stability [2]. Concomitantly, disruption to

Supplementary Information The online version contains supplementary material available at [https://doi.org/10.1007/978-1-0716-2317-6_12].

Dukka B. KC (ed.), *Computational Methods for Predicting Post-Translational Modification Sites*,
Methods in Molecular Biology, vol. 2499, https://doi.org/10.1007/978-1-0716-2317-6_12,

either the amino acid or modification of highly functional PTM sites can contribute to cellular dysfunction and disease [3–5].

The scientific community has witnessed an exponential increase in PTM data over the last 20 years, fueled by high-throughput mass spectrometry that has identified hundreds of different PTM types occurring on nearly all of the 20 common amino acids. However, the rate at which PTM data is generated—a parallel process involving hundreds of independent labs—far surpasses the rate at which it is being curated and/or processed for interpretation—a task undertaken by a much smaller set of labs and institutions [6, 7]. A longstanding question emerging from these efforts is whether all PTMs (detected accurately) are functionally important—a question not easily answered due to the high burden of experimental evidence needed to prove functionality, which involves significant time, cost, and specific expertise for any given protein. These challenges are compounded by unnecessary redundancy in experimental effort and the tendency of most labs not to report nonfunctional results. Underlying challenges such as these tend to promote a perspective that the study of PTMs is risky and quite possibly not worth the effort.

Computational approaches aimed at the prediction of functional PTM sites have made a tangible impact on the discovery of regulatory elements in proteins. They are also rapidly evolving due to the exponential increase in data availability that can be used for their training and validation. In general, two types of models have emerged over the last 10 years, each of which stems from a unique philosophy born from the intent of functional PTM site prediction. Since these differences have not been discussed in the literature previously, we will take this opportunity to define them here as they exist currently. The first type of model, we will call it an *intrinsic feature model*, seeks to approximate biological truth as purely as possible through the use of intrinsic features that approximate the natural features of the PTM site and are assumed to be largely resistant to new information. Intrinsic features have a longstanding history of being used for functional PTM site prediction and include information such as evolutionary conservation [8–10], PTM site colocalization or crosstalk [11–14], proximity to functional structures (e.g., regulatory motifs or catalytic pockets) [15, 16], proximity to disease-causing mutation sites [17–19], enzyme–substrate pairing [9, 20–24], among other natural features that can be sampled experimentally, such as PTM site occupation stoichiometry and PTM dynamics that are currently difficult to capture accurately [9, 25]. The second type of model, a *heuristic model*, also utilizes intrinsic biological features but is additionally informed by extrinsic features that reflect *community wisdom* in the form of literature evidence. Extrinsic features include things like PTM observation frequencies that proxy the reproducibility and abundance of an observed PTM throughout the community at large or information about preestablished functionality of proximal PTM sites [26]. Indeed, experimentalists do not choose to study PTM sites randomly and often consider several intrinsic and extrinsic aspects of the PTM site in their decision making, which a machine learning model can learn.

Intrinsic feature and heuristic models tend to differ not in performance but in ease of training, lifespan, and interpretation. Models built to approximate a system as closely as possible with innate properties tend to rely on hundreds of individual features with low individual impact making them costly to train due to complexity, and prone to local minima errors. However, when done well these models have lifespans that last for years and produce insights into how the underlying system behaves. Heuristic models, in contrast, tend to be easier to train and rely on far fewer features without suffering a cost in performance. These do typically have a shorter lifespan, as the observations they rely on shift and do not inform the user about the mechanics of the system described. Heuristic models that combine both intrinsic and extrinsic features (as described in this chapter) consequently have a longer lifespan while benefitting from decreased complexity and easier training.

Regardless of whether an intrinsic feature or heuristic model is used for functional PTM site prediction, it is clear that no one feature is a sufficient predictor by itself. Indeed, careful inspection reveals that not all functional PTMs are highly conserved, colocalized, near catalytic centers or near other important protein structures that make them otherwise obvious candidates for functional investigation. Cases wherein a PTM site's potential for function is easily predicted by any one of these features alone may be considered "low-hanging fruit." Machine learning models that incorporate multiple PTM site features have shown promise in capturing a larger proportion of the functional PTM population in eukaryotes [17, 27, 28], and have enabled the identification of functional PTMs not readily identifiable through single feature analyses alone [26]. Within this space, two basic strategies have been described to date: rank-based and recommendation-based models [28, 29]. Rank-based models place PTMs in a competitive hierarchy of functional importance wherein PTMs for which functional evidence already exists end up broadly distributed in the scoring regime and with only a small fraction of candidates rising to the top [17, 28]. This strategy limits the utility of rank-based methods for identifying PTM sites of putative function as only the most extreme outliers are confidently predicted. Recommendation-based models seek to reinforce recall (i.e., accurate prediction/capture) of the bulk of functional PTM sites above carefully established scoring thresholds. As such, recommendation-based models provide more concrete criteria for evaluating model performance.

In this chapter we describe the materials and step-by-step methods involved in development of a heuristic recommendation-based machine learning model for functional PTM site prediction. Throughout the chapter a sample PTM dataset is carried through model development, training, and validation. The model, which we refer to as SAPH-ire (Systematic Analysis of PTM Hotspots), incorporates intrinsic and extrinsic feature data from many different PTM types, PTM site features, and functional consequences. We stress that, like any model, there are many *dials* that can be turned to affect the outcome and that the model shown here is used best as an example of the parts that are needed, and how to build and validate a functional PTM site prediction model. Indeed, the model described here is a snapshot in the evolution of a functional PTM site prediction model likely to undergo further refinement over time. In this chapter, we will explore the dials that can be used for model refinement, highlight some of the most important ones to consider, and discuss underlying caveats. Lastly, we will discuss key limitations of functional PTM site prediction models in general, whether or not such limitations can be mitigated, and how those interested in functional PTM site prediction should approach the development and interpretation of such models in the future.

2 Materials

A table of the required resources (datasources, webtools, programs) to carry out the methods described is provided here (Table 1).

2.1 Sources of Experimentally Observed and Functional PTMs

The SAPH-ire method is PTM agnostic and includes 56 different PTM types (PTMtype), the bulk of which correspond to phosphorylation, ubiquitination, acetylation, methylation, N-linked glycosylation, and sumoylation depending on current data availability (*see* **Note 1**). A summary table of PTM types available as of mid-2019 are shown here (Table S1). PTMs are collected from multiple sources including PhosphoSitePlus [30], SysPTM [31], and dbPTM [32], which provide open access to curated PTM data (*see* **Note 2**). Each PTM is mapped to UniProt identifiers (UID) [33], and validated by matching the native position (NP) and residue (res) of the curated PTM to UniProt sequences verified for 100% sequence identity using Basic Local Alignment Search Tool (BLAST) [34]. Isoforms, although rare in the PTM dataset, are also included. The final PTM dataset for training a machine learning model should be sufficiently large and, in the example shared here, contains 390,697 unique PTMs (identified by what we call the PTM-ID: UID-NP-res-PTMtype; Table S2). Functional PTMs are collected from PhosphoSitePlus, which includes several pieces of useful information about the functional knowledge associated with ~12,000+ PTM sites, including literature references that provide evidence in support of such knowledge.

Table 1
Table of resources needed to accomplish the study described here

	Resource	Usage	Access Point	References
1	PhosphoSitePlus	PTMs	https://www.phosphosite.org/home	[68]
2	dbPTM	PTMs	http://dbptm.mbc.nctu.edu.tw/	[32]
3	SysPTM	PTMs	http://lifecenter.sgst.cn/SysPTM/	[31]
4	UniProt	Protein sequences	https://www.uniprot.org	[33]
5	InterPro	Protein families	https://www.ebi.ac.uk/interpro/	[35]
6	Basic Local Alignment Search Tool (BLAST)	PTM position validation	https://blast.ncbi.nlm.nih.gov/Blast.cgi	[34]
7	MUSCLE	Sequence alignment	https://www.drive5.com/muscle/	[36, 37]
8	Iupred	Protein disorder pred.	https://iupred2a.elte.hu	[69]
9	Python and PANDAS	Coding	https://www.python.org https://pandas.pydata.org	[70]
11	TensorFlow	Machine learning	https://www.tensorflow.org	[38]
12	JMP	Data analysis	https://www.jmp.com	[71]
13	R	Data analysis	https://www.r-project.org	[72]
14	ClinVar	Disease-linked SNPs	https://www.ncbi.nlm.nih.gov/clinvar/	[73]
15	ELM Resource	Short Linear Motifs	http://elm.eu.org	[21]
16	LIME	ML feature contributions	https://christophm.github.io/interpretable-ml-book/lime.html	[55]

2.2 Protein Family Associations and Protein Sequence Retrieval

SAPH-ire relies on associating PTMs within groups of closely related proteins, or protein families (Table S3). This information is not generally provided in any of the available PTM repositories and must be acquired independently. We recommend using InterPro as the source for mapping UID entries to whole sequence protein families [35] (*see* **Note 3**). Each UID from the list of modified proteins must be mapped to at least one specific protein family (IPR ID) resulting in a UID-to-IPR mapping table. This map will serve as the basis for further processing of modified protein sequences by alignment (*see* **Note 4**).

Protein sequences for each UID are retrieved directly from UniProt and used for whole sequence protein family alignment that provides anchoring for all further steps in the SAPH-ire method. Importantly, UniProt sequences are properly annotated with methionine as the initiating residue. However, the reader should be aware that PTM repositories are not always annotated properly and in many cases are numbered starting from the second residue (*see* **Note 5**). This necessitates BLAST-based correction of some PTM site data, as described earlier (*see* Subheading 2.1).

2.3 Multiple Sequence Alignment

Once the UID-to-IPR map table is generated, it can be used to guide the alignment of protein sequences systematically. We use MUSCLE with default parameters for multisequence alignment [36, 37] (*see* **Note 6**). In the standing SAPH-ire method, only proteins with experimental evidence of PTM on at least one site in their sequence are used for alignment (*see* **Note 7**). Within the context of the example presented here, a final 390,697 PTMs have been mapped to 8039 families containing 38,231 UIDs representing 763 eukaryotic organisms.

2.4 Model Input Features

A comprehensive list of feature descriptions used in the example are shown here (Table 2). Assembly of these features is described in Subheading 3.

Table 2
Description of PTM site features used in SAPH-ire model development, training, and performance evaluation

Feature	Description
Family	InterPro Family ID (e.g., IPR005419)
AP	Alignment position
MAP	Modified Alignment Position. Defined by Family-AP.
UID	UniProt Identifier (e.g., P50124)
NP	Native Position
PTMtype	Type of PTM(s) at the position in question (e.g., phosphorylation). Not a direct input feature but used in calculating AP_PTc.
PRC_AP[a]	PTM residue conservation; % identity of a modifiable residue within a MAP
Dis[a]	Disorder; Mean short disorder tendency predicted using IUPred
OIDc[a]	Organism ID count; Count of unique taxonomic IDs (taken for protein family and alignment position)
MRc[a]	Modified Residue Count; Number of unique modified UID-NPs in the MAP
NMRc[a]	Neighbor Modified Residue Count (sum of unique modified UID-NPs within maps neighboring but not including reference MAP); +/−7 alignment positions; gaps included

(continued)

Table 2 (continued)

Feature	Description
Nc[a]	Neighbor Count; Neighboring MAP Count within +/−7 alignment positions not including the reference MAP; gaps included
KNc[a]	Known Neighbor Count; Sum of neighboring MAPs that harbor at least one reference supporting known biological PTM function +/−2 alignment positions; gaps included
AP_MEM[a]	The number of UniProt entries in an Alignment Position No GAPs.
PF_MEM[b]	Protein family membership (count of UIDs contributing to the family alignment)
AP_PTc[a]	PTMtype count within the alignment position
OBSrc[a]	Observation source count (from PSP; if no PSP data, then sum of observed PTMs in the AP)
PF_OBSrc[a]	Observation source count normalized to number of members in family
Known[c]	MAP functional status binary classifier. 0 = functional evidence absent; 1 = functional evidence present
PTMc[b]	MAP PTM count. Sum of nonredundant PTMs (tracked by UID-NP-Res-PTMtype)
KFSC[b]	Known Function Source Count. Within a MAP, the sum of PubMed IDs linked to functional data reported for any and all functional PTMs harbored within the MAP.
Cc	PTM cluster count (sum of PTMc for MAP +/− 2 or +/− 7 alignment positions)
KF-PTM	PTM site established experimentally (and published) as having a modification with some functional impact (curated in the PhosphoSitePlus database of regulatory PTMs)
UF-PTM	PTM site with unknown function (either untested or not reported)
KF-MAP	Modified alignment position harboring at least 1 member site classified as KF-PTM
UF-MAP	Modified alignment position lacking any evidence of functional impact (either untested or not reported)

[a] Feature input used in the SAPH-ire machine learning model described in this chapter
[b] Feature used for model performance evaluation but not included in the model
[c] Feature used for model training

2.5 Machine Learning

SAPH-ire is based on an Artificial Neural Network (ANN) model that can be implemented using TensorFlow—an end-to-end open-source platform for machine learning developed by Google [38]. TensorFlow is particularly useful as it is equipped with tutorials for every level of user from beginner to advanced, reducing barriers to implementation and making the method widely accessible.

2.6 Graphical and Statistical Data Analysis

Analysis of model data can be accomplished using any appropriate statistical analysis package capable of handling large datasets. We have opted to use a combination of R, Python (specifically the pandas library), and JMP 14.1 (Table 1).

3 Methods

3.1 Overview and Rationale of PTM Site Function Prediction

The goal of SAPH-ire and other similar models is to provide a method of comparing PTM sites quantitatively so as to allow experimentalists to evaluate each site systematically rather than subjectively, and in so doing, make informed decisions about the *function potential* of any given site. This information can then be used to guide empirical studies aimed at testing PTM site functionality for a given protein of interest, with the aim of relating a given PTM to some molecular and/or cellular phenotype. Machine learning models are ideally suited to this goal as they can be trained to classify PTM sites based on whether they harbor quantitative feature combinations that mimic sites of known function (derived from prior experimental evidence). The SAPH-ire method utilizes an ANN machine learning model in which individual PTMs are first aggregated into Modified Alignment Positions (MAPs) from which model input features are extracted (*see* Subheading 2.4.). A discussion of ANN machine learning is outside the scope of this work but is reviewed elsewhere [39–42]. Support vector machine (SVM) [43], and Random Forrest [44], machine learning models are also sometimes used for PTM site functional prediction, but do not significantly outperform ANN models for this purpose [27, 28].

Quantitative output is essential for comparative PTM site analysis and the SAPH-ire model derives quantitative data for MAPs rather than individual PTM sites (*see* **Note 8**). The features of each MAP are passed through the model to produce a probability score ranging from 0 to 1 (called a *functional site similarity* (FSS) score), where scores closer to 1 are indicative of MAPs with a high degree of similarity to MAPs of known function (Fig. 1). MAPs scored in this way allow systematic and quantitative comparison between MAPs, and therefore, between different positions of modification in a protein family. The methods for deriving such models, validating their performance, and interpreting their meaning are described in detail below.

3.2 Model Feature Extraction and Aggregation

SAPH-ire relies on neural network-based machine learning in which features derived from multiple sequence alignment positions are used for model training (Table 2 contains the list of features and their associated acronyms). The choice of features to use can have a significant impact on model performance. As a heuristic model, SAPH-ire utilizes a set of basic features that capture information about evolutionary conservation of modifiable PTM sites in an alignment position, PTM observation frequencies, neighboring PTM site observations, and alignment diversity (sequences contributing to the alignment position), for example. Extraction and aggregation of this feature data is therefore an essential first step in the process. There are no commercially available tools to

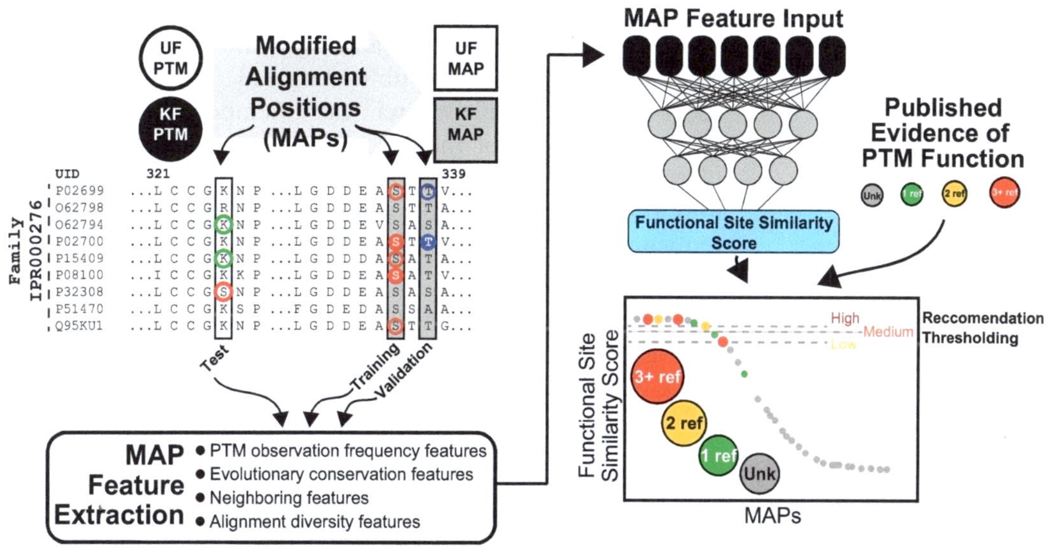

Fig. 1 Schematic diagram of SAPH-ire PTM function prediction workflow. PTMs of both unknown (UF) and known functional (KF) consequence (as determined through curated public record) are organized by full length protein family multiple sequence alignment, creating modified alignment positions (MAPs) of unknown or known function. KF-MAPs are used either for model training and/or model validation (via calculation of model recall) while UF-MAPs represent the test cases for which a functional impact is not currently known for any of the aligned PTMs. Features are extracted from MAP data and then used as inputs into a neural network trained to identify KF-MAPs. At this point the model is blind to whether a MAP is KF or UF. Each MAP (both KF and UF) passes through the model to receive a Functional Site Similarity Score that ranges from 0 to 1, where 1 indicates a MAP that closely resembles a KF-MAP. After scoring, the status of each MAP as KF or UF and the sum of literature sources supporting evidence of function (i.e., the Known Function Source Count; KFSC) is revealed. Model performance is graded and recommendation thresholds are created using recall of KF-MAPs and KFSC as a guide

accomplish this work, which we have done using custom Python scripts and JMP. The workflow proceeds through the creation of a series of tables (*dataframes*) as follows.

1. ***DATAFRAME A***: Create a table of nonredundant experimentally observed PTM sites (nrPTMs) (*see* Subheading 2.1). Each table should be nonredundant for the PTM-ID, defined as (UID-NP-residue-PTMtype), and should also include the total observation count (OBSrc) of each PTM (*see* **Note 9**).
2. ***DATAFRAME B***: Create a table of functional PTM sites that includes a tabulation of the known function source count (KFSC) for each PTM-ID.
3. Map each UID to at least one InterPro (IPR) protein family (*see* Subheading 2.2) (*see* **Note 10**).
4. Retrieve the fasta sequence for each UID in *DATAFRAME A, making sure to filter out redundant sequences* (*see* Subheading 2.3) (*see* **Note 11**).

5. Organize each UID-fasta sequence into a separate InterPro protein family sequence file and align the sequences in each file using MUSCLE (*see* Subheading 2.3). This results in an MSA output (.aln) file wherein each member sequence is now spread the length of the entire alignment and dashes are in place of nonaligning positions in each protein sequence. It is important that dashes be included to ensure each aligned sequence is the same length.
6. ***DATAFRAME C***: Convert each MSA file into columnar format with columns: FamilyID, Alignment Position (AP), UID, Native Position (NP). We call these FAUN files for: Family, AP, UID, NP. Concatenate all FAUN files into a single table.
7. Now, inclusively left-join (on UID-NP) *DATAFRAME C* with *DATAFRAME A*. This final table should have the following columns: Family, AP, UID, NP, PTMtype (*see* **Note 12**).
8. ***DATAFRAME D***: Pass the table from **step 7** through the iupred prediction tool to predict short disorder values (Dis) for every residue in the table.
9. ***DATAFRAME E***: Calculate the PTM residue conservation (PRC) of each modified alignment position (MAP). This is achieved by calculating the percentage of members in the AP that harbor residues modifiable by any of the PTM types found in the AP (*see* **Note 13**). We have a custom script that carries out this function and adds a PRC_AP column to *DATAFRAME D*.
10. ***DATAFRAME F***: Filter out all rows from *DATAFRAME E* except for modified alignment positions. Add a column indicating the organism for each UID in the table. This is the deconvoluted MAPs table from which all further feature data will be extracted.
11. ***DATAFRAME G***: To *DATAFRAME F*, inclusive left-join (on PTM-ID) the following: First, the tabulated sumOBSrc for each PTM-ID (derived from *DATAFRAME A*); Second, the tabulated sumKFSC for each PTM-ID (derived from *DATAFRAME B*) (*see* **Note 14**).
12. ***DATAFRAME H***: Condense *DATAFRAME G* by tabulating aggregate values across each MAP (Family-AP) into what is referred to as the "intermediate table." The columns should include: Family, AP, PTMc (sum), MRc, OBSrc (sum), PF_OBSrc, PF_MEM, AP_MEM (exclude gaps), PF_OIDc, AP_OIDc (exclude gaps), AP_PTc, KFSC (sum), and Known (either 0 = no functional data present or 1 = for functional data present.
13. ***DATAFRAME I:*** *DATAFRAME H* is used to determine the relationship between each MAP and neighboring MAPs. We

use a custom Python script that generates the following additional features based on neighboring alignment positions (gaps included): Nc, NMRc, KNc, and Cc.

DATAFRAME I serves as the input for predictive modeling (an example of *DATAFRAME I* used in the model describe here is provided; Table S4). Some of these columns are used as model feature inputs, others for training, and others for model evaluation (Table 2). Regarding features used as inputs to the SAPH-ire model: AP_PTc, PRC_AP, Dis, and MRc provide the model with an evolutionary conservation-based perspective on the MAP—a feature that has been shown to be effective in the past [8]; NMRc, Nc, and KNc provide information on the local environment of the MAP (not including the MAP in question), the degree to which they are modified and whether any neighbors are known to be functional (neighboring residue context has been shown to be an effective predictive feature in the past by us and others [9, 17, 26]) (*see* Subheading 3.8); OBSrc, PF_OBSrc, and AP_MEM take absolute and relative PTM observation and alignment frequencies into account (*see* **Note 15**).

3.3 Defining Optimal Model Architecture and Model Training

One significant challenge in applying neural network models to machine learning problems is optimizing the network architecture. Adjusting the activation functions, the weight initialization, layer connectivity, and layer count, among other things, are typically done through heuristics built from experience. In recent years, however, automated iterative techniques have solved this problem through iteratively searching through the possible combinations. This is particularly useful as it allows for quick optimization of a model, without the concern of overlooking an optimal combination. One of the simpler optimization techniques is particle swarm optimization (PSO). PSO is an iterative training technique where groups of models, referred to as particles, are generated with random parameters, trained, then evaluated. The best performing particles(s) are then selected to use as the starting point for the next iteration of models. Through this process, the population of models, called a swarm, typically converges to an optimal parameter set. PSO and its modifications are effective at traversing large multidimensional parameter space with few iterations and typically converge [45]. In the optimization of the SAPH-ire model, we manually implemented the logic of PSO as follows.

1. Define the metrics for evaluation. PSO defines the sets of metrics as the fitness; however, in the case of SAPH-ire, standard metrics like accuracy would misrepresent the performance (*see* **Note 16**). In the example discussed here, loss and recall can be used effectively during early model development, but were eventually replaced with a combination of recall and receiver

operating characteristic (ROC) to evaluate swarm fitness, as defined by an *Fzero* metric.

$$Fzero = \frac{2(auroc)(recall)}{(auroc + recall)}$$

where *auroc* is the area under the ROC curve and optimal models are defined as those with the greatest *Fzero* score and *recall* is the proportion of true positives correctly predicted. Taken together, the combined metric emphasizes retaining known data and can tolerate false positives. In contrast to normal applications the metrics for evaluating the SAPH-ire model are not evenly weighted, but instead recall is given a higher weight since maximizing recall is valued at a significantly higher rate than precision in this case.

2. Define the search space of possible parameters. For SAPH-ire this includes the number of layers, the activation function between layers, and the extent of connectivity during training (*see* **Note 17**). In this example, evaluation was initially extended between 1 and 15 layers, but was eventually constrained from 1 to 5 layers as it became apparent that additional layers beyond 5 did not show improvements in any metrics. All activation functions that did not include recurrent terms, were included in the search space. In this example we included: softmax, softsign, tanh, sigmoid, selu, relu, and linear (*see* **Notes 18** and **19**).
3. Randomize the starting swarm based on this search space. Ensure that each parameter is varied with a defined range so that each swarm has a diverse set of solutions. When the parameter space is small (no larger than 29) all possible combinations can be generated programmatically (*see* **Note 20**).
4. Train all models (i.e., particles) in the swarm, using either a fixed number of epochs (training passes) for speed training or until all particles converge if the focus is on a representative perspective of each particle. The training dataset for these models is bootstrapped in order to over-represent the positive class to avoid sample distribution biasing [46], producing a 1: 2 ratio of KF-MAPs to UF-MAPs instead of the basal 1: 84 ratio. This is an important process that allows the model to learn the difference between the classes instead of the class distribution. Models are then trained using a 33% holdback rate, where 33% of the data is only used to evaluate fitness across epochs. Each model within the swarm is trained until their loss has converged using default evaluator trainer settings in TensorFlow.
5. Select which model has the highest fitness to use as the baseline for the next swarm. This can be done automatically; however,

we chose to visualize each iteration on a *Fzero* x *recall* plot and selected the best model respective to both axes. Typically, the model with the highest recall was also the model with the highest *Fzero*, but if not, both models were used to seed the next swarm.

6. Repeat **steps 4–6**, until the parameters and performance converge. We opted to select two models that performed similarly after over 20 iterations, because we had reached a point where the swarm had converged in terms of *Fzero*. (*see* **Note 21**).

In the example shown here, we show sample swarms from an iterative PSO approach used in developing SAPH-ire, along with the final model generated (Fig. 2). Early stages of PSO are geared toward deciding discrete parameters like the number of layers and the activation functions used for those layers, wherein the most important factors are *recall* and training *loss*. From this swarm, a model architecture of 2 fully connected layers with a tanh activation function was selected as the seed for the next generation (Fig. 2a). Later in the PSO approach, Loss had reached a range of overfitting where further reduction would negatively affect fitness (as determined by *Fzero* and *recall*) (Fig. 2b). In the final swarm most of the parameters had become fixed for every model and we then opted to vary different weight initializations (Fig. 2c), which are typically uniform with default TensorFlow settings (*see* **Note 22**). The best model from this epoch (batch 1) is selected for use as the SAPH-ire model. In this example, we joined the batch 1 model with another model (batch 2) trained using the same PSO approach, but with one additional feature added (*see* **Note 23**). Averaging the scores produced by both models together produces the final PTM function site similarity score, which outperforms either model alone (Fig. 2c, d).

3.4 Model Performance Evaluation Using Aggregated Literature Evidence—Known Function Source Count (KFSC)

The premise underlying any PTM function predictor is that not all PTM sites are of equivalent functional significance because some but not all PTM sites are important for proteins (and therefore cells) to function properly. We go further to suggest that not all functional PTMs (i.e., PTMs that have been published as biologically functional in some way) are equivalent either. Indeed, in some cases evidence of PTM site functionality is derived from a single study (most cases) while in others it is founded on several independent studies. For example, two closely spaced phosphorylation sites conserved in the activation loop of mitogen-activated protein kinases (MAPKs) that are essential for catalytic activation [47], have hundreds to thousands of independent studies confirming the functionality of these sites across the MAPK protein family (*see* **Note 24**).

As a heuristic method, we evaluate SAPH-ire performance using aggregated literature evidence (KFSC) as a proxy for

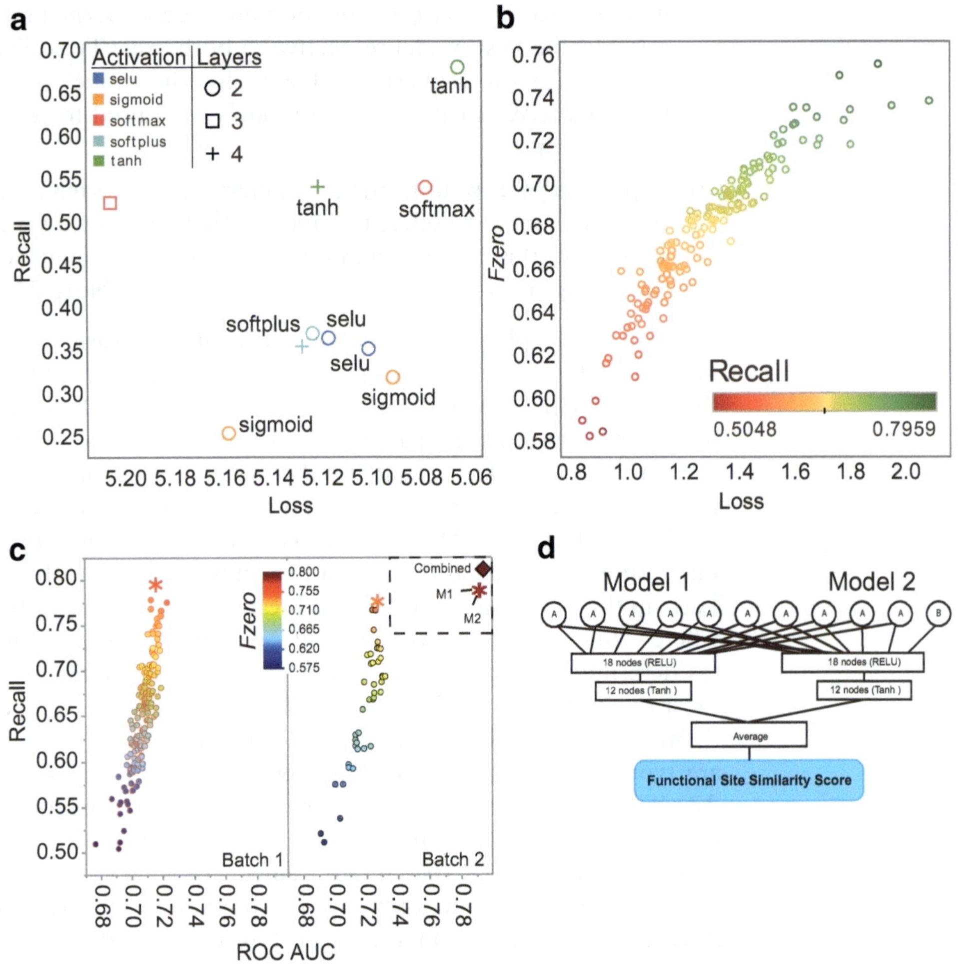

Fig. 2 General overview of PSO training for SAPH-ire and final model architecture. As an application of PSO, the models had their parameters varied to select an architecture and initialization method that produced effective models. (**a**) An early swarm of 10 models with varying number of layers and activation functions were assessed based on loss and recall. An ideal model would have both a high recall and a low loss. From this swarm, a model with 2 tanh activation layer was selected as the progenitor of the next swarm. (**b**) A later swarm showing how models with different random weight initializations with the same architecture produced different *Fzero* metrics and *Recall*. Layer connectivity was varied among this swarm. Despite decreasing loss typically being associated with better performing models, Recall and Fzero were anti correlated at this point of iteration. (**c**) The last swarm of models considered where the weight initialization parameters were varied, and the highest fitness models were selected. Batch 1 and Batch 2 were trained in the same way; however, batch 2 included one additional input feature that was found to have a significant positive effect on fitness. (**d**) Both models from Batch 1 and Batch 2 are averaged together to create the final SAPH-ire model

confidence in the functional status of a MAP and the PTM sites contained therein. Importantly, the SAPH-ire model is trained to call a MAP as either functional (max score = 1) or not (min score = 0) based on features from KF-MAPs that are known to harbor functional PTM sites (Known = 1) and does not include

KFSC. Consequently, KFSC can be used for more detailed model evaluation wherein one should expect that high-scoring KF-MAPs should include those MAPs that have higher KFSC if the model works properly. Furthermore, such an analysis increases confidence that high scoring UF-MAPs represent PTM sites that, by conventional wisdom, are likely functional but await experimental proof to solidify such status (*see* Subheading 3.6 for further detail).

3.5 Optimization of PTM-Agnostic Recommendation Thresholding

Without thresholds that provide a way to interpret model scores, a PTM functional prediction model should be interpreted as a rank-based model and interpretation of such models is difficult because there is no unified quantitative experimental metric for PTM functional importance. Implementing recommendation thresholds can improve interpretation of, but are rarely included in, functional prioritization models. Conversely, thresholds can create boundaries that, if inappropriately placed, can lead to inaccurate predictions. These problems can be addressed by modeling the tradeoff between true and false positive rates using ROC curves. Our philosophy in this goal has been to set a minimum threshold score over which MAPs could be considered having a high chance of functionality. Due to our desire for SAPH-ire scores to be agnostic across different PTM types, we first consider that the selected thresholds must not create a bias in distribution of PTMs occurring above a recommended threshold. To evaluate this, we plot the percent representation of each of the most common PTMs in the dataset relative to the model score (Fig. 3a). In the example shown here, the relative representation of each PTM type deflects significantly above a score of 0.9897 but is stable below this point and above 0.945. Next, we evaluate the ROC curves for the highest confidence true positive MAPs (KFSC >1, >2), and unfurl each curve to reveal the independent rates for true and false positives with respect to the model score. In this example, three thresholds were chosen within the ideal range (0.95, 0.972, and 0.987) that strike a balance between true positive hits and false positive recommendations (Fig. 3b). These recommendation thresholds provide useful landmarks to interpret model scores for a protein or family of interest, as shown here for family IPR000043 (Fig. 3c). These thresholds also allow for the evaluation of SAPH-ire in context of other models.

3.6 Model Validation with External Data Sources

As with any predictive model, time is usually required to determine whether or not the model's predictions are accurate. That said, we have derived several ways in which PTM function prediction models can be evaluated pseudoexperimentally, circumventing the need to wait for the otherwise slow rate of data publication. While model development and training necessarily include holdback validation procedures, we recommend that more rigorous steps be taken to determine the accuracy and utility of the final model. Importantly,

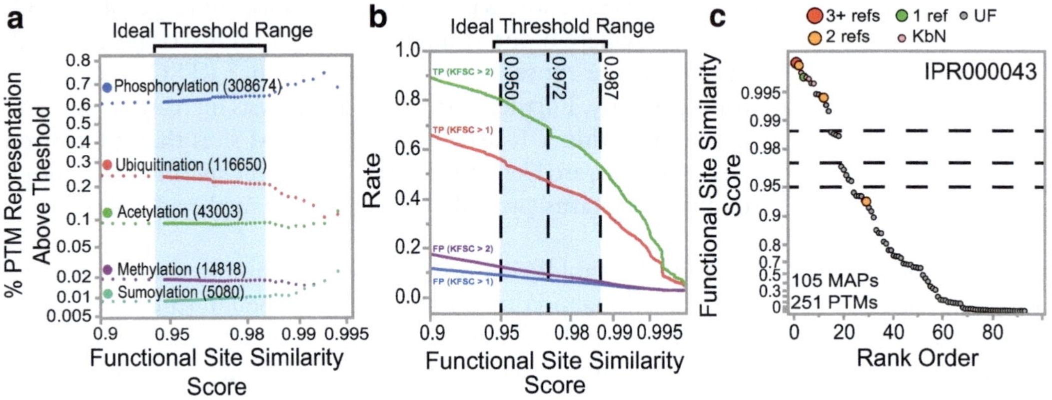

Fig. 3 Derivation of SAPH-ire recommendation thresholds. (**a**) Plot of the percent representation for different PTM types relative to Functional Site Similarity Score, revealing an ideal threshold range inside which no one PTM becomes over or underrepresented. (**b**) Unfurled ROC curves showing true positive (TP) and false positive (FP) rates above given SAPH-ire score. Rates shown are KFSC >1 (lower confidence) or KFSC >2 (higher confidence). Dashed vertical lines represent chosen thresholds where TP and FP rates are as follows: (KFSC >2): 0.95 (TP = 0.82, FP = 0.11); 0.972 (TP = 0.7, FP = 0.07); 0.987 (TP = 0.53, FP = 0.04). (**c**) Representative SAPH-ire score plot for family IPR000043 (Adenosylhomocysteinase-like family) with indicated thresholds shown for reference

these mechanisms should be independent from the data used in model training and holdback validation. They should also be model agnostic.

We have identified several useful mechanisms for external evaluation of SAPH-ire, placed in order of increasing rigor: (1) Evaluation of expanded datasets and functional reference citations; (2) Evaluation of new functional PTM data not included in the original model; (3) Precision/recall analysis of PTM function predictions using site-by-site experimental data; and (4) Direct experimental evaluation. We suggest mechanisms 1–3 as a minimum requirement for estimating the utility of any PTM function predictive model (*see* **Note 25**).

3.6.1 Evaluation of Expanded Datasets

PTM data is being collected at a nearly constant rate, with an average of ~500 publications per year from 2013 to 2019 (https://pubmed.ncbi.nlm.nih.gov). However, data retrieved from online curation sources (e.g., PhosphoSitePlus) are subject to curation rates that are usually slower than the rate of data generation. Thus, a model built to comprehensively analyze the compendium of publicly available PTM sites today is likely not going to be comprehensive after one year's time.

PTM site data curated after the predictive model is built can be used to evaluate robustness of the model. In the example described here, the SAPH-ire model was initially developed on a dataset compiled in late 2018, consisting of 435,750 total PTMs of which 9151 were known functional (the training dataset). We then evaluated model performance on an expanded dataset in which 102,475 PTMs (3233 known functional) that had not been used in training or holdback validation were added to the original set. The entirety of the merged dataset (the expanded dataset; 512,015 total PTMs) was then repassed through SAPH-ire and new predictions established (Fig. 4a). Model performance on the expanded dataset was better than on the training dataset as estimated by area under the receiver operating characteristic (ROC) and recall curves (AUCROC 0.794 and AUCRecall 0.764), suggesting that the addition of new data did not diminish performance (Fig. 4b). As expected, score distributions for MAPs also show increasing enrichment for MAPs with KFSC >1 as well as a decrease in the variance of the prediction with increasing KFSC (Fig. 4c). These trends were evident regardless of whether the data was from the training or expanded dataset (Fig. 4c, dashed lines).

Data can be further evaluated by tracking model scores in terms of the citation frequency of KF-MAPs (citation frequency, as opposed to KFSC, is a measure of the frequency at which the functional PTM literature has been cited elsewhere). In this case, the citation frequency for each functional PMID reference is measured relative to its year of publication (to control for the length of time the work was available for citation). While citation frequency does not inherently validate a study (citations can be both supportive or critical of the publication's conclusions), citation count is commonly and strongly associated with scientific impact and relevance [48]. In the example here, we found that half of all MAPs with a single source of literature evidence supporting PTM functionality (i.e., KFSC 1) have either no or low relative citation (Fig. 4d). This drops to 34% for KFSC 2 and to 10% for KFSC 3+. Inversely, KFSC 3+ MAPs are enriched in publications with high and medium relative citation (90%), followed by KFSC 2 (67%), and KFSC 1 (49%). The SAPH-ire model score scales with the relative citation frequency of a MAP, whereby known functional MAPs with a higher relative frequency fall almost entirely above the recommendation threshold score of 0.95 (Fig. 4e). Despite this, a large proportion of MAPs (>50%) with no citations were also highly scored above this threshold, most of which were published within the last 4 years.

3.6.2 Evaluating Model Performance for PTM Sites That Undergo a Change in Functional Status—Pseudoexperimental Validation

A relatively simple but powerful test of functional PTM site prediction performance can be achieved by asking whether or not the model's predictions become true later in time. This mechanism relies on the continued growth of PTM functional literature that continues to expand over time (albeit more slowly than PTM observation data) and represents a pseudoexperimental method for validating model scoring. We have accomplished such analyses whereby updates to the database of regulatory PTM sites (updated

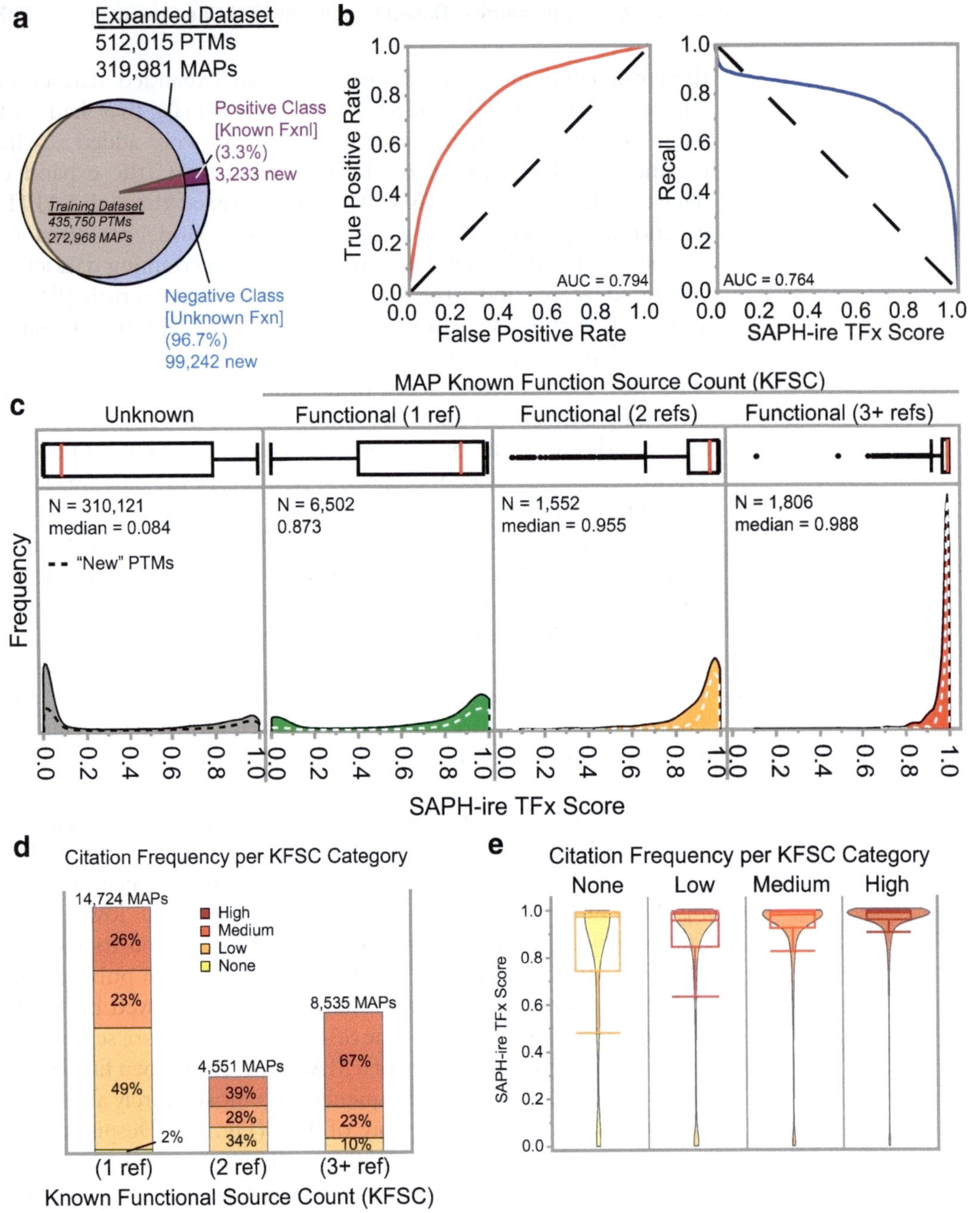

Fig. 4 Model performance on data not used for development. (**a**) Venn diagram showing the relationship between the training and expanded datasets. The expanded dataset contained 102,475 newly curated PTMs. (**b**) ROC and recall curves for SAPH-ire results from the expanded dataset. (**c**) Frequency distribution of functional site similarity scores relative to true positive status in terms of known function source count (KFSC = 0, 1, 2, or 3+ references). Area contained by solid lines corresponds to the total expanded PTM dataset. Area contained by dashed lines corresponds to model output for unexposed PTMs not contained in the original training dataset. (**d**) Bar Charts showing the citation frequency of each KFSC category (27810). Citation Frequency is based on the frequency of citation of a publication relative to other publications published in the same year: Low (lowest 25% Quantile), Medium (25–75% Quantile), High (75% + Quantile). (**e**) Violin plots showing the SAPH-ire functional site similarity score distribution of MAPs with a known function by citation frequency. All statistical data shown is aggregated at the MAP level

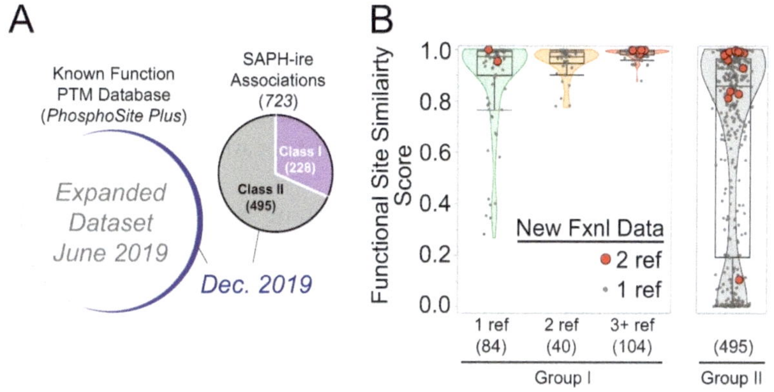

Fig. 5 Performance evaluation of functional site similarity scoring on newly curated functional data. (**a**) Venn diagrams depicting newly curated functional PTM data in comparison to the expanded dataset from Fig. 4. Group I PTMs are new functional PTM data already associated with KF-MAPs in the SAPH-ire model. Group II PTMs are new functional PTM data associated with MAPs previously classified as UF-MAPs in the model. (**b**) Score distributions for new functional PTM data in Groups I and II. Red circles correspond to PTMs with 2 references supporting functional impact of the PTM from the updated dataset. MAP bins (1, 2, 3+ refs) are based on the original classifications from Fig. 4

a few times per year by PhosphoSitePlus) provides new information on PTM sites that were scored previously but not themselves known as functional at the time of scoring. In the example here, such an update resulted in the acquisition of 1066 new functional PTM sites that were originally listed as having unknown function in the model (by virtue of their residence in UF-MAPs). Since scoring is based on alignment position, these new PTM data are coalesced into to 723 MAPs, which we found could be classified into two groups: those new functional sites that were already associated with KF-MAPs (Group I; 228) and those sites that had no previous link to function as members of UF-MAPs (Group II; 495) (Fig. 5a). As we show here, the median score for Group I MAPs was above the 0.95 recommendation threshold, supported by evidence from 1 to 3+ references (Fig. 5b, left). Moreover, new functional PTMs with more than one reference (from the updated dataset) were further enriched above the 0.95 threshold in most cases (red circles). Interestingly, in this study we discovered that some of the associated references for the new functional data were from as recent as 2018, suggesting that experimental redundancy within protein families (i.e., where functional studies have been reported for orthologous PTM sites within a protein family) is common and also probably not always well known to the experimentalist. In Group II, we found a similar trend wherein the bulk density of new data scored near or above the 0.95 recommendation threshold, especially for sites with more than one reference (red circles).

In summary, testing model performance on PTM sites that have undergone a UF to KF status change evaluates the ability of a model to correctly predict sites of functional modification before they have been demonstrated as such experimentally and is therefore an excellent means of model validation.

3.6.3 Precision/Recall Analysis of PTM Function Predictions Using Site-by-Site Experimental Data

Predictive machine learning models are often evaluated by precision/recall analysis. The definitions for precision and recall are as follows.

$$precision = \frac{TP}{TP + FP} \; \& \; recall = \frac{TP}{TP + FN}$$

where *TP* is true positives (e.g., KF-MAP that scores above recommendation threshold); *FP is* false positives (e.g., KF-MAPs that score below the recommendation threshold), and *FN* is false negatives (e.g., UF-MAPs that score above the recommendation threshold).

Problematically, publicly available PTM functional data rarely (*read: almost never*) curate negative data, and furthermore, false positive data are considered potential discoveries in functional PTM site prediction models! Thus, evaluating model precision and recall is inherently difficult for such models. This is complicated further by the fact that the database of functional PTM sites contains a mixture of publications varying in the depth with which they have experimentally proven PTM function. This has led to questions about the utility of KFSC-based model validation. On the contrary, we would argue that there is currently no better way to evaluate PTM function prediction models.

We find that negative PTM functional data is in fact not absent from the public domain, but rather simply is not actively curated. Consequently, such data could be included in the validation of SAPH-ire or other PTM function prediction models but requires manual curation. In the example shown here, we reviewed the Group II dataset (*see* Subheading 3.6.2) that was published within the last 5 years, seeking literature examples where multiple phosphosites in a protein were experimentally evaluated for functional effect and where both positive and negative data were included. A total of 23 peer-reviewed publications examining 90 distinct phosphosites result from this search (*see* **Note 26**). For each site, we review the published data and classify each as either positive or negative for functional impact, making sure to maintain consistency with the interpretations of the publishing authors and excluding 11 cases for which the published experiment was not site-specific (*see* **Note 27**) (Table S5). We then cross-reference these against PTM sites that are scored by SAPH-ire, using the previously established scoring thresholds to evaluate performance of the model. In the example shown here, we show that the model accurately called ~80% of the data based on a 0.95 score threshold (Fig. 6a). Model

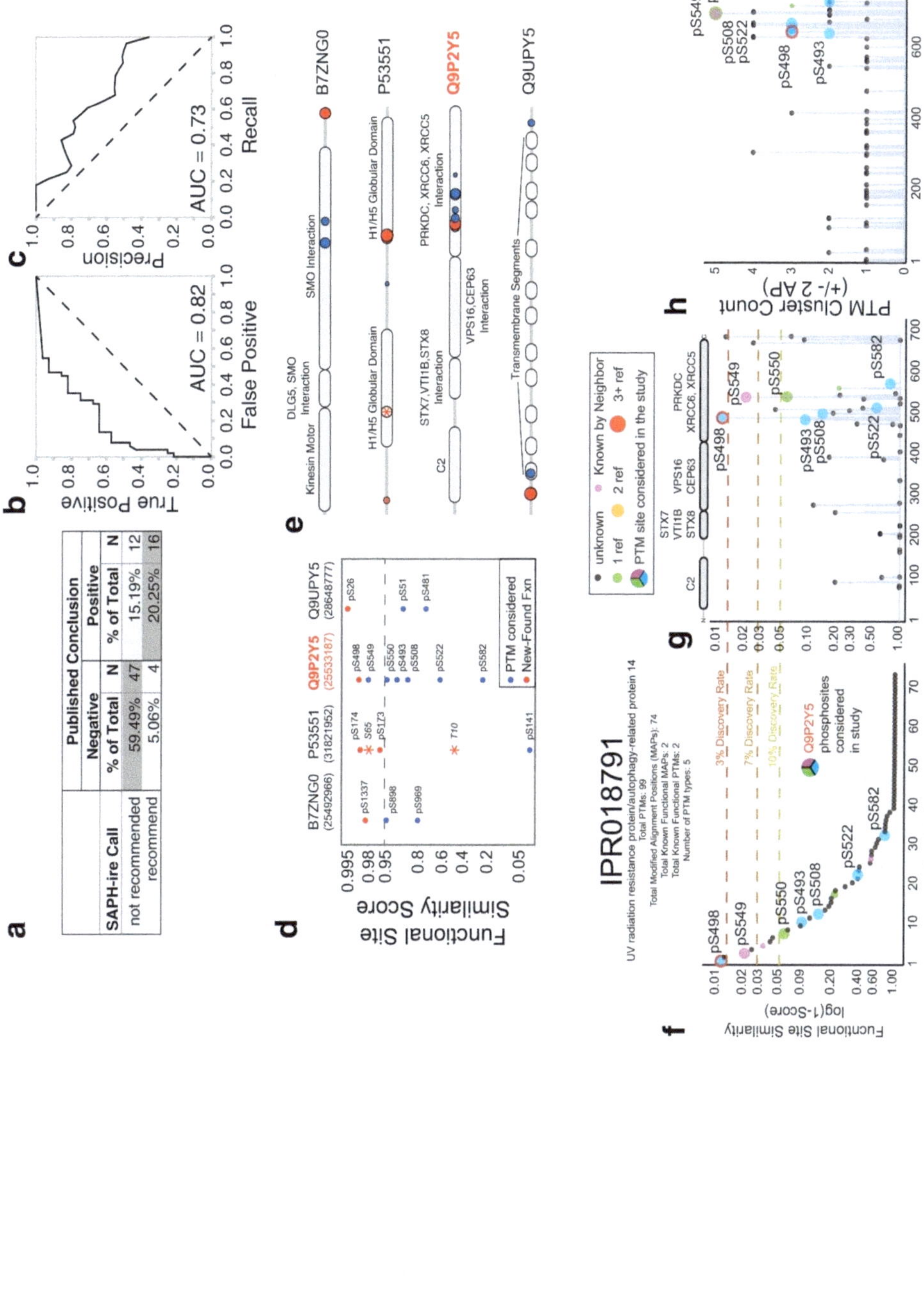

Fig. 6 Survey of model performance on a per-site basis. 79 phosphosites considered in publications from the last 5 years that had not been curated during development of SAPH-ire were manually investigated and the experimental results compared with the model's score. **(a)** Confusion matrix showing model call

Fig. 6 (continued) versus experimental result, where positive corresponds to functional impact as reported in the publication. (**b**) ROC curve and (**c**) precision-recall curve for logistic regression of model score versus the published conclusion on functionality. (**d**) Model scores for phosphosites examined in four example proteins (PMID in parentheses). Protein Q9P2Y5 is further examined in this figure. Asterisks correspond to phosphosites that aligned with positions studied in the protein but were not actually observed as phosphorylated in the protein. (**e**) Domain architecture of each protein from **d**, showing location of the examined phosphosites. Circle size reflects model score. Red-bordered circle corresponds to the MAP harboring the newly discovered function. (**f–h**) Output from https:/pearl.biosci.gatech.edu for protein Q9P2Y5 corresponding to human UVRAG showing: (**f**) rank order of MAPs harboring PTMs investigated in the study (labeled); (**g**) the model score for each PTM observed in the protein relative to its native position; and (**h**) PTM clustering across the family sequence alignment (this represents PTM cluster count described earlier (*see* Table 2)) performance was also deemed excellent by ROC curve analysis (AUCROC 0.82) and by precision-recall analysis (AUCPRC 0.73), which specifically takes into account true negative data (Fig. 6b, c). In general, we have found that successfully recommended (*or rejected*) sites are accurately called due to proper recommendation by SAPH-ire or due to lack of any evidence of modification associated with the phosphosite in question. For example, some sites found to be nonfunctional or sites specifically not chosen for testing by the authors [by rational design] have not been detected previously (by mass spec or other means) and are therefore not found within the SAPH-ire dataset. Interestingly, in some cases, we also find site(s) that are recently published as functional that could have been classified as such by alignment alone, either through neighboring functional residues or direct sequence alignment with known functional PTMs associated with other protein family members (i.e., function association by alignment) (Table S5).

We also recommend evaluating PTM function prediction models on a per-protein basis wherein both positive and negative data are taken into account. This type of analysis is unique from metadata analyses, which do not provide details on performance for individual proteins where regulation by PTMs is more constrained. In this example, we chose four examples to highlight exemplary multisite PTM investigations (i.e., rigorous experimental studies): Mouse kinesin-like protein Kif7 (B7ZNG0), yeast linker histone Hho1 (P53551), human UV radiation–resistant protein UVRAG (Q9P2Y5), and human cysteine/glutamate transporter SLC7A11 (Q9UPY5). Within this set, a total of 18 sites are considered, of which SAPH-ire scores are observably consistent with the published report of functional impact in 16 cases (89%) (Fig. 6d). Evaluated sites were found both in and outside domains within these proteins, and in most cases, the sites were also clustered in relatively close proximity, making it difficult to predict function based on modification density alone (Fig. 6e). This was demonstrated in human UVRAG, in which seven independent phosphosites cluster within the PRKDC/XRCC5/6 interaction domain. Within this cluster, phosphorylation of UVRAG-S498 by MTORC1 regulates autophagy [49]. SAPH-ire reveals that this site falls within the highest scoring MAP within its protein family (IPR018791) (Fig. 6f), as well as the highest scoring PTM in the protein (Fig. 6g). It is also nested within a dense cluster of PTMs that align within the family—further emphasizing this region as a PTM hotspot for other member proteins (Fig. 6h). Interestingly, pS549 also fell above the 0.95 recommendation threshold but was excluded for testing by the authors. This site neighbors pS550 (score 0.943) that was determined to be functionally important for autolysosomal tubulation later that same year [50].

3.6.4 Experimental Validation

A rigorous method for validating functional PTM site prediction models is to study them experimentally (*see* **Note 28**). Phosphorylation, which is readily detectable by a wide variety of methods and which can be functionally studied through phospho-nullifying or mimicking mutations, is the easiest PTM type to study. However, experimental evaluation also requires a robust functional assay that can be used to quantify differential effects of phospho-mimic versus null mutants or an in vitro system to evaluate the effect of site-specific phosphorylation on structure–function relationships. We have directly tested SAPH-ire based predictions of PTM site function for proteins of interest, previously [51]. In this case, we systematically evaluated SAPH-ire-scored phosphosites in the yeast linker histone protein, Hho1. Five distinct phosphosites were tested for their ability to impact DNA double strand break (DSB) repair, which is regulated in part by Hho1 and for which well-established quantitative assays are readily accessible in the yeast model system [51]. Importantly, we evaluated sites that spanned the entire scoring range so as to evaluate scoring accuracy. Three sites were experimentally observed on the yeast protein and two were aligned with phosphosites in other family members. Three of these sites (pS174, pS65, and pS173) were scored above the 0.95 recommendation threshold, each of which was found to negatively affect DNA repair in their phospho-null form (Ser-to-Ala mutation) but not in their phospho-mimic form (Ser-to-Glu) in support of a functional requirement for phosphorylation at these positions. In contrast, phospho-null or mimic mutations to S141 were found to have no impact on repair. The functional effect of phospho-null mutants on DSB repair was also found to correlate well with the functional similarity score in this study (Fig. 7) (*see* **Note 29**).

3.7 Additional Mechanisms for Model Validation

Several additional mechanisms are available for validation of PTM function prediction models, which like those described above, involve the relation of PTM sites and model predictions with independently determined functional data, such as disease-causing mutations or short linear motifs. However, any functional data source that experimentally links protein function or organismal fitness to precise amino acid residues within a protein can be a useful source for prediction model validation. For various reasons, we view these as good but not ideal validation mechanisms and have therefore grouped them separately here.

3.7.1 Disease-Linked Variant Correlation

A growing collection of pathogenic mutations can be identified using free online resources such as ClinVar (Table 1) or the HGV database (https://hgv.figshare.com), among other sources [52, 53]. These resources contain thousands of rigorously curated and validated human disease-linked variants classified by their pathogenicity, which serves as a proxy for the functional importance of any PTM that would be disrupted by such genetic variants (*see*

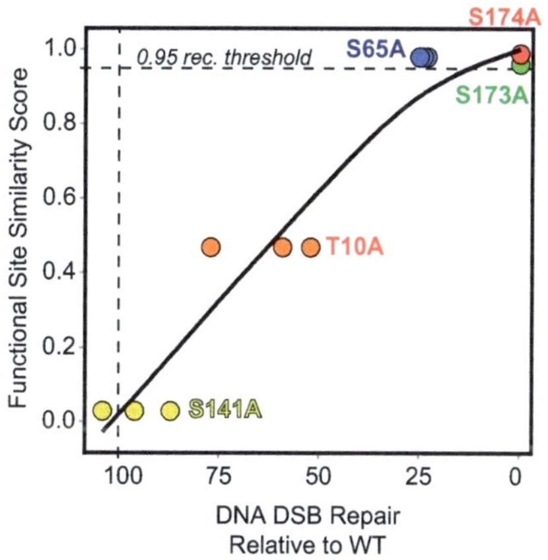

Fig. 7 Experimental evidence reveals correlation between functional prediction scores and DNA repair response for nullified phosphosites in yeast linker histone Hho1 (P53551). Relative DNA double strand break repair responses for five nullified phosphosites in Hho1 (ser/thr to ala) relative to functional site similarity score. Circles represent results from replicate analysis of each site as described in the cited article [51]

Note 30). Consequently, one should expect that their PTM function prediction model would enrich pathogenic but not benign mutations. However, careful considerations regarding the inherent bias contained within these datasets (e.g., arginine is more commonly mutated) should be considered before assumptions are made as to the correlation between model scoring and disease linkage, which have been carefully described elsewhere [18] (see **Note 31**).

3.7.2 SLiM Association

Functional enrichment analyses, as shown for pathogenic mutations (*see* Subheading 3.6.1), may also be carried out with other sources of functional data, such as Short Linear Motifs (SLiMs) curated by the Eukaryotic Linear Motif (ELM) resource for functional sites in proteins (Table 1) [21]. SLiMs are comprised of short protein sequences that have been validated as functional elements across several proteins. ELM contains ~289 motif classes clustered into six motif categories based on functional assessment of 3523 validated instances. However, when filtered for direct motif/PTM correspondence in the SAPH-ire dataset, we have found that very few instances (<300) can be retrieved (*see* **Note 32**). Thus, we recommend using ELM-designated regular expressions to search for additional motif examples within a PTM site dataset. However, this should be done with caution as it can result in significant false positive motif instances. In the example shown here, we took the following steps to identify PTM-linked SLiMs whilst minimizing false positive SLiM instances.

1. Identify the maximum SLiM length and use this length to define a sequence window size surrounding each PTM site in the dataset. A window size of ±18 residues surrounding the PTM site was used in the example described here.
2. Scan each windowed sequence for amino acid sequences matching the regular expression patterns provided by ELM.
3. Filter out any windows where the PTM site does not overlap with the regular expression (*see* **Note 33**).
4. Now, filter the resulting windows by keeping only those instances with greater than 40% occurrence using strict alignment position boundaries—as prescribed by ELM curators previously [54] (*see* **Note 34**). Only the regular expression sequence match is used for this filter and *not* the ±18 residue window.

Once identified, the PTM-associated SLiM data can be evaluated by how effectively the model recommends PTM sites that are part of functional SLiM regular expressions. In the example shown here, we found that the SAPH-ire model recalls 86% of SAPH-ire KF-MAPs that are associated with functional SLiMs (based on the 0.95 recommendation threshold), while 67% of UF-MAPs that are associated with functional SLiMs are recommended as likely functional (Fig. 8).

3.8 Interpreting How Machine Learning Models Arrive at their Predictions—LIME

Most machine learning models (of any type and function) are not designed with interpretability constraints and instead are focused on achieving high accuracy and precision [55]. How such models arrive at their final predictions is therefore often referred to as a "black box," a historical term which is meant to say that humans are

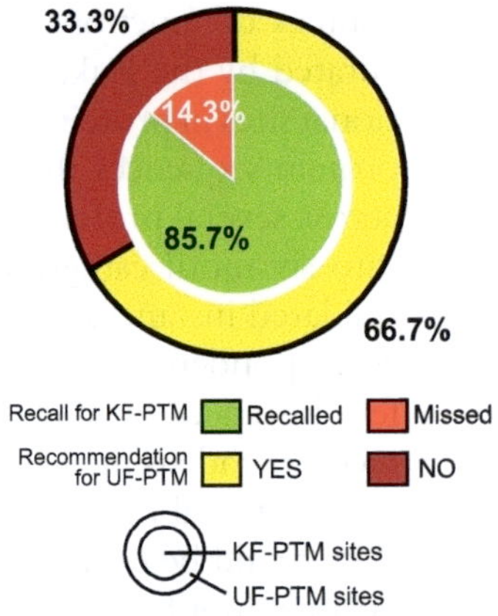

Fig. 8 Model recall and recommendation comparison for PTMs associated with validated functional SLiMs from the ELM resource

not fully aware of how the machine makes the prediction. Moreover, model predictions are necessarily based on combinatorial feature contributions not readily identifiable without a means to establish the importance of each feature in a prediction. However, recent advances geared toward improving machine learning model interpretation have begun to tackle this problem. This is particularly useful to consider with respect to PTM function prediction as it provides a way to glean commonalities in the features of functional versus nonfunctional modifications.

Local Interpretability Model Explanation (LIME) provides a mechanism for determining the relationship between model input features through linear estimates of the weight that each feature contributes to each prediction made by the PTM function prediction model (in this case, SAPH-ire) [56]. To understand how the SAPH-ire model balances quantitative feature content to make predictions about MAPs, and to determine if it is overly reliant on any single feature, we have used LIME as follows.

1. Pass model input feature values and model prediction scores through LIME (Table 1). Retrieve linear estimates for each feature and prediction.
2. Cluster the data using the LIME linear estimates (we recommend using normal mixtures clustering). This can be done using a variety of statistical programs. We use JMP (SAS). Principal components are derived automatically from normal mixtures clustering in JMP.
3. Identify principal components that contribute the greatest separation between the clusters.
4. Deconvolute the principal components into their eigenvector values. These values convey a broad sense of the contributions from each feature that contribute to the distinctions between clusters (*see* **Note 35**) and indicate whether a single or several features contribute to the scoring mechanism.

In the example shown here, we performed LIME analysis of a 29,859 sample of SAPH-ire MAPs that organized into three clusters (Fig. 9a). Notably, cluster 2 is represented mostly by UF-MAPs while clusters 1 and 2 are represented by KF-MAPs. The two first principle components of each cluster are plotted to reveal distinctions between each (Fig. 9b). Recognizing that the greatest variability occurs in PC1, and readily distinguishes clusters 1 and 3 from cluster 2, we deconvoluted the component to reveal how individual features contribute to this variability, revealing that no single feature dominates the prediction field but that several features make large contributions to the prediction (Fig. 9c).

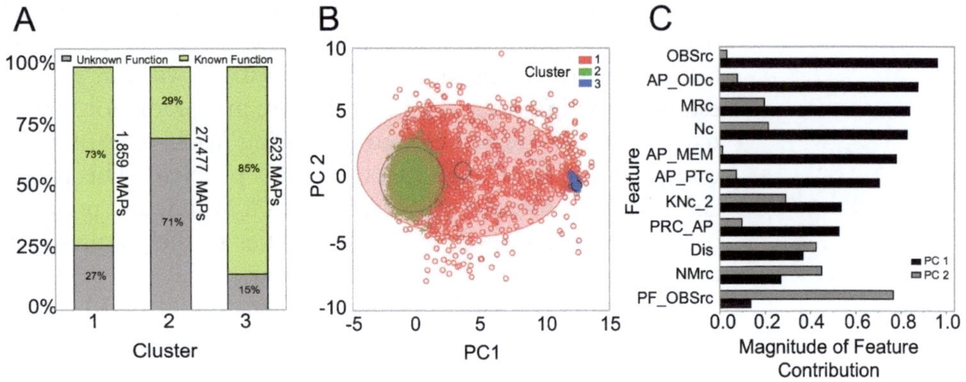

Fig. 9 Exploring the model's interpretation of feature space using LIME. Normal mixtures clustering of LIME analysis data from 29,859 representative MAPs. (**a**) Percentage of MAPs within normal mixture clusters that are known to contain a functional PTM. Clusters 1 and 3 harbor mostly KF-MAPs compared to Cluster 2, which has the fewest. (**b**) Clusters projected onto two of their principal components. (**c**) Magnitude of each feature within the principal components shown in B (*see* Table 2)

3.9 Continuing Challenges and Additional Considerations When Developing PTM Function Prediction Models

3.9.1 What is a Functional PTM?

Ironically, one of the major challenges in the field of PTM function prediction is in defining what is a functional PTM in the first place. Surveying the database of functional PTM literature reveals that what defines a functional PTM depends on the level of the observer. An observer at the cellular level may define a functional PTM site as one that disrupts a measurable cellular phenotype when blocked (e.g., a phospho-null mutation) or precisely introduced (e.g., a phospho-mimic mutation). An observer at the molecular level may define it as one for which disruption produces measurable changes in proteome behavior (e.g., a change in signaling pathway responses measured at the molecular level). An observer at the single molecule level may define it as one for which a protein structural change (from secondary to quaternary) can be detected. While samples of each of these examples can be found in the literature, they are by no means equivalent.

Experimental analysis of PTM function is difficult and context dependent. Consequently, the quality of experimental results supporting the claim of PTM functionality vary widely. In some cases we have found that multiple PTM sites have been classified as functional wherein their mutation together, but not apart, produces an observable effect. However, this type of information is incongruent with site-specific function predictors and will likely impact their accuracy. In other cases, we have found that functional PTM sites have been claimed despite lacking proper control experiments to demonstrate the potential impact of the PTM, specifically. For example, some reports document measurable effects of phospho-null mutation without also attempting to rescue behavior with phospho-mimic mutations. Such studies are inadequate for distinguishing the effect of the amino acid substitution itself versus the effect of possible phosphorylation at the site. In other cases, putative sites of PTM are mutated without any evidence of modification occurring at the site.

We recognize that unique challenges are associated with every PTM functional analysis and that new knowledge provided by a wide range of experiments at different levels of observation can be valuable. We suggest that systematically encoding the limitations/quality of curated function studies would likely improve the quality of function prediction models (*see* **Note 36**).

3.9.2 Inherent Biases

A predictive model can only be as good as the data used in its creation. PTM function prediction is a relatively new and rapidly evolving endeavor that is subject to several limits and unavoidable biases that should be considered during their development, application, and interpretation. First, PTMs that are easier to detect are more frequently observed and curated than those that are not. Phosphorylation is by far the most commonly observed PTM due to several factors including its evolved importance in controlling proteome/cellular behavior but also because it is small and methods for its analysis by high throughput mass spectrometry are highly accessible and relatively simple to accomplish [57]. Second, the population size of a given PTM type can change within the time frame of a single publication. For example, before 2011 ubiquitination was listed as the eighth most populous PTM by dbPTM. However, technology advances in ubiquitination site detection by mass spec revealed thousands of newly discovered ubiquitination sites and the PTM quickly jumped to second place in the table of PTM frequencies [58]. Similar trends have been observed for lysine acetylation [59], N- an O-glycosylation [60], among other PTMs wherein improved methods of enrichment and detection have dramatically changed our view of the PTM landscape. Thus, observation frequency of certain PTMs will change rapidly over short periods of time.

A third form of bias common in PTM function prediction comes from the fact that functional studies are often initiated on PTMs that represent "low hanging fruit" of the modified metaproteome, wherein a PTM's biological impact is indicated by a single predictive feature that does not necessarily require a predictive model. Such low hanging fruit is represented by PTMs localized within catalytic centers [16], that exhibit high observation frequencies, or are neighboring PTM sites that have already been established as functional [17, 26]. The tendency to study PTMs with high observation frequency has been propagated in-part by resources such as PhosphoSitePlus, for example, which utilizes observation frequency as a primary ranking feature for each PTM in their dataset. Consequently, the database of functional PTM sites becomes overpopulated with PTMs exhibiting this feature. Indeed, we have found that observation frequency is 6.8-fold greater for PTMs with known function than for those without evidence of function. As a heuristic model that includes a broad range of intrinsic as well as community wisdom features like observation frequency, SAPH-ire does show bias toward PTM sites with higher observation frequency; however, we have also found that high observation frequency is not a required feature for high scoring predictions [26]. This is likely due to the fact that community wisdom likely also entails evolutionary conservation and other sequence features that guide experimentalists toward functional PTM sites. Thus, the inclusion of a broad scope of input features in heuristic models can enable predictions based on community wisdom or intrinsic features, but should be carefully interpreted since community wisdom can and does change over time (Fig. 9).

We also find that there is considerable redundancy in the experimental analysis of PTMs such that if one site is shown to be functional, additional publications emerge showing the functionality of neighboring sites in the same protein or aligning PTM sites in different proteins (i.e., comparison of human and mouse versions of the same protein) (*see* Subheading 3.6.3). The SAPH-ire model reveals such redundancy by virtue of its dependence on protein family sequence alignment. The model also actively attempts to exploit this feature (KNc) by inclusion in the prediction. Indeed, such an inclusion ensures that nearby SAPH-ire MAPs harboring evidence of functionality drive predictions higher so as to highlight the functional importance of local protein residues and to provide indicators of the potential for redundancy in functional analysis. As we have shown here (Fig. 9) and previously [26], inclusion of this positive data is not required for a positive recommendation, nor is it even the most impactful feature in the prediction (Fig. 9c).

3.9.3 Negative Data

Improvements to PTM function prediction would likely also be gained through systematized documentation and curation of negative data. Indeed, one of the major limitations in model training and performance evaluation stems from the lack of negative data (i.e., PTM sites that are nonfunctional). This is not a trivial challenge to overcome since negative data are rarely published and, as the saying goes: "absence of evidence is not evidence of absence" (i.e., just because a PTM site appears to have no functional importance under one condition does not mean that it will be irrelevant under all conditions). Despite these challenges and concomitant with the systematic labeling of PTM functional data, we suggest that improving the mechanisms for reporting and curating negative functional evidence is just as important.

3.9.4 PTM Stoichiometry and Dynamics

PTMs are inherently sub-stoichiometric, meaning that for every 100 protein copies in a cell, a site-specific PTM may only occur on two of them. Moreover, PTM stoichiometry can change in response to different cellular conditions. For example, the stoichiometry of negative feedback phosphorylation on serine-7 of the yeast G protein gamma subunit, Ste18, increases by ten-fold from ~8% to ~80% upon stimulation of the yeast pheromone pathway [61]. Modification stoichiometry and dynamics are suggested to be important indicators of functional PTM sites [25]. However, neither form of data is readily or routinely accessible currently, largely because they are laborious to properly quantify experimentally [62], and the data is difficult to curate systematically. Furthermore, many proteomics studies do not included controls that would be necessary to estimate stoichiometric PTM changes by mass spectrometry. Thus, improved curation and quality control mechanisms for PTM stoichiometry and dynamics data curation will likely allow the inclusion of such features and the accuracy benefits they could add to PTM function predictive models.

3.9.5 PTM and Enzyme Interactions—Functional Association

In some cases, it has been shown that PTMs on different sites coordinately drive functionality. The so-called functional association between PTMs has been observed in many cases [11, 13], but is perhaps best represented by a well-known interaction between phosphorylation and ubiquitination. Phosphorylation sites (serine, threonine, or tyrosine) occurring within five residues of ubiquitination sites (occurring on lysine) are more likely to be functionally associated than sites beyond the 5 residue distance [9]. This is inherently tied to their roles in phosphorylation-mediated ubiquitination and protein degradation—a common mechanism of protein abundance regulation in eukaryotes that is responsible for cell cycle and other forms of cellular control [63]. Another well described example is the functional association between site-specific epigenetic modifications on histone N-terminal tails that control chromatin remodeling and gene expression [64]. Models that exploit PTM functional association data for prediction continue to evolve and, with increasing data availability and in combination with improvements in data curation and experimental analysis, will likely have a major impact on our understanding of how to identify functional PTM sites.

Longstanding evidence also indicates that the likelihood of PTM site functionality is higher for sites that can be clearly associated with specific modifying enzymes [9]. Indeed, in and of itself, this relationship could be used as a predictive feature. Thus, it is also possible that predictions of enzyme substrate relationships associated with a given PTM site may also enhance functional prediction accuracy [22–24].

4 Notes

1. While the SAPH-ire model described here is PTM agnostic, it should be noted that restriction of the method to a single PTMtype that is highly abundant (e.g., phosphorylation) is acceptable and may enhance accuracy. However, the method will likely not work if the data is restricted to PTMtypes of lower abundance as this would compromise model accuracy. Model development and training must occur after the choice of PTMs to include is made.
2. Currently there is no other refined source of curated PTM data, which is a major limitation. PTM data curation requires expertise in mass spec data curation and automated efforts to curate PTM data from mass spec data do not currently exist.
3. SAPH-ire can be extended to the analysis of domain families, although the reader should be aware that additional challenges arise in this case due to the repetition of domains within single proteins and the breadth of protein types contained within a single domain family (due to the existence of the domain in several different protein families). To date, SAPH-ire has only been published for whole protein families.
4. Not all modified proteins are necessarily linked to a protein family in InterPro. If a protein is missing it is likely not yet been assigned a family.
5. Methionine is often cleaved from many proteins and we have found through experience with PTM repositories like PhosphoSitePlus that the native position is not accurately reported in all cases. A simple BLAST-based screen of PTM sites that validates the accuracy of residue identity for each PTM site will overcome this limitation.
6. Improved methods of alignment and alignment validation continue to emerge, and could positively impact alignment-based models like SAPH-ire [65, 66].
7. Multiple sequence alignment is a cornerstone of functional PTM site prediction algorithms and the alignment approach should be chosen carefully at the onset of model development as changes in alignment will result in changes in model training and scoring. In considering the approach to take, one should consider recent evidence that more accurate alignment of a protein family (i.e., an alignment that reflects the true phylogenetic relationship between amino acids in each protein family member) is best achieved through alignment of most if not all of its corresponding members and through ensemble methods that utilize more than one alignment algorithm [65]. However, one should also recognize the hidden caveats of such

approaches, such as the inclusion of nonreviewed protein sequences (sequences that are not necessarily verified open reading frames), the tremendous size of the alignments (often thousands upon thousands of member sequences), and much longer compute times. For larger families, the membership of whole sequence protein families is also often extremely diverse and leads to major gaps in the alignment that introduce significant noise in alignment space, which may be avoidable through analysis of subfamilies. Domain families, which are more well-conserved on a per alignment position basis compared to whole sequence protein families, can mitigate some of these problems but also produce unique challenges, such as the exclusion of PTM sites not found within domains. This is a rapidly evolving methodology and care should be taken to investigate the state-of-the-art approaches at the time of study. As a compromise to these caveats and as a heuristic model, the SAPH-ire method described here aligns only modified sequences. Further work exploring the limiting returns of full protein family alignments and the "sweet spot" for model accuracy is an ongoing endeavor in the field.

8. Feature data from individual PTM sites is highly limiting and therefore aggregate data (as in MAPs) provide greater numbers and dynamic range of feature information that can be used for prediction. For example, PTM sites by themselves harbor no information about the conservation of the modification site across species – a feature found to be correlated with functional impact [8, 67].

9. OBSrc corresponds to the number of literature and mass spec observations associated with the PTM. These data are available from PhosphoSitePlus or are otherwise calculated for other PTMs collected from alternative PTM site resources if available. The minimum value is 1.

10. In some cases, a single UID may map to more than one protein family (e.g., local and super protein families may result in nested data). Our current perspective is that this is ok as each family will be different—composed of some overlapping but largely unique sets of protein sequences. Consequently, predictive scoring will be slightly different and only lend additional perspective about the function potential of the PTM site.

11. It is imperative that redundancy be eliminated to avoid inaccurate feature estimates in downstream steps.

12. Some PTM sites will be lost in this join step. Thus, only PTMs that can be aligned in a protein family with at least two modified members will make it through. Despite this, we have found that the bulk of PTM site data generally make it through this step. Also, note that some UID-NPs will have redundancy

due to different PTM type observation of the same residue (e.g., ubiquitination, acetylation, methylation, citrullination of lysine). Finally, we verify that the residue identity matches with the type of PTM called for the site (e.g., phosphorylation is allowed on S,T,Y).

13. As an example, for an alignment position in which ubiquitination and phosphorylation are both found to occur (meaning that both phosphorylatable (S/T/Y) and ubiquitinatable (K) residues are present in the AP) the PRC-AP is the sum of positions harboring S/T/Y/K relative to the total number of members in the alignment (i.e., contributors to the AP). Dashes (empty cells in the AP) are included in this calculation. Also note that PRC is only calculated for APs that harbor at least one PTM site (most APs are unmodified and filtered out in the next step).

14. Recall that nrPTMs table will have some redundancy for PTM-IDs (e.g., lysine modifications), which necessitates that you first tabulate a sumOBSrc for each PTM-ID before joining to the deconvoluted MAPs table.

15. Absolute observation source count (a measure of PTM observation frequency) is directly dependent on how many members fall into the family alignment and normalization is important. Nonnormalized data is also taken as it reflects an absolute scaling of observation that is also viewed as important in the context of the collective dataset. Also note that any of the features described herein can be excluded if so desired, although we have found that inclusion of all the described features provides the overall best performance [26].

16. Accuracy is the rate of correctly labeled true positives and true negatives. If we take KF-MAPs to be true positives and UF-MAPs to be true negatives, then the SAPH-ire dataset described in this chapter would have a 1:84 ratio of true positives to true negatives and the model would be biased to label more MAPs as false since the distribution itself is weighted in that direction (during training).

17. Tensorflow is a robust toolkit for the implementation of neural networks. One can easily customize the input function batching, cost function, the backpropagation algorithm, the weight matrix initialization, and what metrics are reported between runs. However, when working through a PSO it is important to start with the few parameters that will have the most significant impact on model performance, then fine tune the others.

18. These activation functions were implemented via the layers API in the high level keras interface of TensorFlow. Each activation function has its own parameters that can be adjusted and throughout the iterative process in the development of

SAPH-ire many of these parameters were adjusted. Since model creation, the interface has been updated and more than a dozen new activation functions have been introduced. They can be found here: https://www.tensorflow.org/api_docs/python/tf/keras/activations.

19. The possible parameters to vary over PSO is vast. Throughout our iterations we varied: the number of layers, the activation functions, the parameters that define the shape of those activation functions, the cost function, the backpropagation method, and finally the weight initialization method. As TensorFlow has become more refined, additional parameters that can be varied have been introduced, and we believe it is worth exploring them before beginning a PSO approach.

20. Deciding what level of granularity to vary the parameters with is its own challenge. As shown in Fig. 1a, simply selecting the type of activation functions is one level of granularity; however, the terms of the activation function can also be varied for increased depth. Progressively expanding the search space while keeping the number of models constrained is useful for efficient training, which is why we decided on a soft cap of 512 models.

21. Deciding when to stop further iterations depends on the perceived benefit of adjusting certain parameters. TensorFlow holds many of the low-level parameters (weight initialization, backpropagation method, and weight connectivity) to standard defaults as they have been shown to be effective in most situations. These are parameters that can be modified as part of a PSO; however, they typically have less of an impact than the more accessible parameters like activation functions, cost function, and layer count.

22. Dozens of different options are available for weight initialization of neural network layers using TensorFlow (https://www.tensorflow.org/api_docs/python/tf/keras/initializers). For this example, we sampled Normal, Uniform, Glorot Normal, Glorot Uniform, RandomNormal, Random Uniform, Zeros and Ones. All implemented with the default parameters assigned by TensorFlow. For Glorot and Random initializations this meant the seed for each particle varied.

23. The feature set for the two models differed in that, Model 1 (from batch 1) takes in 10 features without preprocessing, allowing it to use the network architecture to scale the inputs globally. In contrast, Model 2 (from batch 2) uses 11 features wherein OBSrc is normalized by PF_MEM for each MAP, serving as the additional feature while the other 10 features normalized globally. The output of both models is averaged together to give the final model score.

24. The list of publications is too long to present here. For reference, antibodies that recognize the activation sites in MAPKs Erk1 and Erk2 have been used in nearly 4400 different papers at the time of this publication (p44/42 antibody; *cellsignal.com*).
25. Even despite these several validation efforts, the reader should realize that no model is perfect and some bias is inevitable in the building of PTM function prediction models (*see* Subheading 3.6).
26. Phosphosite functional data is by far the most commonly encountered and is particularly useful because it can be studied genetically through phospho-mimic versus phospho-null mutations for which opposite functional effects are expected.
27. In some cases we found that functional evidence was convoluted across multiple modification sites that were mutated in combination. While all of the sites in these cases may in fact be necessary for the functional effect, we have eliminated them from precision/recall analysis described here since SAPH-ire does not consider the function potential of more than one position at a time.
28. It is arguable that more systematic surveys across several proteins whose PTM regulation is newly discovered is actually better than any one experimental study since it stems from several independent sources and is therefore unbiased toward any one protein or functional assay.
29. It should not be expected that a functional PTM site prediction model correlate with the quantitative degree of functionality observed for the protein in question, although if the functional assay and the biological phenotype in question can capture a linear relationship between site and functional effect, this is certainly an ideal system to use for model evaluation.
30. Considering the complexity of human disease and its link to genetic variants, it is often difficult to say whether a single site variant is truly the cause of a disease in all cases. If there are in fact additional mutations that must be coupled with a variant in question to achieve the disease state, this information (which is not available), can easily skew the correlation between PTM function prediction scores and pathogenic variants. This is a major reason why correlating PTM function predictions with pathogenic mutation is considered a less-rigorous validation mechanism.
31. Correlation should be restricted to SNPs only to minimize the possibility (as could be seen for some indel mutations) of relating model scores (which are based on a single protein residue) to diseases caused by gross gene mutations that may affect more than the PTM site in question.

32. Restriction to only ELM-validated functional sequences results in a small number of associations (<300), which is why this mechanism of validation is considered less than ideal. However, expansion of this list through regular expressions can provide greater survey space.
33. We allow for the PTM to occur at any position within the SLiM. The rationale for this is that SLiMs, by definition, are evolutionarily conserved and just as mutation of such an element could disrupt its function, so too could PTM of any residue within the element.
34. Sequence alignment, if not already built into the PTM prediction tool, should be carried out before this step as it is a requirement for filtering SLiM regular expressions as prescribed by the ELM resource.
35. We have observed that clusters fall into groups that are enriched either in KF-MAPs (i.e., positive class) or UF-MAPs (largely a negative class) and this natural clustering outcome has allowed us to detect feature contributions that tend to contribute to accurate model predictions.
36. PhosphoSitePlus does in fact curate some details about the type of functional effect observed for reported PTM sites, including categories such as: protein function, process, protein interact, other interaction. These can be very helpful tags when the field is populated.

Acknowledgments

Special thanks to Henry Dewhurst who was the first to encode the SAPH-ire method for broad application. Special thanks to Dr. Peng Qiu of Georgia Institute of Technology for critical review of the manuscript and contributions to model design. This work was funded by National Institutes of Health R01 GM117400 and to M.T.

References

1. Prabakaran S, Lippens G, Steen H, Gunawardena J (2012) Post-translational modification: nature's escape from genetic imprisonment and the basis for dynamic information encoding. Wiley interdisciplinary reviews. Syst Biol Med 4:565–583
2. Csizmok V, Forman-Kay JD (2018) Complex regulatory mechanisms mediated by the interplay of multiple post-translational modifications. Curr Opin Struct Biol 48:58–67
3. Gibson DG, Glass JI, Lartigue C, Noskov VN, Chuang R-Y, Algire M et al (2010) Creation of a bacterial cell controlled by a chemically synthesized genome. Science (New York, NY) 329:52–56
4. Reimand J, Bader GD (2013) Systematic analysis of somatic mutations in phosphorylation signaling predicts novel cancer drivers. Mol Syst Biol 9:637
5. Reimand J, Wagih O, Bader GD (2015) Evolutionary constraint and disease associations of post-translational modification sites in human genomes. PLoS Genet 11:e1004919

6. Chen C, Huang H, Wu CH (2017) Protein Bioinformatics Databases and Resources. Methods Mol Biol 1558:3–39
7. Pascovici D, Wu JX, McKay MJ, Joseph C, Noor Z, Kamath K, Wu Y, Ranganathan S, Gupta V, Mirzaei M (2018) Clinically relevant post-translational modification analyses-maturing workflows and bioinformatics tools. Int J Mol Sci 20(1):16. https://doi.org/10.3390/ijms20010016
8. Landry CR, Levy ED, Michnick SW (2009) Weak functional constraints on phosphoproteomes. Trends Genet 25:193–197
9. Beltrao P, Albanèse V, Kenner LR, Swaney DL, Burlingame A, Villén J, Lim W, Fraser JS, Frydman J, Krogan NJ (2012) Systematic functional prioritization of protein posttranslational modifications. Cell 150:413–425
10. Strumillo MJ, Oplová M, Viéitez C, Ochoa D, Shahraz M, Busby BP, Sopko R, Studer RA, Perrimon N, Panse VG, Beltrao P (2019) Conserved phosphorylation hotspots in eukaryotic protein domain families. Nat Commun 10: 1977
11. Minguez P, Parca L, Diella F, Mende DR, Kumar R, Helmer-Citterich M, Gavin A-C, van Noort V, Bork P (2012) Deciphering a global network of functionally associated post-translational modifications. Mol Syst Biol 8:599
12. Minguez P, Letunic I, Parca L, Bork P (2013) PTMcode: a database of known and predicted functional associations between post-translational modifications in proteins. Nucleic Acids Res 41:D306–D311
13. Minguez P, Letunic I, Parca L, Garcia-Alonso-L, Dopazo J, Huerta-Cepas J, Bork P (2015) PTMcode v2: a resource for functional associations of post-translational modifications within and between proteins. Nucleic Acids Res 43: D494–D502
14. Swaney DL, Beltrao P, Starita L, Guo A, Rush J, Fields S, Krogan NJ, Villén J (2013) Global analysis of phosphorylation and ubiquitylation cross-talk in protein degradation. Nat Methods 10:676–682
15. Dewhurst HM, Choudhury S, Torres MP (2015) Structural analysis of ptm hotspots (SAPH-ire)--A quantitative informatics method enabling the discovery of novel regulatory elements in protein families. Mol Cell Proteomics 14:2285–2297
16. Johnson JR, Santos SD, Johnson T, Pieper U, Strumillo M, Wagih O, Sali A, Krogan NJ, Beltrao P (2015) Prediction of Functionally Important Phospho-Regulatory Events in Xenopus laevis Oocytes. PLoS Comput Biol 11:e1004362
17. Torres MP, Dewhurst H, Sundararaman N (2016) Proteome-wide Structural Analysis of PTM Hotspots Reveals Regulatory Elements Predicted to Impact Biological Function and Disease. Mol Cell Proteomics 15:3513–3528
18. Holehouse AS, Naegle KM (2015) Reproducible analysis of post-translational modifications in proteomes--application to human mutations. PLoS One 10:e0144692
19. Li S, Iakoucheva LM, Mooney SD, Radivojac P (2010) Loss of post-translational modification sites in disease. Pac Symp Biocomput:337–347. https://doi.org/10.1142/9789814295291_0036
20. Xue B, Jordan B, Rizvi S, Naegle KM (2021) KinPred: A unified and sustainable approach for harnessing proteome-level human kinase-substrate predictions. PLoS Comput Biol 17: 1–20
21. Kumar M, Gouw M, Michael S, Sámano-Sánchez H, Pancsa R, Glavina J, Diakogianni A, Valverde JA, Bukirova D, Čalyševa J, Palopoli N, Davey NE, Chemes LB, Gibson TJ (2020) ELM-the eukaryotic linear motif resource in 2020. Nucleic Acids Res 48:D296–D306
22. Patrick R, le Cao KA, Kobe B, Boden M (2015) PhosphoPICK: Modelling cellular context to map kinase-substrate phosphorylation events. Bioinformatics 31:382–389
23. Linding R, Jensen LJ, Pasculescu A, Olhovsky M, Colwill K, Bork P, Yaffe MB, Pawson T (2008) NetworKIN: a resource for exploring cellular phosphorylation networks. Nucleic Acids Res 36:D695–D699
24. Wang C, Xu H, Lin S, Deng W, Zhou J, Zhang Y, Shi Y, Peng D, Xue Y (2020) GPS 5.0: An Update on the Prediction of Kinase-specific Phosphorylation Sites in Proteins. Genomics Proteomics Bioinformatics 18: 72–80
25. Prus G, Hoegl A, Weinert BT, Choudhary C (2019) Analysis and interpretation of protein post-translational modification site stoichiometry. Trends Biochem Sci 44:943–960
26. Dewhurst HM, Torres MP (2017) Systematic analysis of non-structural protein features for the prediction of PTM function potential by artificial neural networks. PLoS One 12: e0172572
27. Xiao Q, Miao B, Bi J, Wang Z, Li Y (2016) Prioritizing functional phosphorylation sites based on multiple feature integration. Sci Rep 6:24735
28. Ochoa D, Jarnuczak AF, Viéitez C, Gehre M, Soucheray M, Mateus A, Kleefeldt AA, Hill A, Garcia-Alonso L, Stein F, Krogan NJ, Savitski MM, Swaney DL, Vizcaíno JA, Noh K-M,

Beltrao P (2020) The functional landscape of the human phosphoproteome. Nat Biotechnol 38:365–373

29. English N, Torres M (2019) SAPH-ire TFx: A Machine Learning Recommendation Method and Webtool for the Prediction of Functional Post-Translational Modifications. bioRxiv. https://doi.org/10.1101/731026
30. Hornbeck P, Kornhauser JM, Latham V, Murray B, Nandhikonda V, Nord A, Skrzypek E, Wheeler T, Zhang B, Gnad F (2019) 15 years of PhosphoSitePlus®: integrating post-translationally modified sites, disease variants and isoforms. Nucleic Acids Res 47:D433–D441
31. Li J, Jia J, Li H, Yu J, Sun H, He Y, Lv D, Yang X, Glocker MO, Ma L, Yang J, Li L, Li W, Zhang G, Liu Q, Li Y, Xie L (2014) SysPTM 2.0: an updated systematic resource for post-translational modification. Database 2014: bau025
32. Huang K-Y, Lee T-Y, Kao H-J, Ma C-T, Lee C-C, Lin T-H, Chang W-C, Huang H-D (2019) dbPTM in 2019: exploring disease association and cross-talk of post-translational modifications. Nucleic Acids Res 47: D298–D308
33. The UniProt Consortium (2017) UniProt: the universal protein knowledgebase. Nucleic Acids Res 45:D158–D169
34. Altschul SFF, Gish W, Miller W, Myers EWW, Lipman DJJ (1990) Basic local alignment search tool. J Mol Biol 215:403–410
35. Mitchell A, Chang H, Daugherty L, Fraser M, Hunter S, Lopez R, Mcanulla C, Mcmenamin C, Nuka G, Pesseat S, Sangrador-vegas A, Scheremetjew M, Rato C, Yong S, Bateman A, Punta M, Attwood TK, Sigrist CJ, Redaschi N, Rivoire C, Xenarios I, Kahn D, Guyot D, Bork P, Letunic I, Gough J, Oates M, Haft D, Huang H, Natale DA, Wu CH, Orengo C, Sillitoe I, Mi H, Thomas PD, Finn RD (2015) The InterPro protein families database : the classification resource after 15 years. Nucleic Acids Res 43:D213–D221
36. Edgar RC (2004) MUSCLE: multiple sequence alignment with high accuracy and high throughput. Nucleic Acids Res 32: 1792–1797
37. Edgar RC (2004) MUSCLE: a multiple sequence alignment method with reduced time and space complexity. BMC Bioinformatics 5:113
38. Abadi M, Barham P, Chen J, Chen Z, Davis A, Dean J, Devin M, Ghemawat S, Irving G, Isard M, Kudlur M, Levenberg J, Monga R, Moore S, Murray DG, Steiner B, Tucker P, Vasudevan V, Warden P, Wicke M, Yu Y, Zheng X (2016) TensorFlow: A system for large-scale machine learning. Proceedings of the 12th USENIX Symposium on Operating Systems Design and Implementation (OSDI '16)
39. Cheng B, Titterington DM (1994) Neural Networks: A Review from a Statistical Perspective. Stat Sci 9:667–678
40. Lancashire LJ, Lemetre C, Ball GR (2009) An introduction to artificial neural networks in bioinformatics--application to complex microarray and mass spectrometry datasets in cancer studies. Brief Bioinform 10:315–329
41. Schmidhuber J (2015) Deep learning in neural networks: an overview. Neural Networks 61: 85–117
42. Almeida JS (2002) Predictive non-linear modeling of complex data by artificial neural networks. Curr Opin Biotechnol 13:72–76
43. Akbani R, Kwek S, Japkowicz N (2004) Applying Support Vector Machines to Imbalanced Datasets. Lnai 3201:39–50
44. Couronné R, Probst P, Boulesteix A-L (2018) Random forest versus logistic regression: a large-scale benchmark experiment. BMC Bioinformatics. 19:270
45. Kennedy J, Eberhart R Particle swarm optimization. in Proceedings of ICNN'95 - International Conference on Neural Networks, pp. 1942–1948, IEEE, 4, 1942–1948
46. Dupret G, Koda M (2001) Bootstrap re-sampling for unbalanced data in supervised learning. Eur J Oper Res 134:141–156
47. Cargnello M, Roux PP (2011) Activation and function of the MAPKs and their substrates, the MAPK-activated protein kinases. Microbiol Mol Biol Rev 75:50–83
48. Aksnes DW, Langfeldt L, Wouters P (2019) Citations, Citation Indicators, and Research Quality: An Overview of Basic Concepts and Theories. SAGE Open 9:2158244019829575
49. Kim Y-M, Jung CH, Seo M, Kim EK, Park J-M, Bae SS, Kim D-H (2015) mTORC1 phosphorylates UVRAG to negatively regulate autophagosome and endosome maturation. Mol Cell 57:207–218
50. Munson MJ, Allen GF, Toth R, Campbell DG, Lucocq JM, Ganley IG (2015) mTOR activates the VPS34-UVRAG complex to regulate autolysosomal tubulation and cell survival. EMBO J 34:2272–2290
51. Mukherjee K, English N, Meers C, Kim H, Jonke A, Storici F, Torres M (2019) Systematic analysis of linker histone PTM hotspots reveals phosphorylation sites that modulate homologous recombination and DSB repair. DNA Repair 86:102763
52. Stenson PD, Mort M, Ball E, Shaw K, Phillips A, Cooper DN (2014) The Human

Gene Mutation Database: building a comprehensive mutation repository for clinical and molecular genetics, diagnostic testing and personalized genomic medicine. Hum Genet 133: 1–9

53. Wheeler DL, Barrett T, Benson DA, Bryant SH, Canese K, Chetvernin V, Church DM, DiCuccio M, Edgar R, Federhen S, Geer LY, Kapustin Y, Khovayko O, Landsman D, Lipman DJ, Madden TL, Maglott DR, Ostell J, Miller V, Pruitt KD, Schuler GD, Sequeira E, Sherry ST, Sirotkin K, Souvorov A, Starchenko G, Tatusov RL, Tatusova TA, Wagner L, Yaschenko E (2007) Database resources of the National Center for Biotechnology Information. Nucleic Acids Res 35: D5–D12
54. Gibson TJ, Dinkel H, Van Roey K, Diella F (2015) Experimental Detection of Short Regulatory Motifs in Eukaryotic Proteins: Tips for Good Practice as Well as for Bad. Cell Commun. Signal 13(1):42.
55. Rudin C, Radin J (2019) Why Are We Using Black Box Models in AI When We Don't Need To? A Lesson From An Explainable AI Competition. Harvard Data Sci Rev 1:1–9
56. Ribeiro MT, Singh S, Guestrin C (2016) "Why Should I Trust You?": Explaining the Predictions of Any Classifier. arXiv
57. Mertins P, Tang LC, Krug K, Clark DJ, Gritsenko MA, Chen L, Clauser KR, Clauss TR, Shah P, Gillette MA, Petyuk VA, Thomas SN, Mani DR, Mundt F, Moore RJ, Hu Y, Zhao R, Schnaubelt M, Keshishian H, Monroe ME, Zhang Z, Udeshi ND, Mani D, Davies SR, Townsend RR, Chan DW, Smith RD, Zhang H, Liu T, Carr SA (2018) Reproducible workflow for multiplexed deep-scale proteome and phosphoproteome analysis of tumor tissues by liquid chromatography-mass spectrometry. Nat Protoc 13:1632–1661
58. Kim W, Bennett EJ, Huttlin EL, Guo A, Li J, Possemato A, Sowa ME, Rad R, Rush J, Comb MJ, Harper JW, Gygi SP (2011) Systematic and quantitative assessment of the ubiquitin-modified proteome. Mol Cell 44:325–340
59. Smith KT, Workman JL (2009) Introducing the acetylome. Nat Biotechnol 27:917–919
60. Suttapitugsakul S, Sun F, Wu R (2019) Recent Advances in Glycoproteomic Analysis by Mass Spectrometry. Anal Chem 92(1):267–291. https://doi.org/10.1021/acs.analchem.9b04651
61. Choudhury S, Baradaran-Mashinchi P, Torres MP (2018) Negative Feedback Phosphorylation of Gγ Subunit Ste18 and the Ste5 Scaffold Synergistically Regulates MAPK Activation in Yeast. Cell Rep 23:1504–1515
62. Wu R, Haas W, Dephoure N, Huttlin EL, Zhai B, Sowa ME, Gygi SP (2011) A large-scale method to measure absolute protein phosphorylation stoichiometries. Nat Methods 8:677–683
63. Vodermaier HC (2004) APC/C and SCF: controlling each other and the cell cycle. Curr Biol 14:R787–R796
64. Strahl B, Allis C (2000) The language of covalent histone modifications. Nature 403:41–45
65. Sloutsky R, Naegle KM (2019) ASPEN, a methodology for reconstructing protein evolution with improved accuracy using ensemble models. eLife 8:1–27
66. Chang JM, di Tommaso P, Lefort V, Gascuel O, Notredame C (2015) TCS: A web server for multiple sequence alignment evaluation and phylogenetic reconstruction. Nucleic Acids Res 43:W3–W6
67. Landry CR, Freschi L, Zarin T, Moses AM (2014) Turnover of protein phosphorylation evolving under stabilizing selection. Front Genet 5:245
68. Hornbeck PV, Zhang B, Murray B, Kornhauser JM, Latham V, Skrzypek E (2015) PhosphoSitePlus, 2014: mutations, PTMs and recalibrations. Nucleic Acids Res 43:D512–D520
69. Dosztányi Z, Csizmok V, Tompa P, Simon I (2005) IUPred: Web server for the prediction of intrinsically unstructured regions of proteins based on estimated energy content. Bioinformatics 21:3433–3434
70. McKinney W (2010) Data Structures for Statistical Computing in Python. PROC. OF THE 9th PYTHON IN SCIENCE CONF
71. Inc., S. A. S. I. JMP Pro 12 JMP®, Version 15. SAS Institute Inc., Cary, NC, 1989–2021. https://www.jmp.com/support/notes/35/282.html
72. R Core Team (2013) R: A language and environment for statistical computing
73. Landrum MJ, Lee JM, Benson M, Brown GR, Chao C, Chitipiralla S, Gu B, Hart J, Hoffman D, Jang W, Karapetyan K, Katz K, Liu C, Maddipatla Z, Malheiro A, McDaniel K, Ovetsky M, Riley G, Zhou G, Holmes JB, Kattman BL, Maglott DR (2018) ClinVar: improving access to variant interpretations and supporting evidence. Nucleic Acids Res 46:D1062–D1067

Chapter 13

Exploration of Protein Posttranslational Modification Landscape and Cross Talk with CrossTalkMapper

Arthur Grimaud, Frederik Haugaard Holck, Louise Marie Buur, Rebecca Kirsch, and Veit Schwämmle

Abstract

Post-translational modifications (PTMs) of proteins play crucial roles in defining protein function. They often do not occur alone, leading to a large variety of proteoforms that correspond to different combinations of multiple PTMs simultaneously decorating a protein. Changes of these proteoforms can be quantified via middle-down and top-down mass spectrometry experiments where the simultaneous PTM settings are obtained by measuring long peptides or entire proteins. Data from such experiments poses big challenges in identifying relevant features of biological and clinical importance. Generally, multiple data layers need to be considered such as proteoforms, individual PTMs, and PTM types. Therein, visualization methods are a crucial part of data analysis as they provide, if applied correctly, insights into both general behaviors as well as a deep view into fine-grained behavior. Here, we present a workflow to visualize histone proteins and their myriad of PTMs based on different R visualization modules applied to data from quantitative middle-down experiments. The procedure can be adapted to diverse experimental designs and is applicable to different proteins and PTMs.

Key words Epigenomics, Histones, Posttranslational modifications, Cross talk, Middle-down

1 Introduction

Classic bottom-up mass spectrometry, despite its high sensitivity and high throughput, does not allow for the study of distant co-occurring posttranslational modifications (PTMs) because of the commonly short length of the tryptic digests [1]. For this reason, and supported by recent increases in mass spectrometer precision, middle-down and top-down mass spectrometry approaches are gaining popularity in the study of protein modifications. In these approaches, researchers analyze long peptides resulting from the digestion of proteins by different proteases

Dukka B. KC (ed.), *Computational Methods for Predicting Post-Translational Modification Sites*, Methods in Molecular Biology, vol. 2499, https://doi.org/10.1007/978-1-0716-2317-6_13,

(middle-down) or even unfragmented proteins (top-down) [2]. These methods allow the identification of the full set of concurrent PTMs on long peptides or even entire proteins, paving the road to the association of fully characterized proteoforms and their different biological functions. These two mass spectrometry approaches are becoming powerful platforms in the large-scale and quantitative study of combinatorial PTMs and their cross talk [3], which are highly relevant to determine protein activity and their regulatory functions [4].

The information-rich spectra obtained from the analysis of long peptides in middle-down and top-down experiments has raised new challenges in both the identification and quantification of proteoforms and in the interpretation of the resulting quantification profiles. The analysis of long peptidoforms or proteoforms by tandem mass spectrometry leads to the acquisition of complex spectra where the mass differences between different peptidoforms are often very low. The identification of such spectra by database search engines often necessitates identifying multiple species per spectrum [5].

Moreover, interpretation of quantitative profiles of a multitude of proteoforms becomes demanding due to the consideration of multiple data layers such as proteins, proteoforms, single PTMs or even PTM cross talk, in addition to experimental designs where proteoforms are quantified across different conditions, timepoints, and replicates [6]. We introduced CrosstalkDB [7] to provide a platform that collects and statistically assesses such data. The platform is used to store and analyze datasets containing quantitative information about long peptidoforms and proteoforms.

With the challenge to further condensate and visualize information on PTM changes and PTM cross talk comparing multiple experimental conditions, we developed a new form of visualization with PTM-CrossTalkMapper [8]. This statistical and visualization framework was created to further explore and visualize the PTM quantification data from CrosstalkDB or other sources. PTM-CrossTalkMapper calculates the abundances of individual PTMs and the co-occurrence of PTM pairs from proteoform data. PTM cross talk is calculated via the interplay score, which allows determining positive or negative cross talk between two PTMs. The different data layers of single PTM abundance, co-occurrence, and cross talk are mapped onto so-called PTM cross talk maps. These visually combine the complex information from multiple conditions at once. Since a toolkit of different visualizations is often needed to explore such multilayered datasets of quantified modifications, PTM-CrossTalkMapper now also provides additional visualization methods as shown in this tutorial.

2 Materials

2.1 Resources

To illustrate the functionalities of PTM-CrossTalkMapper, we will analyze a dataset of quantified histone H3 proteoforms from mouse embryonic stem cells (mESCs) [9]. In this dataset, proteoform quantification has been performed in wild-type mESCs as well as in three mESCs with knockouts in different key components known to be involved in the regulation of the histone PTM landscape. Specifically, the cell lines mESC *Ring1A/B*$^{-/-}$ and *Suz12*$^{-/-}$ have knockouts of genes coding for proteins of the polycomb repressive complex 1 and 2, respectively. mESCs *Dmnt1/3a/3b*$^{-/-}$ have three knockouts of genes coding for DNA methyltransferases. The data contains abundance profiles of the modified isoforms of canonical histone H3 (variants H3.1 and H3.2) and H3.3.

This dataset is available in the tutorial folder of the GitHub repository "https://github.com/veitveit/CrossTalkMapper". It is also possible to download this dataset from the CrosstalkDB platform by following the below-described procedure.

Downloading "MouseStemCells.csv" dataset from CrosstalkDB

1. Open the page http://crosstalkdb.bmb.sdu.dk/ in your browser (*see* **Note 1**).
2. In "Available data sets" filter datasets with description = 'mouse' and paper reference: '27302890'.
3. Select the four datasets matching the special identifiers CrDB000038, CrDB000039, CrDB000041, and CrDB000043 (*see* **Note 2**).
4. Then click on "Start" in the "Show dataset(s)" frame.
5. Download the data by clicking on the CSV icon in the bottom left frame named "Export entire table," and rename the file to **MouseStemCell_CrDB.csv** (*see* **Note 3**).

2.2 Software

CrossTalkMapper and the R script for this tutorial can be downloaded from the GitHub repository (https://github.com/veitveit/CrossTalkMapper/archive/refs/heads/master.zip). To run the program, the R environment (3. 6. 3 or later) must be installed on your computer. Additionally, we recommend using the RStudio IDE (https://www.rstudio.com/products/rstudio/download/) to execute the different tasks of the tutorial script step by step. The R code ("Tutorial_ES_Cells.R") and the dataset ("MouseStemCells.csv") used to perform the analysis are available in the "tutorial" folder of the GitHub repository. The source code of PTM-CrossTalkMapper with the functions used in this tutorial can be found in the file "ptm-crosstalkmapper.R" in the folder "ctm-functions"'of the GitHub repository. Here, we do not show all the R code used for the analysis but refer to the corresponding parts of the tutorial R script. These are marked with tags of the form **Task X.Y**.

2.3 Preparation of the R Environment for CrossTalkMapper

In the first part of the tutorial script "Tutorial_ES_Cells.R" (**Task 1.1**), you will find the necessary commands to set up your R environment for the analysis. Several R packages are required to run PTM-CrossTalkMapper. A list of the required packages is stored in the vector **packages** and can be loaded by running the first line of **Task 1.1** (*see* **Note 4**). If some or all of the necessary packages have not been installed, run the commented-out lines calling the **update.packages() and install.packages()** functions in **Task 1.1** of the tutorial script. Then, load the packages into your R environment, by executing the following line.

```
invisible(lapply(packages, library, character.only = TRUE)).
```

Run the code in **Task 1.2** to define your working directory as the directory in which the tutorial script is located. Then, a subfolder ***plots*** in which the generated plots will be stored is created (*see* **Note 5**). Finally, import the functions from ***ptm-crosstalkmapper.R*** via the **source()** function. If you changed the relative path of the tutorial script to ***ptm-crosstalkmapper.R***, you will have to specify its path in the **source()** function.

3 Methods

The purpose of this section is to demonstrate the different analysis modules of CrossTalkMapper through the analysis of histone H3 modifications across mESC knockouts.

3.1 Data Preparation

In **Task 2.1**, we show how to import and prepare the dataset previously downloaded from CrosstalkDB. If the mESC file has not been downloaded from CrosstalkDB, it can be found in the tutorial folder under the name **MouseStemCell.csv**.

First, load this dataset as an R data frame named **mytable** by carrying out the function **read.csv()** in which you specified the path to your dataset. To compute the PTM abundances and interplay scores, the CrossTalkMapper's function **calcPTMab()** requires some specific values and column names in the input data (*see* **Note 6**). For that, several modifications are made to the imported dataset as follows:

1. A column "timepoint" with values equal to "None" is appended to the data frame.
2. The column "cell.type...tissue" is renamed to "tissue".
3. The resulting data frame is saved as csv file "t.csv" in the current working directory.
4. The function **prepPTMab()** of CrossTalkMapper is finally used to remove columns not necessary for the calculation. If

the argument **histvars** is set to TRUE the quantification values will be normalized according to histone variants. On the contrary, if this argument is set to FALSE, individual histone variants will be ignored (*see* **Note 7**). In the case where the dataset contains multiple replicates and you want to perform a comparison between replicates, you can set the argument **avrepls** to FALSE. By default this argument is set to TRUE, and replicates will be averaged.

PTM abundances and interplay scores can now be computed from this dataset using the function **calcPTMab()** and the result stored in a data frame named **ptm_ab** (**Task 2.2**). Detailed information on the content of this data frame can be found in **Note 8**. At this stage, we are reordering the factor in the column "tissue" to display the control condition first in the plots.

3.2 Visualization

The computation of interplay scores for each pair of PTMs on a protein often results in a dataset containing thousands of PTM pairs (about 3400 in our case). We therefore need to restrict our analysis to several specific PTM modification types and positions. This can be achieved on the basis of prior knowledge or by looking into the dataset as a whole. In this tutorial, we present two visualization methods that can be used to determine PTMs of interest that will be further investigated later.

3.2.1 Overview of the PTM Landscape

Bar Plot

The first overview of the entire dataset is to investigate the abundances of individual modifications on the two histone variants. This is achieved by creating a grouped bar plot of individual PTMs for each condition. In the R script of this tutorial, we create such a bar plot using the "ggplot" R package. To create this bar plot, a subset with the columns "hist," "tissue," "timepoint," "repl," "mi," and "mj" is created from the data frame **ptm_ab**. Then PTMs with abundance inferior to 0.1 are filtered out (*see* **Note 9**). Finally, two grouped bar plots corresponding to the two histone variants H3.1 and H3.3 are created from this filtered data frame and stored as a pdf file in the plots folder (**Task 3.1**).

This bar plot allows us to quickly determine which modifications are the most abundant in the dataset and if a modification varies greatly in its abundance between conditions. In our example (Fig. 1), we can observe that dimethylations of lysine (K) 27 (K27me2) are highly abundant in both histone variants. However, the abundance of K27me2 remarkably decreases in mESCs Suz12−/− (*see* **Note 10**).

Heatmaps

A more adequate way to display variations between conditions is to use heatmaps in combination with hierarchical clustering. CrossTalkMapper includes the function **heatmap_all()** to generate such a heatmap with PTM profiles which are clustered according to their

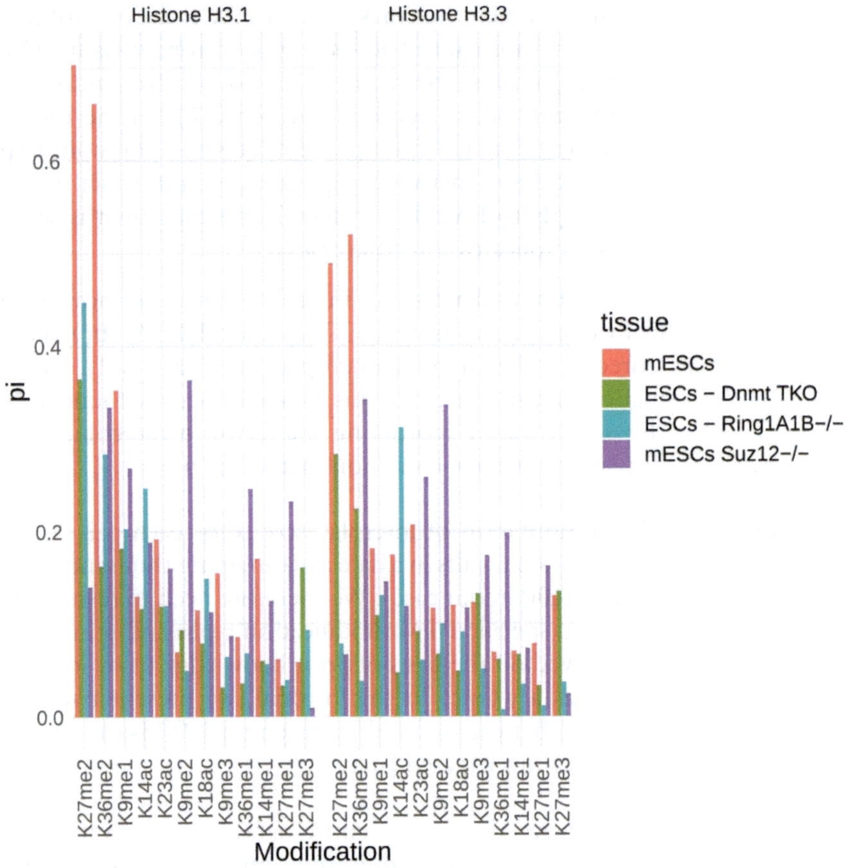

Fig. 1 Grouped bar plot of single PTM abundances on histone H3.1 and H3.3. Only PTMs with abundance superior to 0.1 are shown

similarity. In this tutorial, we present how to create heatmaps of single PTMs, co-occurring modification frequencies, interplay scores, and peptidoform/proteoform abundances. The code to create these four heatmaps is presented in **Task 3.2**.

Heatmaps of single PTM abundances display the same information as the bar plot previously created but also provide information on single PTMs with similar variations in abundance across the different conditions. Heatmaps of co-occurring modification frequency and interplay score (Fig. 2) are both used to visualize interactions between PTM pairs and show common trends on this data level. Peptidoform and proteoform abundance heatmaps provide insight into the behavior of proteoforms.

3.2.2 In-Depth Visualization of PTM Cross Talk and Abundance Variations

The previous tasks showed how to create bar plots and heatmaps, as well as to use them to determine the PTMs of interest in a dataset. Based on that information, we can generate cross talk maps for given PTM pairs that provide a detailed view of their behavior. In our example, we are interested in the cross talk between the three methylation states of H3 K27 with PTMs on other lysines.

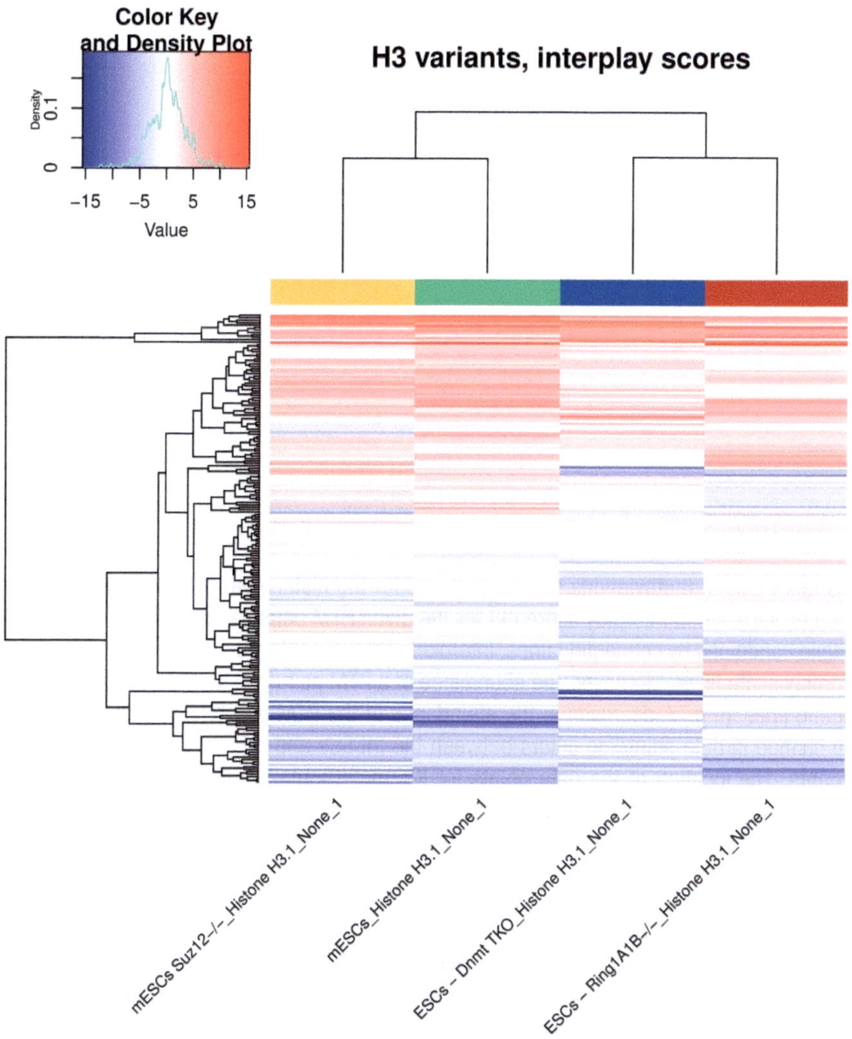

Fig. 2 Heatmap of all PTM interplay scores of histone H3.1. Blue heatmap cells indicate negative cross talk between PTMs, that is, these PTMs tend to exclude each other, while red indicates positive cross talk between PTMs that are more frequently observed on the same protein. Side labels with the names of the PTM pairs are omitted due to their high number

Cross Talk Maps

Cross talk maps allow for a combined view of the different data layers. An extensive description of how to interpret cross talk maps can be found in [8]. To create a cross talk map, a data frame of PTM abundances is passed to the **CrossTalkMap()** function. The characteristics of the cross talk maps can be tuned through the multiple arguments of this function that are detailed in the PTM-CrossTalkMapper documentation (*see* **Note 11**).

In **Task 3.3** of the tutorial script, we create a cross talk map of different methylation states on K27 and K36. First, rows of the **ptm_ab** data frame containing the term "K27me" in the column "mi" and "K36me" in the "mj" column are retained using the

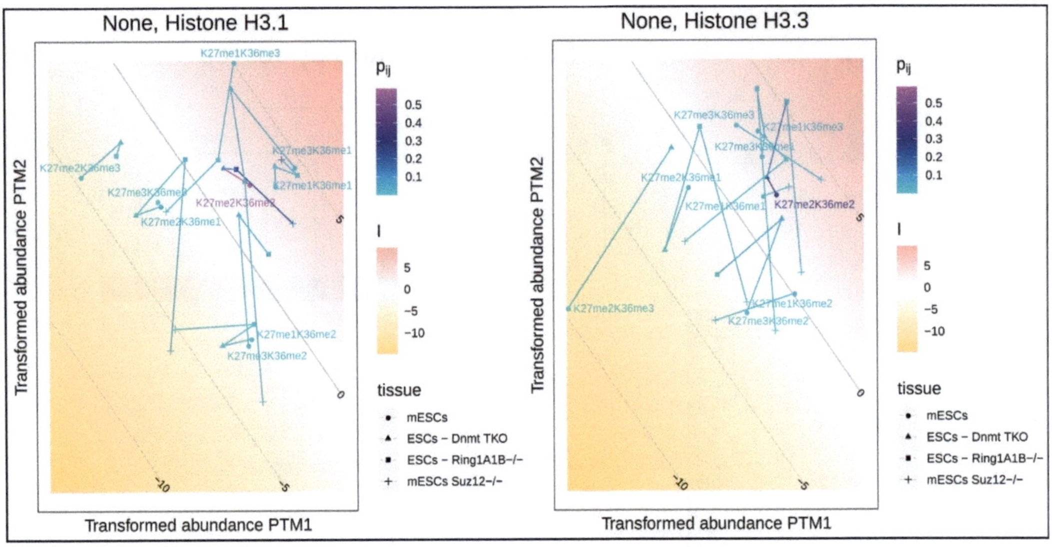

Fig. 3 Cross talk maps of methylations on K27 and K36 for the histone variants H3.1 and H3.3. In these two cross talk maps, variation in interplay score ('I') for the given PTM pairs (labels of the form 'mimj') are shown across multiple conditions by connecting lines. Interplay score is displayed with the background color gradient and the interplay score lines. Moreover, the abundance of a PTM pair ('pij') is color-encoded in the data points and connecting lines. (e.g., A decrease in the abundance of the PTM pair K27me2K36me2 can be observed through the change in data point/line colors between the mESCs and mESCs Suz12−/− conditions for both histone variants. The location of the data points for K27me2K36me2 in H3.1 [left] remain along the same interplay score line, indicating that the cross talk between the two PTMs of the pair is stable across the conditions, while the interplay score for the same pair changes in H3.3 [right])

grepl() function. The cross talk map is then created from the resulting subset **ptm_ab_pos** (Fig. 3). Data points corresponding to the same PTM pair are visually connected across the conditions by setting the parameter **connected** to "tissue."

Further Investigation of PTM Cross Talk with Automatic Generation of Cross Talk Maps and Line Plots

Task 3.4 presents how to generate multiple cross talk maps for a given set of PTM pairs and create line plots of abundance and interplay score for each of the PTM pairs visualized in the cross talk maps. For that, a list of vectors **pos_combs** is created in which the PTM combinations of interest are specified (*see* **Note 12**). In our example, we are interested in observing interactions of mono-, di-, and trimethylation of K27 with modifications on other lysines. Once our pairs of modifications of interest are defined, the following procedure is applied to create the cross talk maps and line plots.

1. Looping through the PTM pairs in the **pos_combs** list, a subset data frame, **ptm_ab_pos_mod,** is created from the **ptm_ab** data frame for matching **mi** and **mj** modifications.
2. The function **CrossTalkMap()** is then called to plot interplay score and PTM abundances in a cross talk map from the

ptm_ab_pos_mod filtered dataset (*see* **Note 13**). The resulting cross talk maps can be found in the "plot" directory (file names ending in "_all.pdf").

3. In order to obtain a clearer picture of PTM interplay, a subset data frame **ptm_ab_pos_mod_top** is subsetted from **ptm_ab_pos_mod** by retaining the top 10 PTM pairs with the most abundant PTM **mj**.
4. As cross talk maps can quickly appear crowded, we filter the dataset to only retain the conditions *mESCs* and *mESCs Suz12−/−* by using the **grepl** function on the column 'tissue' of the **ptm_ab_pos_mod_top**. This filtered dataset is stored under the name **ptm_ab_pos_mod_top_cond.**
5. A second cross talk map is generated for the top 10 PTM pairs and combinations thereof (Fig. 4, file names ending in "_top10.pdf").

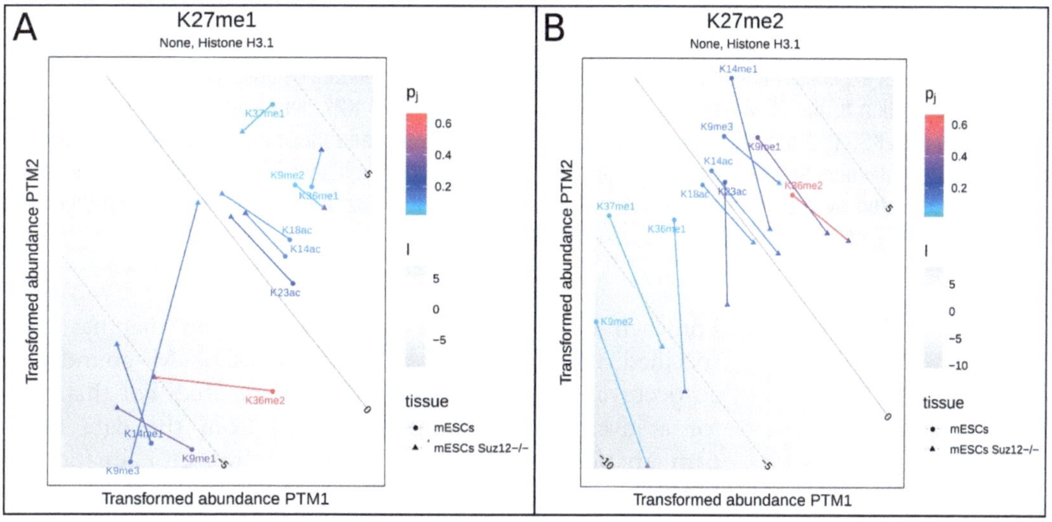

Fig. 4 Visualization of dynamic changes and interplay of K27 methylations in Suz12−/− mEScells compared to wild-type mouse stem cells. Cross talk maps have been generated for the top 10 most abundant PTMs *mj* in PTM pairs with monomethylated and dimethylated K27 (*mi*). Only modifications of the histone H3.1 for wild-type mESCs and Suz12−/− mESCs are displayed here. Note that the abundance *pj* of each K27 methylation is constant across all selected pairs in the cross talk map, which makes it possible to derive the change of co-occurrence from the change of the individual abundance *pi* and the change of interplay score [8]. All PTM pairs with dimethylation (**b**) show the same decrease in PTM co-occurrence in Suz12−/− denoted by the transition of the data points in the direction of the bottom right corner of the map. This result is expected as SUZ12 is a key component of the protein complex associated with the dimethylation of K27. In contrast, PTM pairs with K27me1 (**a**) display the opposite trend in Suz12−/− that could be linked to the decrease in K27me2 abundance

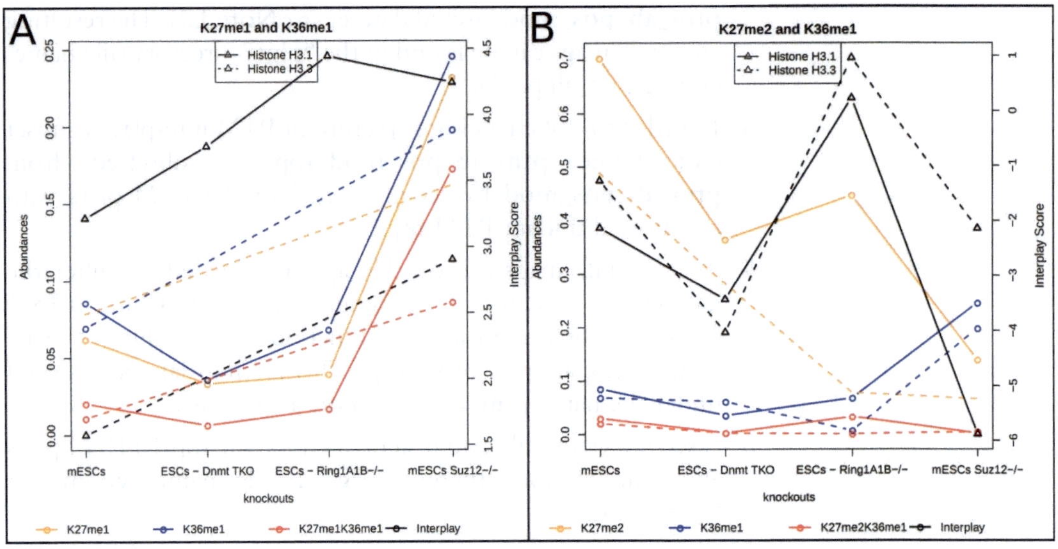

Fig. 5 Change of abundance, co-occurrence, and interplay score for K27me (1,2)K36me1 PTM pairs. The line plots display the detailed variations of abundances, co-occurrence, and interplay score for a specific PTM pair, for the two histone variants. Here we focus on the changes between mESCs and mESCs Suz12−/−. (**a**) In wild-type mESCs, positive cross talk was found between K27me1 and K36me1 on both H3 variants. This positive cross talk explains the simultaneous increase in both PTMs abundances observed in mECSs Suz12−/−. (**b**) In opposition to the PTM pair K27me1K36me1, the interplay of K27 dimethylation and trimethylation (not shown) with K36 monomethylation show negative cross talk between these modifications. This negative interplay score indicates that these modifications tend to be mutually exclusive. The computed negative cross talk is corroborated by the increase in K36me1 abundance in the Suz12−/− cells (blue lines) when the abundance of K27me2 or K27me3 decreases (yellow lines)

6. For each PTM pair in a given combination that has been specified in **pos_comb**, a line plot of PTM abundance, co-occurrence, and interplay score is generated. For that, data for a given PTM pair are extracted from the data frame **ptm_ab_pos_mod** and passed to the **line_ct()** function of CrossTalkMapper (Fig. 5). The condition to be displayed on the *x* axis, "tissue" in our example, is defined with the argument **connected** (*see* **Note 11**).

4 Notes

1. The CrosstalkDB database presents a collection of mass spectrometry-based datasets of quantified long peptides of histone proteins. Statistical and visualization tools are integrated within this platform and can be applied to the database content. The database allows uploading and analyzing data from experiments that yield quantifications of different proteoforms or

long multiply modified peptides. Every dataset of this database can be downloaded and further explored using CrossTalkMapper as described in this tutorial.

2. Each sample of CrosstalkDB is assigned a unique identifier in the format "CrDBXXXXXX."
3. Data from CrosstalkDB can be downloaded in the csv format where each line corresponds to a unique proteoform in a given sample. Proteoforms are defined by their corresponding protein ("protein name"), their amino acid sequence ("protein" or "peptide sequence") and the modification they carry ("modification") in BRNO nomenclature [10]. For each proteoform the relative abundance is stored in the column "quantification."
4. To run code in RStudio, place your cursor on the line of the R script you want to run and press ctrl+enter. The command is executed and its output can be seen in the terminal in the bottom left panel of RStudio. Alternatively, you can select multiple lines and press ctrl+enter to run them.
5. The command used to create the folder will display a warning if the folder already exists in the current working directory.
6. A data frame with the following column names must be passed to the function **calcPTMab()**:

   "protein.name", "tissue", "modifications", "quantification", "timepoint", "biological.replicate".
7. Proteoform quantification is normalized by dividing by the total sum of proteoform abundances of a given histone variant. Quantification values for a protein like histone H3 is achieved by summing up all values of the proteoforms derived from the corresponding histone variants.
8. The data frame obtained from the function **calcPTMab()** is organized such that each row represents a unique PTM pair. The position and type of the PTMs of a pair are specified in BRNO nomenclature, in the columns *'mi' and 'mj'*. For each of these pairs, the abundance of the individual PTMs, 'pi' and 'pj', is stored in the respective columns, and 'pij' is the quantified co-occurrence of both PTMs. *'pi_hat'* and *'pj_hat'* are the transformed *'pi'* and *'pj'* abundances from which the interplay score *'I'* is computed.
9. The data must be filtered as the high number of PTMs will cause the plot to become too crowded. In general, this is true for all the visualizations in this tutorial where data must be subsetted to obtain exploitable plots.
10. SUZ12 is a protein of the Polycomb repressive complex 2. This protein complex is known to have a histone methyltransferase

activity on K9 and K27 of histone H3, explaining the decrease in K27me2 in *Suz12*$^{-/-}$ KO [11].

11. Extensive documentation of PTM-CrossTalkMapper's functions can be found in the section "Documentation" of the "README.md" file at "https://github.com/veitveit/CrossTalkMapper".

12. The pos_combs list must be of the following form "(miA, mjA), (miB, mjB)", where miX and mjX are terms to filter the pairs of PTMs specified in the columns "mi" and "mj" of a PTM abundance data frame (e.g., If the pair of terms "('K4ac', 'K14me')" is specified, pairs of PTMs with acetylation of K4 in the column 'mi' and all methylation states of K14 in column 'mj' are retained).

13. When generating a cross talk map from a PTM abundance data frame containing infinite values, a warning message like "Transformation introduced infinite values in continuous y-axis" might be returned. This can be observed for special cases for which the interplay score is infinite, for example when one PTM does never occur without the other, that is, the abundance of one PTM *pi* is equal to the co-occurrence of the combinatorial PTM "pij." Please note that data points with infinite interplay score are plotted at the margin of the map and that the described warning messages do not alter the creation of the correct cross talk map.

Acknowledgments

F. Holck was supported by the Independent Research Fund Denmark (DFF-FSS). A. Grimaud acknowledges the Marie Curie ITN PROTrEIN for financial support. R. Kirsch was supported by a postdoctoral fellowship from the VILLUM Foundation.

References

1. Arnaudo AM, Molden RC, Garcia BA (2011) Revealing histone variant induced changes via quantitative proteomics. Crit Rev Biochem Mol Biol 46:284–294
2. Moradian A, Kalli A, Sweredoski MJ et al (2014) The top-down, middle-down, and bottom-up mass spectrometry approaches for characterization of histone variants and their post-translational modifications. Proteomics 14:489–497
3. Tian Z, Tolić N, Zhao R et al (2012) Enhanced top-down characterization of histone post-translational modifications. Genome Biol 13: R86
4. Venne AS, Kollipara L, Zahedi RP (2014) The next level of complexity: crosstalk of posttranslational modifications. Proteomics 14:513–524
5. Houel S, Abernathy R, Renganathan K et al (2010) Quantifying the impact of chimera MS/MS spectra on peptide identification in large-scale proteomics studies. J Proteome Res 9:4152–4160
6. Tvardovskiy A, Schwämmle V, Kempf SJ et al (2017) Accumulation of histone variant H3.3 with age is associated with profound changes in the histone methylation landscape. Nucleic Acids Res 45:9272–9289

7. Schwämmle V, Aspalter C-M, Sidoli S et al (2014) Large scale analysis of co-existing post-translational modifications in histone tails reveals global fine structure of cross-talk. Mol Cell Proteomics 13:1855–1865
8. Kirsch R, Jensen ON, Schwämmle V (2020) Visualization of the dynamics of histone modifications and their crosstalk using PTM-CrossTalkMapper. Methods 184:78–85
9. Schwämmle V, Sidoli S, Ruminowicz C et al (2016) Systems level analysis of histone H3 post-translational modifications (PTMs) reveals features of PTM crosstalk in chromatin regulation. Mol Cell Proteomics 15:2715–2729
10. Turner BM (2005) Reading signals on the nucleosome with a new nomenclature for modified histones. Nat Struct Mol Biol 12:110–112
11. Pasini D, Bracken AP, Jensen MR et al (2004) Suz12 is essential for mouse development and for EZH2 histone methyltransferase activity. EMBO J 23:4061–4071

Chapter 14

PTM-X: Prediction of Post-Translational Modification Crosstalk Within and Across Proteins

Yuxuan Li, Yuanhua Huang, and Tingting Li

Abstract

Posttranslational modifications (PTMs), which are processes of adding covalent groups in protein amino acids after the translation, play an important role in regulating proteins' localization, degradation, and functions. Different PTMs both within a single protein and across multiple proteins can work together or regulate reciprocally, known as PTM cross talk. However, high-throughput experimental identifications of PTM cross talk are lack due to technical limitations. In this chapter, we review in silico prediction approaches and illustrate the usage of PTM-X, a suite of recently proposed machine learning methods to predict both intra- and interprotein PTM cross talk.

Key words Posttranslational modifications cross talk, Machine learning prediction, Protein sequence coevolution, Protein structure

1 Introduction

Modifications of cellular proteins have been of great interest in biological research for a long time since the introduction of the center dogma of molecular biology by Crick in 1970s. Posttranslational modifications (PTMs) are covalent additions (e.g., phosphate, methyl, and acetyl) to special residues in protein amino acids, happening after the proteins are translated. PTMs play important roles in multiple cellular processes such as protein folding, activation, translocation, and degradation [1], and mis-PTMs are found to be key regulators of diseases or tumorigenesis, for example, through TP53 [2].

In the past two decades, high-throughput measurement of PTMs has become possible and attracted much attention thanks to the developments in mass spectrometry [3]. With the accumulated PTM data, increasing evidence suggests that PTMs do not function alone but act in an orchestrated manner. Multiple PTMs within one protein or among different proteins are identified and

Dukka B. KC (ed.), *Computational Methods for Predicting Post-Translational Modification Sites*,
Methods in Molecular Biology, vol. 2499, https://doi.org/10.1007/978-1-0716-2317-6_14,

these PTMs can work together in the same cellular events, which is called PTM cross talk [4]. More than 10 PTMs in TP53 work precisely to suppress tumorigenesis [5]. Lysine 4 in tyrosine kinase PYK2, when SUMOylated, promotes PYK2 autophosphorylation in Tyrosine 402 [6]. Moreover, PTM cross talk contributes to the regulations of DNA damage repair [7], gene expression [8] and protein stability [9]. Besides within a single protein, PTM cross talk events between different proteins have also been reported. Phosphorylation of histone-lysine N-methyltransferase EZH2 in Ser-21 decreases Lys-27 trimethylation on histone H3 due to the suppression of methyltransferase activity of EZH2 [10].

Despite the important function the PTM cross talk presents, experimental identification of PTM cross talk on a large scale has been bottlenecked and time-consuming. It has only been applied to small- or medium-scale organisms, for example, in *Mycoplasma pneumonia* for genome-wide cross talk between phosphorylation and acetylation [11], and in *Saccharomyces cerevisiae* for global interaction between phosphorylation and ubiquitylation [21]. On the other hand, some studies are restricted to several proteins or specific environments, for example cross talk between on histone modifications in *S. cerevisiae* [12].

At the same time, computationally analysis of PTM cross talk has also gained more attentions. Coevolution features of PTM sites are used for measuring the relationship among different PTM types [13]. Mutation simulations of acetylation sites were used to discover the relationship between acetylation and any other PTMs [14]. However, prediction of PTM cross talk between different proteins still needs to be further studied because of their large protein pair space and lacking of gold standard dataset. In our previous studies, we established PTM-X, a platform to predict intraprotein as well as interprotein PTM cross talk. For the intraprotein PTM cross talk predictions, based on 193 pairs of PTM sites in 77 human proteins, PTM-X integrated sequence motifs, structural proximity, and coevolution to train a naïve Bayes model. For the interprotein PTM cross talk predictions, based on 199 interprotein PTM cross talk pairs, PTM-X integrated coevolution and comodification features to train a random forest model [15]. Recently, the dataset curated from PTM-X project has accelerated the further development of new algorithms for prediction of PTM cross talk, for example PCTpred [16]. Meanwhile, the community of protein structure prediction has recently embraced a breakthrough by introducing AlphaFold [17] and its release of predicted structure on almost all human proteins. We anticipate that these 3D structural information will further enhance the PTM cross talk in near future.

In this chapter, we will give a brief introduction about PTM-X and demonstrate its usage for predicting intra- or interprotein PTM cross talk at a proteome-wide scale on a local machine or a small scale via our web server.

2 Materials

In this chapter we will use PTM-X, an open-source software developed in Python 2.7, for prediction of both inter- and intraprotein PTM cross talk. Here we provide a handful of PTM cross talk samples as example to illustrate the function of PTM-X in PTM cross talk prediction. The web server, Python package, and our compiled datasets are freely accessible at following links.

1. Web server of PTM-X: http://bioinfo.bjmu.edu.cn/ptm-x/
2. PTM-X packages: https://github.com/huangyh09/PTM-X
3. PTM cross talk data: https://sourceforge.net/projects/ptm-crosstalk/

3 Methods

3.1 Description of PTM-X

Large-scale quantification and analysis of PTM cross talk have been implemented experimentally and computationally. For experimental quantification, technical shortages restrict the scale to a small group of proteins or specific perturbations. Computational analysis of PTM cross talk raise the attention for discovering distinct patterns of sequence features in different PTMs when they process the interplay. Recent studies have shown more details in sequence or motif features contributing to the cross talk between two PTMs. Minguez et al. presented a measurement of correlation among different PTM types using coevolution level of PTM sites [18]. Schwammle et al. found some coexistence patterns of PTM sites [19]. However, lack of gold-standard dataset and suitable model has been the bottleneck of the in silico identification of PTM cross talk. The PTM-X projects started with curation of experimentally validated samples of PTM cross talk at both intra-protein and interprotein levels, followed by identifying a set of predictive features and fitting prediction model.

The intraprotein part using three features including tertiary structure distance, residue coevolution and modification evolution, as Fig. 1 shows. Tertiary distance is obtained from PDB database and it is defined as distance between 2 α-carbon atoms adjacent to the carboxyl group of amino acids [20]. Residue coevolution is measured by multiple sequence alignment across 50 vertebrates in eggNOG v4.5. Comodification level of PTM sites is represented by

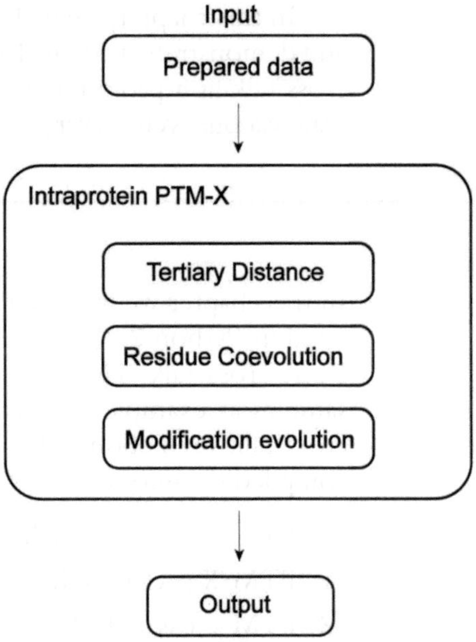

Fig. 1 Flowchart of intraprotein PTM cross talk prediction of PTM-X. Given the input data (a pair of PTMs) for query, three types of features will be fetched. Then, PTM-X gives the prediction score according to these three features

the comparison of PTM modification capabilities among rat, mouse, and human. Features are integrated using a naïve Bayes classifier with kernel density estimation. 193 pairs of PTM sites in 77 human proteins experimentally supported are collected for training.

The interprotein model uses four features including sequence residue coevolution, sequence motif coevolution, comodification across different species and comodification across different conditions, as shown in Fig. 2. Sequence residue coevolution level is defined as two amino acids across multiple vertebrates and the sequence motif coevolution level is defined as the ±3 surrounding amino acids of the PTM sites. Residue coevolution is measured by 1-Hamming loss after sequence alignments using eggNOG v4.5. As for motif coevolution, the score is calculated by a mean of two vectors based on motif conservation. Comodification across different species refers to a comparison among rat, mouse and human and data here is downloaded from PhophoSitePlus. The last feature in this algorithm is the comodification across different conditions, which is defined as the co-occurrence between the two PTMs across 88 tissue, disease and cell line conditions. Based on these four features and the full-collected dataset, PTM-X is able to predict PTM cross talk precisely.

3.2 Data Preparation for PTM-X

PTM-X processes two main functions: intraprotein cross talk prediction and interprotein prediction. The prediction process can be

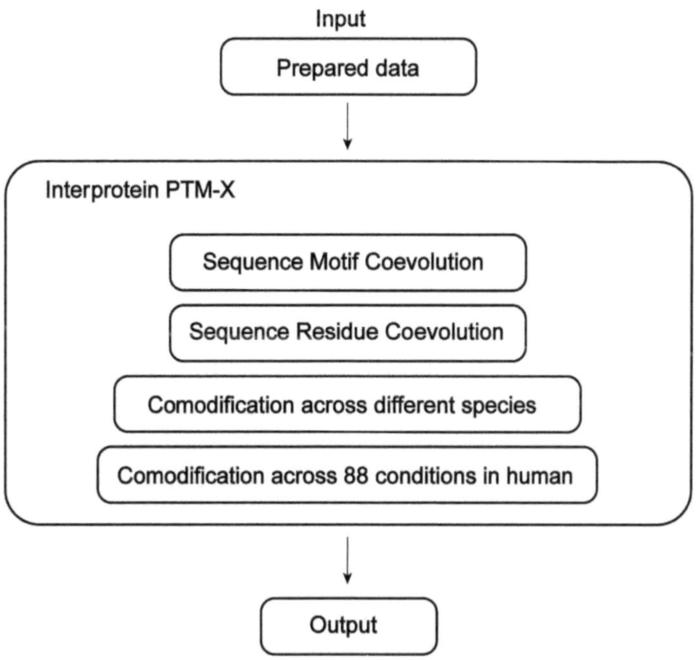

Fig. 2 Flowchart of interprotein PTM cross talk prediction of PTM-X: Given the input data (a pair of PTMs) for query, three types of features will be fetched. Then, PTM-X gives the prediction score according to these four features

carried out by web server for small scale datasets or by local machine with Python package supporting larger scale of datasets.

Intraprotein cross talk is defined as interplay of different modification sites within one protein. Data format in this function of PTM-X must arrange as follows.

P04637 S15 phosphorylation T18 phosphorylation

1. The first column is the UniProt accession number.
2. The second column is the position of first candidate modification site.
3. The third column is the PTM type of the first position.
4. The forth column is the position of second modification site.
5. The fifth column is the PTM type of the second position.

Columns here are separated by Tab, and unknown PTM type can be replaced by nan. Example data can be downloaded here: http://bioinfo.bjmu.edu.cn/files/ptm_input.txt.

Interprotein cross talk is defined as interplay of different PTM sites between two different proteins. Data format here should be prepared like this.

P12814 Y12 phosphorylation Q05397 Y397 phosphorylation

1. The first column is the UniProt accession number of the first protein.
2. The second column is the position of first candidate modification site in the first protein.
3. The third column is the PTM type of the first position.
4. The forth column is the UniProt accession number of the second protein.
5. The fifth column is the position of modification site in the second protein.
6. The sixth column is the PTM type of the second modification.

Columns here are separated by Tab, and unknown PTM type can be replaced by nan. Example data can be downloaded here: http://bioinfo.bjmu.edu.cn/files/ptm_ppi_input.txt.

3.3 Usage of PTM-X Software Locally

Here, we introduce to how to setup the computing environment for using PTM-X locally. The PTM-X software needs an environment of Python 2.7 to support all packages. As a common choice, we illustrate using conda through Anaconda (https://anaconda.org) to set up the environment.

3.3.1 Part 1: Setup Computing Environment

1. Step 1, create a conda environment with Python 2.7 and activate it as follows. See note on Python versions in Note 1.

```
conda create -name ptmxPy2 python=2.7 numpy scipy \
 scikit-learn joblib==0.11
 conda activate ptmxPy2
```

2. Step 2: install the PTM-X package through the source code from GitHub (https://github.com/huangyh09/PTM-X), which can be done with the following command line directly.

```
pip install -U git+https://github.com/huangyh09/PTM-X
```

3. Step 3: to fetch the above four features, download the data from http://ufpr.dl.sourceforge.net/project/ptm-crosstalk/PTM-X_data_v2.2.zip. Unzip it and set the directory as the according parameter in the command line below.

3.3.2 Part 2: Using PTM-X to Feature Features and Perform Prediction

Once the local computing environment set up, users can run the PTM-X software on a local machine. It contains two separate steps: fetching features and predicting the cross talk states.

1. Step 4: fetch the four features with the following command line by input the samples by "-i" and specifying the directory of downloaded data in step 3.

```
PTMX-feature -i SAMPLE_FILE -d DATA_DIR -o FEATURE_FILE
```

protein	site1	PTM1	site2	PTM2	distance_1d	distance_3d	disorder	residue_co	modify_co	prediction_score
P04637	S15	phosphorylation	T18	phosphorylation	3	7.254	0	0.740	1.000	0.982

Prot_id1	Residue1	PTM1	Prot_id2	Residue2	PTM2	Seq_residue	Seq_motif	PTM_conditions	prediction_score
P62805	K17	acetylation	Q71DI3	K5	dimethy/trimethy	0.750	0.964	nan	0.973

Fig. 3 Example output of PTM-X for intraprotein PTM cross talk (**a**) and interprotein cross talk (**b**). Output chart of PTM-X shows the input query table (column 1–5 for intraprotein and 1–6 for interprotein), followed by fetched features (5 columns for intraprotein and 3 columns for interprotein) and predicted score (last column)

Then this command line will output a table by adding the fetched features as additional columns. For detailed instructions, one can use the command line "`PTMX-feature -h`".

2. Step 5: with step 4, you can obtain the features for your query samples (PTM pairs) and our curated positive and negative samples. Now you can use the following command line to perform prediction.

```
PTMX-predict -i QUERY_FEATURE_FILE -o PREDICT_FILE \
 --positive POSITIVE_FEATUR_FILE \
 --negative NEGATIVE_FEATUR_FILE
```

Then this command line will output a table by adding the prediction score as an additional column to the input feature table. For detailed instructions, one can use the command line "`PTMX-predict -h`".

3.4 Direct Usage of PTM-X Web Server

With the format mentioned in Subheading 3.2, the query data can be directly input to the web server for immediate use. Example items, including their fetched features and prediction scores, are shown in Fig. 3. Here, you can also set a more stringent or lenient cutoff, as discussed in Note 2.

4 Notes

PTM-X provides a reliable prediction of modification cross talk in both intraprotein and interproteins. Different four features are used to predict PTM cross talk via a random forest classifier. In this section, we provide a brief introduction of PTM-X. PTM-X is still a strong method to predict PTM cross talk so far, here we give some tips on its usage:

1. As Python3 is more general for daily use, there will be a tiny problem to run the software in Python3. The first consideration for its usage is that this python packages contains many tools developed in Python2. If PTM-X is failed in one environment, the simplest way to solve the problem is to setup a specific environment based

on Python2.7 as told in Subheading 3.3. Conda will be a great way to manage your environments(https://conda.io/projects/conda/en/latest/user-guide/tasks/manage-environments.html).

2. The interprotein PTM cross talk data is summarized by the developer and it contains 199 PTM cross talk from 82 human proteins. Your input PTM pairs can be taken as potential cross talk pairs if their prediction scores are higher than given threshold. To get more sensitive prediction, a loose threshold can be used. In web server(http://bioinfo.bjmu.edu.cn/ptm-x/) the ROC curve from the ten-fold cross-validation is displayed after clicking on the prediction score and related false positive and true positive rate are also presented immediately.

In summary, the PTM-X is useful tool for both in inter- and intra-PTM cross talk prediction. The web server allows for quick prediction on a handful of PTM cross talk candidates and the local installation allows a much larger scale prediction.

References

1. Olzscha H (2019) Posttranslational modifications and proteinopathies: how guardians of the proteome are defeated. Biol Chem 400: 895–915
2. Bode AM, Dong Z (2004) Post-translational modification of p53 in tumorigenesis. Nat Rev Cancer 4:793–805
3. Witze ES, Old WM, Resing KA, Ahn NG (2007) Mapping protein post-translational modifications with mass spectrometry. Nat Methods 4:798–806
4. Beltrao P, Bork P, Krogan NJ, van Noort V (2013) Evolution and functional cross-talk of protein post-translational modifications. Mol Syst Biol 9:714
5. Dai C, Gu W (2010) p53 post-translational modification: deregulated in tumorigenesis. Trends Mol Med 16:528–536
6. Uzoma I, Hu JF, Cox E, Xia SL, Zhou JY, Rho HS, Guzzo C, Paul C, Ajala O, Goodwin CR, Jeong J, Moore C, Zhang H, Meluh P, Blackshaw S, Matunis M, Qian J, Zhu H (2018) Global identification of small ubiquitin-related modifier (SUMO) substrates reveals crosstalk between SUMOylation and phosphorylation promotes cell migration. Mol Cell Proteomics 17:871–888
7. Ivanov GS, Ivanova T, Kurash J, Ivanov A, Chuikov S, Gizatullin F, Herrera-Medina EM, Rauscher F 3rd, Reinberg D, Barlev NA (2007) Methylation-acetylation interplay activates p53 in response to DNA damage. Mol Cell Biol 27: 6756–6769
8. Khidekel N, Hsieh-Wilson LC (2004) A 'molecular switchboard' - covalent modifications to proteins and their impact on transcription. Org Biomol Chem 2:1–7
9. Zhang J, Bu X, Wang H, Zhu Y, Geng Y, Nihira NT, Tan Y, Ci Y, Wu F, Dai X, Guo J, Huang YH, Fan C, Ren S, Sun Y, Freeman GJ, Sicinski P, Wei W (2018) Cyclin D-CDK4 kinase destabilizes PD-L1 via cullin 3-SPOP to control cancer immune surveillance. Nature 553:91–95
10. Cha TL, Zhou BP, Xia W, Wu Y, Yang CC, Chen CT, Ping B, Otte AP, Hung MC (2005) Akt-mediated phosphorylation of EZH2 suppresses methylation of lysine 27 in histone H3. Science 310:306–310
11. van Noort V, Seebacher J, Bader S, Mohammed S, Vonkova I, Betts MJ, Kuhner S, Kumar R, Maier T, O'Flaherty M, Rybin V, Schmeisky A, Yus E, Stulke J, Serrano L, Russell RB, Heck AJR, Bork P, Gavin AC (2012) Cross-talk between phosphorylation and lysine acetylation in a genome-reduced bacterium. Mol Syst Biol 8: 571
12. Guan X, Rastogi N, Parthun MR, Freitas MA (2013) Discovery of histone modification crosstalk networks by stable isotope labeling of amino acids in cell culture mass spectrometry (SILAC MS). Mol Cell Proteomics 12: 2048–2059
13. Beltrao P, Albanese V, Kenner LR, Swaney DL, Burlingame A, Villen J, Lim WA, Fraser JS,

Frydman J, Krogan NJ (2012) Systematic functional prioritization of protein posttranslational modifications. Cell 150:413–425

14. Lu Z, Cheng Z, Zhao Y, Volchenboum SL (2011) Bioinformatic analysis and post-translational modification crosstalk prediction of lysine acetylation. PLoS One 6:e28228
15. Huang R, Huang Y, Guo Y, Ji S, Lu M, Li T (2019) Systematic characterization and prediction of post-translational modification cross-talk between proteins. Bioinformatics 35: 2626–2633
16. Liu HF, Liu R (2020) Structure-based prediction of post-translational modification cross-talk within proteins using complementary residue- and residue pair-based features. Brief Bioinform 21:609–620
17. Jumper J, Evans R, Pritzel A, Green T, Figurnov M, Ronneberger O, Tunyasuvunakool K, Bates R, Zidek A, Potapenko A, Bridgland A, Meyer C, Kohl SAA, Ballard AJ, Cowie A, Romera-Paredes B, Nikolov S, Jain R, Adler J, Back T, Petersen S, Reiman D, Clancy E, Zielinski M, Steinegger M, Pacholska M, Berghammer T, Bodenstein S, Silver D, Vinyals O, Senior AW, Kavukcuoglu K, Kohli P, Hassabis D (2021) Highly accurate protein structure prediction with AlphaFold. Nature 596:583–589
18. Minguez P, Parca L, Diella F, Mende DR, Kumar R, Helmer-Citterich M, Gavin AC, van Noort V, Bork P (2012) Deciphering a global network of functionally associated post-translational modifications. Mol Syst Biol 8:599
19. Schwammle V, Aspalter CM, Sidoli S, Jensen ON (2014) Large scale analysis of co-existing post-translational modifications in histone tails reveals global fine structure of cross-talk. Mol Cell Proteomics 13:1855–1865
20. Huang Y, Xu B, Zhou X, Li Y, Lu M, Jiang R, Li T (2015) Systematic characterization and prediction of post-translational modification cross-talk. Mol Cell Proteomics 14:761–770
21. Danielle L, Swaney Pedro, Beltrao Lea, Starita Ailan, Guo John, Rush Stanley, Fields Nevan J, Krogan Judit, Villén (2013) Global analysis of phosphorylation and ubiquitylation cross-talk in protein degradation. Nature Methods 10(7) 676-682 10.1038/nmeth.2519

Chapter 15

Deep Learning–Based Advances In Protein Posttranslational Modification Site and Protein Cleavage Prediction

Subash C. Pakhrin, Suresh Pokharel, Hiroto Saigo, and Dukka B. KC

Abstract

Posttranslational modification (PTM) is a ubiquitous phenomenon in both eukaryotes and prokaryotes which gives rise to enormous proteomic diversity. PTM mostly comes in two flavors: covalent modification to polypeptide chain and proteolytic cleavage. Understanding and characterization of PTM is a fundamental step toward understanding the underpinning of biology. Recent advances in experimental approaches, mainly mass-spectrometry-based approaches, have immensely helped in obtaining and characterizing PTMs. However, experimental approaches are not enough to understand and characterize more than 450 different types of PTMs and complementary computational approaches are becoming popular. Recently, due to the various advancements in the field of Deep Learning (DL), along with the explosion of applications of DL to various fields, the field of computational prediction of PTM has also witnessed the development of a plethora of deep learning (DL)-based approaches. In this book chapter, we first review some recent DL-based approaches in the field of PTM site prediction. In addition, we also review the recent advances in the not-so-studied PTM, that is, proteolytic cleavage predictions. We describe advances in PTM prediction by highlighting the Deep learning architecture, feature encoding, novelty of the approaches, and availability of the tools/approaches. Finally, we provide an outlook and possible future research directions for DL-based approaches for PTM prediction.

Key words Deep learning, Posttranslational modification site, Proteolytic cleavage, Phosphorylation, Machine learning

1 Introduction

The living cell extends the chemical repertoire and information content of the protein by attachment of covalent modification to polypeptide chain or proteolytic cleavage [1, 2]. Most of these modifications occur after the synthesis of the polypeptide chain and thus they are called "posttranslational modification" (PTM). Proteolytic cleavage is usually an irreversible modification, while

Subash C. Pakhrin and Suresh Pokharel contributed equally with all other contributors.

Dukka B. KC (ed.), *Computational Methods for Predicting Post-Translational Modification Sites*,
Methods in Molecular Biology, vol. 2499, https://doi.org/10.1007/978-1-0716-2317-6_15,

covalent modifications may be reversible, for example, in the case of protein phosphorylation. There are more than 450 unique protein posttranslational modifications including the most common and extensively studied phosphorylation. Over the past decade, the number and variety of protein posttranslational modifications that have been detected and characterized have rapidly increased.

PTMs are involved in most cellular processes including the maintenance of protein structure and integrity, regulation of metabolism and defense processes, and cellular recognition events and morphology changes. There are several challenges involved in the analysis and understanding of PTMs, and efficient and sensitive methods for the detection of PTMs are indispensable. Traditionally, PTMs have been identified by Edman degradation, amino acid analysis, isotopic labeling, or immunochemistry. More recently, mass spectrometry (MS) has proven to be extremely useful in PTM discovery [3]. Mass spectrometry-based characterization of PTMs is based on the principle that covalent modifications in proteins affect the molecular weight of the modified amino acids, and the change in mass (molecular weight) can be detected by MS. Please *see* Chapter 1 for details about advances in MS-based proteomics technologies to maximize PTM coverage. Although MS provides a good platform for experimental determination of PTM sites, given the cost of instruments and specialized expertise required, computational approaches provide a promising strategy for identifying and understanding of PTMs.

1.1 Computational Prediction of PTM Sites

During the last 30 years, several computational methods have been developed for both PTM site prediction and proteolytic cleavage prediction. To our knowledge, one of the first computational predictions of PTM sites using primary sequences was the work of Nakai and Kanehisa [4] in 1988 that uses consensus sequence pattern (motif)–based approach combined with discriminant analysis. In 2004, a Neural Network approach by Blom et al. [5] to predict phosphorylation sites was published. In 2005, an SVM-based approach for the prediction of phosphorylation sites was published by Plewczynski et al. [6]. These events led to the opening of the flood gate for machine learning and deep learning–based approaches for various PTM predictions.

Similarly, the work of Blom et al. in 2004 [2] was one of the earliest review papers for summarizing the computational prediction of PTM sites that reviewed approaches for phosphorylation and glycosylation site prediction. Subsequently, numerous computational methods were developed. In 2010, Eisehnaber and Eisenhaber [7] reviewed the existing computational methods for predicting cleavage of targeting signals, lipid PTMs, phosphorylation, and glycosylation. The field witnessed an even greater number of computational approaches followed by some excellent review papers that summarized the existing computational approaches to study/predict PTMs. In 2011, Trost and Kusalik [8] summarized various prediction tools by taking a comparative approach to examine issues and challenges with the prediction of eukaryotic phosphorylation sites from 1999 to 2011 focusing on the ML approach (or computational approach), the number of residues in the window, whether the input is the primary sequence or tertiary structure and so on.

Recently in 2017, Audagnotto and Dal Peraro [9] summarized the available online platforms for various PTMs. In addition, in 2020 Ramazi and Zahiri [10] reviewed the major online databases and related tools for posttranslational modifications in proteins focusing on the 10 most studied PTMs. Most recently, Wen et al. [11], while summarizing DL-based approaches for various aspects of proteomics, very briefly touched on a few DL-based approaches in PTM prediction. To our knowledge, there is no comprehensive review paper that summarizes the most recent DL-based approaches for various PTM site predictions.

1.2 Proteolytic Processing/Cleavage

Another type of PTMs is proteolytic processing or proteolysis which involves limited and highly specific enzymatic hydrolysis of a peptide bond in a peptide or protein substrate by a family of specialized enzymes called proteases [12]. Proteolysis is an irreversible posttranslational modification and is often regarded as a mere degradative mechanism in the destruction of proteins. Recent research has led to the recognition that proteolytic cleavage events by a wide repertoire of proteases are pivotal regulators of most, if not all, physiological and pathological processes, and isopeptide bonds of a protein by a protease [13]. Proteolytic cleavage is crucial for cellular health and homeostasis and is also involved in the development and progression of various diseases including cancer, inflammation, autoimmune, cardiovascular, and metabolic disorders [13]. This cleavage generates shorter protein chains displaying neo-N and -C termini with new or modified biological activities [14]. This ubiquitous and irreversible PTM can result in the activation, inactivation, completely altered protein function, and regulate a vast array of biological processes. Proteases are primarily responsible for proteolytic cleavage of the proteins and there is a large repertoire of proteases: in humans, there are 588 proteases [15] that can be grouped into five catalytic classes. In the absence of a protease, peptide bonds are highly stable however are hydrolyzed within milliseconds in protease-assisted biochemical reactions. Proteases recognize and bind their substrates mainly through hydrophobic and electrostatic interactions with the substrate amino acid residue side chains. The knowledge of protease cellular and physiological functions, as well as their substrates and cleavage preferences, are crucial in designing molecules for therapeutic modulation of protease activity. Several mass-spectrometry-based proteomics techniques to systematically study protease–substrate relationships have been developed including the recent "high-throughput protease screen" [16]. These experimental technologies have aided in obtaining repertoires of proteinase-specific cleavages which has helped in the development of computational tools to identify protease substrates and cleavage sites.

1.2.1 Computational Prediction of Proteolytic Cleavage

As discussed above, experimentally various MS-based approaches are utilized to decipher the mechanism of proteolytic cleavage. Due to the success of these experimental approaches in proteolytic cleavage identification, various databases have been available like MEROPS [17] and CutDB [18] that have enabled ML-based approaches for the prediction of proteolytic cleavage. Recently, several computational approaches have been developed for the prediction of protease-specific substrates and the position of proteolytic cleavage sites. Existing methods for protein cleavage prediction can be broadly classified as ones based on sequence-scoring functions (CaSPredictor [19], SitePrediction [20], etc.) and the others based on Machine Learning (CASVM [21], Cascleave [22, 23], PROSPERous [24], etc.).

With the growing number of computational approaches for proteolytic cleavage prediction available, several review articles [25, 26] discussing the computational tools for the prediction of the proteolytic cleavage site and proteases have been published. Among the various proteases, Caspases are an important protease family. Recently, Bao et al. [27] summarized methods, tools, and features for the prediction of caspase cleavage sites and also benchmarked existing 12 state-of-the-art approaches for the prediction of caspases cleavage prediction. It was concluded from this review that CasCleave [22, 23] and SitePrediction [20] achieved the overall best results for predicting caspase substrates in different species. Recently, Li et al. [28] meticulously reviewed various computational bioinformatics approaches for the prediction of protease-specific substrates and cleavage site prediction and benchmarked 19 existing tools based on approaches, features, performance analysis, and software availability. In addition, the authors created an independent test set that is comprised of 3641 cleavage sites for 10 proteases and assessed the performance of each of these tools. For additional details, please refer to this excellent review [28]. More recently, some deep learning–based approaches have been developed for the prediction of proteolytic cleavage. Thus, in this chapter, we will also review some of the recent deep learning–based approaches for proteolytic cleavage prediction.

To our knowledge, compared with the existing review articles that have been published on computational prediction of posttranslational modification tools, our chapter is different as it specifically focuses on the (recent) deep learning–based advances for various posttranslational modification site prediction and also highlights the recent advances in proteolytic cleavage site prediction.

Subheading 2 summarizes deep learning–based advances for (single) posttranslational modification site prediction. Subheading 3 discusses recent deep learning–based advances for multiple posttranslational modification site prediction. Subheading 4 discusses recent development including deep learning–based approaches for proteolytic cleavage prediction (often ignored in comparison to covalent-modification) and finally, Subheading 5 discusses the challenges, future outlook, and concluding remarks for the field.

2 Deep-Learning Based Advances for (Single) Posttranslational Modification Site Prediction

Recently, several DL-based approaches have been reported for the prediction of various types of PTM sites. In this section, we review some of the recent DL-based approaches for (single) posttranslational modification site prediction.

2.1 Deep-Learning Based Advances for Acetylation Site Prediction

Acetylation is one of the important and ubiquitous posttranslational modifications which plays an important role in protein functioning. The identification of Lysine acetylation in a protein sequence is essential to study its roles in various diseases and other biological processes. Many machine learning–based tools like PSKAcePred [29], LAceP [30], and N-Ace [31] have been proposed to predict acetylation sites in given protein sequences using SVM, Bayesian discrimination, logistic regression models to create a binary classifier to identify the presence of acetylation sites in an input protein sequence. Recently, a few deep learning–based acetylation prediction tools have been developed. Here, we describe some of them.

2.1.1 DeepACEt

DeepACEt [32] is a Multilayer Perceptron (MLP) based acetylation prediction tool trained using six different feature encoding schemes: one-hot, BLOSUM62 matrix, a composition of K-space amino acid pairs (CKSAAP), information gain, physicochemical properties (AAIndex), and a position-specific scoring matrix (PSSM). To prepare the dataset for training, 29,923 human lysine-acetylated sites were extracted from the CPLM database and their corresponding proteins from UniProt. After applying preprocessing techniques to remove redundancy between peptides, 16,107 positive and 57,443 negative peptides were obtained. The negative dataset set was undersampled to match the size of the positive dataset to avoid class imbalance problems. The optimal length of the window was found to be 31 from various experiments and 2199 optimal features were selected from each sequence using F-Score based feature selection technique. The finalized dataset was divided in the proportion of 6:4 as a training and independent test set. After completion of data preparation processes, a six-layer MLP was developed and trained using 60% of the total dataset (training dataset). The trained model was tested using the remaining 40% of the total data (independent test set). DeepACEt achieved an accuracy of 84.87%, AUC of 0.8407, specificity of 83.46%, sensitivity of 86.28%, and MCC of 0.6977, which was found to be better compared to other existing tools. The model is available as a web server at https://deepsig.biocomp.unibo.it/welcome/default/index.

2.1.2 DNNAce

DNNAce [33] is an acetylation site prediction tool based on a deep learning framework trained on nine different acetylation sites datasets of prokaryotes collected by Chen et al. [34]. After removing the duplicates and similar sequences, 7288 lysine-acetylated and 41,638 lysine nonacetylated peptides were collected. DNNAce uses sequence information, physicochemical information, and evolution information of amino acid residues as the features. Specifically, the used features are binary encoding (BE), pseudo–amino acid composition (PseAAC), AAindex, normalized Moreau-Broto autocorrelation (NMBroto), encoding based on grouped weight (EBGW), multivariate mutual information (MMI), BLOSUM62, and k-nearest neighbor (KNN). Furthermore, the Group Lasso technique was implemented to pick only relevant features that are important for classification purposes. A six-layer deep neural network (DNN) is trained by varying hyperparameters like activation functions, learning rate, and optimization algorithms. Comparing the results from ten-fold cross-validation and test on independent test datasets (10% of total available data), DNNAce performs well compared to existing machine learning–based methods. The source code and data set of DNNAce are available at https://github.com/QUST-AIBBDRC/DNNAce/.

2.1.3 Deep-PLA

Deep-PLA [35] is another deep learning–based method for identifying histone acetyltransferase (HAT)/histone deacetylase (HDAC)-specific acetylation. The enzyme-specific data sets containing several HATs and HDACs were collected from the PubMed literature database. The final dataset contained 930 unique HAT-specific acetylation sites in 304 proteins and 401 unique HDAC-specific deacetylation sites in 190 proteins. Among them, sites in human proteins were treated as the training data set and the sites in other organisms were treated as the testing data set. Four different feature encoding techniques: Amino acid composition (AAC), composition of k-spaced amino acid pairs (CKSAAP), position-specific scoring matrix (PSSM), and binary encoding profiles (BE) were used. The concept of transfer learning was used to improve the learning performance of the convolutional neural network. The model was initially trained using large datasets (CREBBP, EP300, KAT2B, HDACs, and SIRT1) and pretrained hyperparameters were used before training on the actual dataset. While comparing performance with other existing tools, Deep-PLA performs well based on fivefold, tenfold cross-validation, and independent test set. This work is publicly available as a web server at http://deeppla.cancerbio.info/.

2.2 Deep Learning–Based Approaches for Protein Glycosylation Site Predictor

Protein glycosylation is one of the most important posttranslational modifications. It is known to affect the expression, localization, folding, and half-life of secreted and cell surface proteins, and plays an important role in biological processes such as cell adhesion and trafficking, and receptor binding and activation. N-linked glycosylation generally occurs at N-X-[T/S) sequon where X can be any amino acid except proline, whereas O-linked glycosylation occurs at T/S residue of the protein sequence.

2.2.1 SPRINT-Gly

Taherzadeh et al. developed a deep neural network-based protein N and O linked glycosylation site prediction method called SPRINT-Gly [36]. Sprint-Gly is described in detail in Chapter 9 of this book. In this work, the positive N-linked glycosylation was constrained to N-X-[T/S] sequon where X can be any amino acid except proline, however, the negative sequence was not constrained to such sequon. They intercepted 5 window size positive N-linked peptides which follow N-X-[T/S] sequon from 2369 proteins and 9 window size positive O-linked peptides from 211 O-linked proteins which have middle residue T/S. Intercepted O-linked and N-linked peptides were encoded with ASA, SS, Disordered region, PSSM, HSE, Physicochemical properties, Amino acid sequence (AAS), and those extracted features were sent into a deep neural network for analysis. Based on benchmarking results, SPRINT-Gly is shown to be 18% and 50% higher in MCC than the next best method compared in the prediction of N-linked and O-linked sites, respectively. The web server of SPRINT-Gly is available at https://sparks-lab.org/server/sprint-gly/.

2.3 Deep Learning–Based Approaches for Hydroxylation Site Prediction

Hydroxylation is one of the common posttranslational modifications found in the human body that plays a vital role in different diseases. Some of the traditional ML-based hydroxylation site predictors are RF-Hydroxysite [37], iHyd-PseAAC [38, 39], iHyd-PseCp [40], and so on.

Haixia et al. [41] recently developed a hybrid-deep learning (two different deep learning models, convolutional neural network (CNN) and long short-term memory network (LSTM) integrated together) approach to predict the presence of hydroxylation sites in a given protein sequence. As CNN is commonly used to capture higher-level features and LSTM is used to capture long-term dependencies between amino acids, the combined model captures characteristics of CNN and LSTM to achieve better performance. Two feature encoding techniques, namely, pseudo amino acid composition (PseAAC) and position-specific scoring matrix (PSSM) are used to represent the features of the protein sequences. And the hybrid model is implemented and trained in the open-source deep learning software framework MXNET. The benchmark dataset consists of hydroxylation sites from 164 hydroxyproline proteins and 33 hydroxylysine proteins. While comparing the five-fold cross-validation results against other existing predictors, the hybrid deep learning model performs better than other methods (namely, iHyd-PseAAC [38] and iHyd-PseCp [40].

2.4 Deep Learning–Based Approaches for Malonylation Site Prediction

Malonylation, which occurs in both eukaryotic and prokaryotic cells, involves a covalent linkage between the ε-amino group of lysine and a malonyl moiety. It is another common type of post-translational modification that regulates diverse biological activities and has been implicated in several pervasive disorders. Therefore, the accurate identification of malonylation sites is crucial in the understanding of this important PTM.

2.4.1 DeepMal

DeepMal [42] is a tool for predicting malonylation sites in the given input protein sequence and uses a deep neural network architecture to make a binary classifier that distinguishes whether a given amino acid is malonylation site or not. DeepMal uses five different feature encoding techniques: enhanced amino acid composition (EAAC), enhanced grouped amino acid composition (EGAAC), dipeptide deviation from the expected mean (DDE), K nearest neighbors (KNN), and BLOSUM62 matrix. Then, a linear convolutional neural network is used to determine only the relevant features that are useful in the identification of malonylation sites. The dataset used in this research is taken from the work of Zhang et al. [43]. After removing redundant and similar sequences, 7644 positive and the same number of negative peptides of window size 25 are taken as the training dataset. Similarly, 1000 positive and 1000 negative peptides are taken as independent test sets which are completely isolated from the training dataset. The architecture of DeepMal consists of a convolution layer, pooling layer, dropout layer, and a fully connected layer. Finally, the model is trained using a tenfold cross-validation method on the training dataset and tested using the independent dataset. Compared with other existing prediction models, DeepMal performs relatively well. DeepMal is available from the author's GitHub repository at https://github.com/QUST-AIBBDRC/DeepMal.

2.4.2 DL-MaloSite

DL-Malosite (and RF-Malosite) [44] are malonylation site prediction tools based on random forest and deep learning algorithms respectively. DL-MaloSite only takes primary amino acid sequence as input, whereas RF-MaloSite takes several handcrafted biochemical, physiochemical, and sequence-based features. The 3978 positive training sites and 988 positive training sites are curated from experimentally verified malonylation sites from mice and humans. The same number of negative data is randomly selected to avoid the class imbalance problem. The 4246 features were extracted from Features Extraction from Protein Sequence (FEPS) tool. Details about FEPS can be found in Chapter 2 of this book. Furthermore, these features are combined with other features from amino acid properties (AAP), binary encoding (BE), and enhanced encoding features (EAAC). The gini importance feature selection algorithm is used to select only 104 relevant features for the actual classification purpose. After preparing the dataset, the RF-MaloSite model is trained using the extracted features. On the other hand, the DL-MaloSite model is constructed using CNN architecture with an embedding layer in the input. To overcome the performance degradation problem of deep learning in the limited dataset for Human malonylation sites, transfer learning was used to transfer the initial weights from mouse malonylation sites. Finally, the independent test set was tested on both RF-MaloSite and DL-MaloSite models and the results were compared with existing tools including SPRINT-Mal [46]. DL-MaloSite and RF-MaloSite achieved better results in terms of MCC. For RF-MaloSite and DL-MaloSite, MCC were 54% and 38% higher than those achieved by iLMS, respectively and 90% and 71% higher than those achieved by SPRINT-Mal, respectively). Supplementary Data and source code of this project are publicly available in the author's GitHub link at https://github.com/dukkakc/DL-MaloSite-and-RF-MaloSite.

2.4.3 Kmalo

Kmalo [47] is a tool developed for locating protein malonylation sites in mammals based on a hybrid model of different convolutional neural networks (CNNs) trained on physicochemical properties, evolutionary information, and sequenced-based features. Since the existing studies were only done in animal malonylation sites until the publication of this study, this work extends the study to plants (wheat) in addition to mammals. To study malonylation sites in plants, multiple random forest models and CNN models were merged together into a secondary modeling phase using a support vector machine (SVM). The feature encoding techniques AAC, PAAC, PSSM, One Hot Encoding, and AAindex were used to convert protein peptides into feature vectors. Several experiments were performed to determine the best feature set and the optimal window size of the peptides to maximize the generalization characteristics of the classifiers. The combination of AAINDEX, PSSM, and One hot encoding was found to result in the best Area Under Curve (AUC) value. Inside the Kmalo tool, two separate models were trained for mammals and plants. While testing the models on an independent dataset, the competitive performance was achieved while compared to the results of existing tools. This tool can be publicly accessed at https://fdblab.csie.ncu.edu.tw/kmalo/home.html.

2.5 Deep-Learning Based Approaches for Phosphorylation Site Prediction

Phosphorylation is one of the most studied covalent posttranslational modifications. In that regard, there are myriad computational tools for the prediction of protein phosphorylation sites. For a detailed review on computational tools for (eukaryotic) phosphorylation site prediction until 2011, please refer to Trost and Kusalik [8]. Some of the recent machine learning–based phosphorylation tools are PhosphoSVM [48], RF-Phos [49], PhosphoPredict [50], and so on. Here, we review some of the recent deep learning–based phosphorylation site prediction tools.

2.5.1 MusiteDeep

MusiteDeep [51] developed at Xu Lab is the first deep-learning framework for predicting general and kinase-specific phosphorylation sites. It predicts phosphorylation sites from the raw protein sequence avoiding feature engineering and uses one-of-K encoding commonly known as one-hot encoding to encode 33-residue protein fragments centered at the prediction site. The architecture implemented in MusiteDeep is a stacked convolutional neural network with a two-dimensional attention mechanism. The convolutional neural network (CNN) automatically discovers the complex representations needed for classification.

For the kinase-specific phosphorylation site prediction, where there is only a small sample of training data MusiteDeep uses the notion of transfer learning which learns from the general phosphorylation data, and then all the weights learned are transferred (except the last output layer of the network) to the kinase-specific model. Finally, the whole network is fine-tuned using the kinase-specific data. MusiteDeep achieved impressive results for general phosphorylation sites and obtained competitive results for kinase-specific phosphorylation prediction. MusiteDeep is available at https://github**Error! Bookmark not defined.**.com/duolinwang/MusiteDeep.

2.5.2 DeepPhos

DeepPhos [45] is another DL-based protein phosphorylation site predictor that predicts both types (general and kinase-specific) of phosphorylation sites with the primary sequence as an input. In this method, protein sequences are encoded via one-hot encoding, corresponding encoded features are learned by multilayer one-dimensional convolutional neural networks. DeepPhos uses inter- and intrablock concatenation (functional architecture) layers which help to transfer the abstract information of previous layers with different levels to the current layer as the network deepens. DeepPhos outperforms other contemporary sequential deep learning architecture like CNN, LSTM, RNN, Fully Connected Network (FCN) at prediction of phosphorylation Serine/Threonine (S/T) and Tyrosine (Y). On a benchmark comparison, it outperformed general sites prediction when the independent datasets were compared with other widely available predictors like NetPhos3.0 [2], PPSP [52], Musite [53], MusiteDeep [51]. In addition, it did better for kinase-specific phosphorylation site prediction when compared to other kinase-specific predictors like GPS 2.0 [54] and PPSP [52]. Furthermore, DeepPhos is available at https://github.com/USTCHIlab/DeepPhos.

2.5.3 Chlamy-EnPhosSite

Chlamy-EnPhosSite [55] is a DL-based organism-specific phosphorylation site predictor for *Chlamydomonas reinhardtii* (*C. reinhardtii*), which is a single-cell green alga about 10 mm that swim with two flagella. Two contemporary architectures, convolutional neural networks (CNN) and long short-term memory (LSTM) are combined to receive the best phosphorylation site prediction performance in *C. reinhardtii*. Similar to DeepPhos [45] primary protein sequences are used as the input. However, a character-level embedding technique is used for the encoding of the sequence. Subsequently, the power of stacked 2D convolutional neural networks and stacked long short-term memory were combined to get the best prediction performance in S, S/T, T, Y phosphorylation sites. In addition, the authors also proposed Chlamy-MwPhosSite which concatenates the features extracted by 5 CNN models trained on 5 different window sizes. Chlamy-MwPhosSite alleviates the need to choose best performing window sizes. Chlamy-EnPhosSite is available at https://github.com/dukkakc/Chlamy-EnPhosSite.

2.5.4 DeepPSP

DeepPSP [56] engendered the idea of combining local information and global information around potential phosphorylation sites to enhance the phosphorylation site prediction. In that regard, DeepPSP consists of two modules, namely, local and global modules. The local window (length = 51) and global window (length = 2000) are encoded by one-hot encoder and then the encoded primary protein sequences are fed to the sequence of Convolution, two squeeze-and-excitation (SENet) blocks, and Bi-LSTM blocks. SENet estimates the residue contribution at each position around the potential site as well as improves the prediction performance. This paper elaborates that global information may be linked to the interaction between widely separated modification sites, which may influence function and enable complex cellular information processing. Moreover, they also showed that one-hot encoding is more effective than the other sequence information-based encoding like position-specific scoring matrix (PSSM) and secondary structures (SS) for phosphorylation sites prediction. In addition, to relieve the problem of small data sets occurring in specific-kinase phosphorylation site prediction, DeepPSP uses transfer learning to overcome the overfitting problem commonly associated with small data sets. Furthermore, DeepPSP is implemented at https://github.com/gankLei-X/DeepPSP.

2.5.5 PROSPECT

PROSPECT [57] is a histidine (H) phosphorylation site (pHis) predictor for bacteria. The dataset under study was composed of 244 pHis sites (positive samples) and 1420 non-pHis sites (negative samples) from 172 proteins. For each phosphorylation site, a sequence window that contains the central histidine (H) and its ±13 flanking residues were extracted. This window size was optimized through tenfold cross-validation in the benchmark dataset. Furthermore, the histidine (H) phosphorylated site is predicted from sequence information alone. PROSPECT combines two CNN (one CNN (one-of-k) fed with one-of-K encoding and another CNN fed with enhanced grouped amino acids content (EGAAC) encoding) and a Random Forest with 1000 trees fed with composition of k-spaced amino acid group pairs (CKSAAGP) encoding with optimal weight combination of 0.7, 0.05, and 0.25, respectively. The comparative results show that PROSPECT is better than other pHis predictors in both the running speed and prediction accuracy. The method is available as a web server at http://PROSPECT.erc.monash.edu/.

2.5.6 DTL-DephosSite

The extensive study of phosphorylation sites has resulted in a comparatively large dataset of phosphosites—140K, while there is still a relatively sparse dataset of dephosphorylation. DTL-DephosSite [58] developed at KC lab is a predictor for Dephosphorylation sites that exploit the fact that Phosphorylation and dephosphorylation are closely related to one another at the molecular level. DTL-DephosSite has two predictors: a deep learning–based S/T dephosphorylation site predictor and a deep learning–based Y dephosphorylation site predictor, obtained by transfer learning to predict ST, and Y dephosphorylation sites. In DTL-DephosSite, a model is learned on the phosphorylation sites using a Bi-LSTM deep learning architecture and subsequently, it is applied to learn on the dephosphorylation dataset (ComDephos [dephosphorylation sites from DEPOD and additional dephosphorylation dataset obtained from literature mining]) using transfer learning. Specifically, the Bi-LSTM model is pretrained with available phosphorylation data from MusiteDeep, and during the fine-tuning step, the weights learned by the pretrained Phos-model were transferred to a new instance of the Bi-LSTM architecture. The model was then trained on the Dephos data containing the S/T residues in the center, thus obtaining a transfer-learned Dephos model for S/T residues. A Similar method was followed for the Tyrosine (Y) dephosphorylation sites. The best window fragment was found to be 33 through fivefold cross-validation. These amino acids in the 33-length window are integer encoded and character encoded by the embedding layer of the Keras module. The Independent test results show that DTL-DephosSite [47] achieved a better result than other widely available predictors in terms of phosphoserine/phosphothreonine and phosphotyrosine residue dephosphorylation site PTM prediction. DTL-DephosSite is available at https://github.com/dukkakc/DTLDephos.

2.6 Deep-Learning Based Approaches for Succinylation Site Prediction

Succinylation is one of the common posttranslational modifications (PTM) that occurs on lysine residues and is thought to be playing a role in the interaction between cellular metabolism and important cellular functions. Because of its critical role in different cellular processes, succinylation site prediction has become an important research topic in the field of bioinformatics. The ML-based tools like iSuc-PseAAC [59], iSuc-PseOpt [60], and pSuc-Lys [61] have been developed for locating succinylated sites on proteins. In recent years, deep learning–based models have shown promising results in identifying succinylated sites on proteins.

2.6.1 CNN-SuccSite

CNN-SuccSite [62] is a deep learning–based succinylated site prediction tool developed by Huang et al. The different feature encoding techniques—position-specific amino acid composition, the composition of k-spaced amino acid pairs (CKSAAP), and position-specific scoring matrix (PSSM)—were used to convert the protein sequences into corresponding feature vectors. A data clustering algorithm called maximal dependence decomposition (MDD) was implemented to detect the conserved pattern of lysine succinylation sites. After preparing data, a Convolutional Neural Network (CNN), with two convolution layers, two max-pooling layers, and two fully connected layers, was implemented and trained. The model was tested on an independent test of 218 positive and 2621 negative instances. The accuracy of 86.79% and an MCC value of 0.489 were achieved in an independent test set. This tool is publicly available at https://openebench.bsc.es/tool/cnn-succsite.

2.6.2 DeepSuccinylSite

DeepSuccinylSite [63] developed at KC Lab is another deep learning–based prediction tool developed to identify succinylation sites in proteins based on their primary sequence. DeepSuccinylSite uses a dataset of experimentally derived lysine succinylation sites curated from UniProtKB/Swiss-Prot database and NCBI protein database as presented by Hasan et al. [64] and Ning et al. [65]. After removing redundant and similar sequences, 4755 succinylation sites and 50,565 nonsuccinylation sites were taken as training sets and 254 succinylation sites and 2977 nonsuccinylation sites were taken as the independent test. Since the number of samples in the positive and the negative dataset is highly imbalanced, data was balanced using random undersampling technique. Unlike existing machine learning approaches, DeepSuccinylSite does not use handcrafted feature extraction methods and rather uses a one-hot encoding approach in addition to an embedded encoding technique where each amino acid is assigned a numeric value to generate a feature vector. This embedding technique is found to be a promising alternative to the tedious process of extracting handcrafted features. The CCN-based architecture of DeepSuccinylSite consists of two 2D-convolution layers, a max-pooling layer, and two fully connected layers. The original training data was further divided into training and validation set in the proportion of 4:1. The optimal window size of the training peptides was 33 which was determined by different experiments and cross-validation methods. The trained model was tested on an independent dataset and compared with the results of existing tools and it was observed that it produces better MCC compared to previously developed succinylation site predictors. The source code and model of this work are publicly available at https://github.com/dukkakc/DeepSuccinylSite.

2.6.3 MDCAN-Lys

MDCAN-Lys [66] is a deep learning–based binary classifier tool to predict succinylation sites that uses a multilane dense convolutional attention network. Dense convolutional blocks are implemented to extract sequence information, physicochemical properties, and structural properties of the sequences. Dense convolutional blocks are capable of representing the low-level and high-level features of the sequences more precisely as compared to other previously used networks like CNN and LSTM. To overcome the problem of information loss in the max-pooling layer used in CNN, Convolutional Block Attention Module (CBAM) is implemented which improves the representability of the network to identify more succinylation sites.

The 3085 protein sequences used in this study are extracted from Protein Lysine Modification Database (PLMD). The total available sequences were divided into training and testing sequences in the proportion of 9:1. Then 5885 positive and 64,140 negative peptides were extracted from training sequences and considered as the training dataset. Similarly, 684 positive and 6709 negative peptides were extracted from testing sequences and taken as the negative dataset. Since the number of positive and negative samples is different, the class weighting method is used to avoid the class imbalance problem. The experimental results show that, compared with existing methods, the proposed model can effectively learn the abstract pattern of succinylation and identify more succinylation sites. Finally, a multilane dense convolutional attention network is trained using positive and negative training samples. The performance of the model is evaluated by using an independent test set. While comparing with the existing tools like iSuc-PseAAC [59], SuccinSite [67], pSuc-Lys [61], HybridSucc [68], DeepSuccinylSite [63], and iSuc-PseOpt [60], the results of MDCAN-Lys are better in terms of prediction accuracy.

2.7 Deep Learning–Based Approaches for Ubiquitination Site Prediction

Protein ubiquitination occurs when the ubiquitin protein binds to a target protein residue of lysine (K), and it is an important regulator of many cellular functions, such as signal transduction, cell division, and immune response, in eukaryotes. The ubiquitination system is responsible for many aspects of cellular molecular function, such as protein localization, metabolism, regulation, and degradation. There are several machine learning–based approaches for the prediction of ubiquitination sites. Here, we summarize some of the DL-based approaches for protein ubiquitination site prediction.

2.7.1 DeepUbi

DeebUbi [69] is a ubiquitination site prediction approach based on convolutional neural networks similar to the inception neural network. In DeepUbi, four different types of features are adopted to encode the sequence fragments: one-hot encoding, informative physicochemical properties (IPCP), Compositions of K-spaced amino acid pairs (CKSAAP), and Pseudo Amino Acid Composition (PseAAC). The steps of DeepUbi are input a protein fragment, one-hot encode the fragments and embed it with embedding layer, build multi-convolution-layers with different filter size, max pool the convolved output, add features from pooling layers, construct fully connected layers, and classify the ubiquitination or not ubiquitination site with the output layer. In this study, the ubiquitination data was collected from the PLMD (v3.0, June. 2017) database, which is the biggest online data repository designed for protein lysine (K) modification. The database contains 121,742 ubiquitination sites from 25,103 proteins. Empirical approaches like CD-HIT were used to remove the redundant protein sequences; the researchers obtained 12,053 different nonredundant proteins. A sliding window with a length of 31 is used to represent the protein sequences with lysine (K) residues in the center. After eliminating redundant segment fragments a training dataset containing 53,999 ubiquitination and 50,315 nonubiquitination sites were utilized for further study. The hybrid of One-Hot and CKSAAP exhibited the best area under the receiver operating characteristic curve (AUC) of 0.9066 and an MCC of 0.78. DeepUbi is implemented at https://github.com/Sunmile/DeepUbi.

2.7.2 DeepUbiquitylation

DeepUbiquitylation [70] is another deep learning–based approach for the prediction of ubiquitination sites that uses multimodal deep learning architecture. Initially, polypeptide fragments of window size 49 centered around the site of interrogation, that is, ubiquitination and nonubiquitination sites from the respective proteins were extracted. In one modality, these windows are one hot encoded and then sent into four layers of a 1D convolutional network and to three dense layers. In the second modality, these polypeptide fragments of 49 lengths are encoded with top 13 physicochemical properties that have been validated to be strongly related to ubiquitination. Subsequently, these features (49×13) are reshaped for each sample and the matrix is flattened and passed into deep neural network (DNN) with three dense layers for further analysis. Finally, in the third modality important evolutionary feature group PSSM is generated by the Basic Local Alignment Search Tool (BLAST). The PSSM fragment corresponds to the window size 49 from the PSSM matrix file from the whole protein sequence, which records the position-specific evolutionary profiles of the protein fragment. The PSSM vectors are passed into three layers of 1D CNN to obtain deep evolutionary characterization among different sequence positions; furthermore, a similar process was used for trans-positioned PSSM vectors. The 1D convolved output of PSSM and trans-positioned PSSM vector were merged and sent into three layered deep neural networks. The generative deep representations corresponding to three modalities were merged and sent to two neuron dense layers for classification. This model was evaluated on the largest ubiquitination site database PLMD, and it achieved 66.43% accuracy, and 0.221 MCC. DeepUbiquitylation outperformed several popular protein ubiquitination site prediction tools. Finally, it is implemented at https://github.com/jiagenlee/deepUbiquitylation.

2.7.3 UbiComb

Siraj et al. developed a deep learning–based approach for identifying plant-specific ubiquitination sites called UbiComb [71]. UbiComb uses two modules corresponding to two encoding schemes: embedding encoding and physicochemical properties-based encoding. The embedding encoding module extracts the feature by using long short-term memory (LSTM) followed by max-pooling layer, whereas in the second module, the latent five-dimensional feature of physicochemical properties is extracted using 1D convolutional layers followed by a max-pooling layer. The results in terms of feature vectors of these modules are concatenated and passed as an input to dense layers for deeper feature extraction. The embedding encoding is similar to index-based integers encoding ranging from 1 to 22, whereas for the physicochemical based encoding a small five-dimensional latent vector which corresponds to meaningful major principal components: hydrophobicity, size, number of degenerated triplet codons, preference of amino acid residues in a beta-strand, and frequency of occurrence of amino acid residues in a beta-strand were generated from 237 physicochemical properties for each residue of amino acid in 31 window fragment. This work considers local window fragments of 31 where ubiquitination site (K) residue is at the middle of the window fragment. The training data for UbiComb constitutes 7000 protein fragments constructed from plant subset data containing 3500 positive and 3500 negative fragments, which were selected randomly. From this, 5500 were used for training, and 1500 were used for independent testing. UbiComb demonstrated a better generalization capability than other widely available ubiquitination predictors. The web server for UbiComb can be found at http://nsclbio.jbnu.ac.kr/tools/UbiComb/.

2.7.4 DL-plant-ubsites

Wang et al. [72] proposed a novel transfer learning-based word embedding model for plant ubiquitination site prediction. Three sets of ubiquitination sites: plant subset, animal subset, and fungus subset were collected from PLMD database. After redundancy removal, 7000 sites for plant, 60,000 for animal and 17,000 for fungus subset were obtained. Based on the sequence feature analysis of these sites, it was observed that feature patterns were different and plant ubiquitination site was chosen as the target. The original plant protein sequence was used for pretraining of the word2vec network through the skip-gram model, then the optimized parameters were transferred as the initial weights of the embedding layer, and the fine-tuning with the subsequent layers were done. The multilayer one-dimensional CNN was employed which predicts the label of the fragment whose lysine in the center position can be ubiquitylated or not. While comparing the performance of the proposed method with other popular plant ubiquitination prediction tools on the independent dataset, this method achieved better overall performance, which is mainly because the word-embedding scheme is more effective to extract the sequence features. It is implemented at https://github.com/wang-hong-fei/DL-plant-ubsites-prediction.

2.7.5 DeepTL-Ubi

Liu et al. [73] proposed a deep learning–based approach for predicting ubiquitination sites of multiple species called DeepTL-Ubi. Essentially, the authors collected 62,324, 15,495, 7012, 2452, 4598, 1173, 848, and 616 positive ubiquitination Lysine (K) sites for *H. sapiens*, *M. musculus*, *S. cerevisiae*, *R. norvegicus*, *A. nidulans*, *A. thaliana*, *T. gondii*, and *O. sativa*, respectively, from various sources including PhosphoSitePlus, mUbiSida, and PLMD. From the corresponding proteins, a 31-length fragment centered at ubiquitinated lysine and nonubiquitinated lysine site was extracted. Those protein fragments were one hot encoded and transformed into a 31×21 two-dimensional raw sequence feature. One of the novelties of DeepTL-Ubi is the species-specific prediction method which trains prediction models with fine-tuning strategies that take weights of a pretrained human prediction model and use it to initialize a new model for the target species. DeepTL-Ubi employs a densely connected convolutional neural network (DCCNN) as a deep feature extractor, which can efficiently learn the joint feature space for ubiquitination sites of both a source species (i.e., *Homo sapiens*) and a target species (e.g., *Saccharomyces cerevisiae*) by simultaneously minimizing both species transfer

(ST) loss and classification loss. Test results illustrate that the DeepTL-Ubi method improves the performance of species-specific ubiquitination site prediction compared to the deep network designed with the fine-tuning strategy. It also obtains significant performance improvement compared with existing species-specific and general ubiquitination site prediction tools. It is implemented at https://github.com/USTC-HIlab/DeepTL-Ubi.

3 Deep-Learning Based Approaches for Multiple PTM Site Prediction

One of the interesting questions in the prediction of PTM sites is whether the features that are important for one PTM are sharable with other PTMs. In addition, what about the shareability of the deep learning architecture among various PTMs. Recently there have been a few methods to predict multiple PTM-sites exploiting the fact that the features (or feature encoding) and the deep learning architecture are sharable among various PTMs. Here, we briefly review some of these methods.

3.1 MusiteDeep (Web Server)

Wang et al. introduced a new web server, MusiteDeep [74] which provides a deep-learning framework for multiple protein PTM site prediction and visualization. The predictor only uses protein sequence as input and no complex features are needed, which results in a real-time prediction for many proteins. It predicts multiple PTM like phosphorylation, N-linked glycosylation, O-linked glycosylation, N6-actetyllysine, Methylarginine, Methyl-lysine, S-palmitoylation-cysteine, Pyrrolidone-carboxylic-acid, Ubiquitination, SUMOylation, Hydroxylysine, and Hydroxyproline. For training and testing the model, the protein fragment of window length 33 was used and it was one-hot encoded and fed into the ensemble of Multi Convolutional Neural Network (MultiCNN) and Capsule Network (CapsNet) for the prediction of various PTM sites. The MultiCNN model is constructed by three one-dimensional (1D) CNN layers, two-dimensional (2D) attention layers, and two fully connected layers; the CapsNet is constructed by two 1D CNN layers, a PrimaryCaps layer, and a PTMCaps layer. Both the networks were trained separately, and in the prediction procedure, a final prediction score was calculated by averaging the prediction scores obtained by the two independent networks. Furthermore, researchers used Transfer learning to improve the prediction performance of various PTM that has a smaller number of training samples. For example, for phosphotyrosine PTM prediction, researchers trained the base network on the phosphoserine and phosphotyrosine data; then the pretrained weights of the base network were initialized to the weights for the phosphotyrosine predictor. Finally, they fine-tuned the weights of the predictor using the phosphotyrosine data. Independent test set prediction results suggest that this method can be used for several types of PTM prediction. MuSiteDeep web server is available at https://github.com/duolinwang/MusiteDeep_web.

3.2 pCysMod

Some amino acid residues (e.g., cysteines) can undergo multiple pTMs. pCysMod [75] is a deep learning–based tool that predicts multiple protein cysteine modifications like S-nitrosylation, S-palmitoylation, S-sulfenylation, S-sulfhydration, and S-sulfinylation. Authors collected experimentally verified sites by searching the literature from PubMed as well as several databases and ended up with 23,041 S-nitrosylation sites in 10,671 proteins, 2766 S-palmitoylation sites in 1413 proteins, 4978 S-sulfenylation sites in 3288 proteins, 2721 S-sulfhydration sites in 1707 proteins, and 742 S-sulfinylation sites in 538 proteins as the final training dataset. After removing redundant sequences and balancing the datasets, the authors obtained 19,335 S-nitrosylation positive sites, 2766 S-palmitoylation positive sites, 4746 S-sulfenylation-positive sites, 2535 S-sulfhydration positive sites, and 655 S-sulfinylation positive sites. The sequence window for feature extraction consists of a cysteine in the middle and 15 upstream and downstream amino acids at both ends. Protein sequence features such as Binary Encoding Profiles, Amino Acid Composition, Position-Specific Scoring Matrix, Composition of k-Spaced Amino Acid Paris were extracted and united into a deep learning model, and the hyperparameters were optimized by particle swarm optimization algorithms. For each of the five types of modification, the benchmark peptide dataset was encoded by four feature extraction methods. Although each modification type had a special model, they have analogous model architecture. The model received the numerical transferred sequences in the input layer, which consists of four independent DNN submodules to train four input features. Then the four submodules were merged and flattened into a fully connected layer after sufficiently learning the features. Cross-validations indicated pCysMod model showed excellent robustness and outperformed existing tools for S-nitrosylation, S-palmitoylation, S-sulfenylation, S-sulfhydration, and S-sulfinylation. It has a working web server at http://pcysmod.omicsbio.info/.

3.3 MUscADEL

Multiple Scalable Accurate Deep Learner for lysine PTMs (MUscADEL) framework, based on a bidirectional LSTM-RNN algorithm is also another approach to predict multiple PTMs: eight lysine PTMs which includes acetylation, glutarylation, glycation, malonylation, methylation, succinylation, sumoylation, and ubiquitination. There is also a book chapter in this book about MUscADEL (Chapter 11). Here, we very briefly review the method. The dataset for MUscADEl consists of eight (types of) lysine PTMs for two mammalian species (i.e., *Homo sapiens* and *Mus musculus*) which were derived from the PhosphoSitePlus database. Two predictive models, that is, the full-sequence and the sequence-fragment model are constructed in MUscADEL. The full sequence model considers both contributions of the full-length sequences and motifs surrounding the lysine PTM sites, while the fragment model only extracts the motif patterns with the lysine (K) PTM sites. For the fragment model, the window size was set to 27 residues. Both models contain five layers, including the input layer, the word embedding layer, the recurrent layer, the fully connected layer, and the output layer. The full sequence model outperformed the fragment model for predicting lysine acetylation, glycation, malonylation, methylation, sumoylation, succinylation, and ubiquitination, whereas for predicting glutarylation, the fragment model generally outperformed the full-sequence model. Extensive benchmarking tests show that MUscADEL [76] outperforms current methods for lysine PTM characterization, demonstrating the potential and power of deep learning techniques in protein PTM prediction. The web server for MUscADEL is available at https://muscadel.erc.monash.edu.

3.4 MultiLyGAN

Yang et al. proposed a multiclassification machine learning pipeline MultiLyGAN [77] to identify seven types of lysine modified sites. The authors collected 18 kinds of lysine modification samples from the CPLM2.0 database involving a total of 284,780 modification sites from 53,501 proteins. Peptide fragments were obtained through a sliding window technique, with a length of 8 in the upper and lower lysine amino acids (window size L = 17). After data integration and deletion of fragments with length less than 17, researchers came up with Ubiquitination (3185), Acetylation (3114), Succinylation (1645), Glycation (1399), Malonylation (1224), Sumoylation (1174), and Methylation (1147), seven different types of lysine modified sites to be predicted. They evaluated their methods with eight different sequential features including AAindex, CKSAAP, PWM, Reduced Alphabet, FoldAmyloid, Binary Encoding, PC-PseAAC, SC-PseAAC, and five structural features like Accessible Surface Area (ASA), secondary structures (SS), backbone torsion angles, Half Sphere Exposure (HSE), and Contact Number (CN). When all the features were fused then they were passed to the Pearson correlation coefficient filtering module which gave forth 1497 valid features. Since there is a data imbalance between positive sample number and negative sample number that problem was sorted out with the help of the Conditional Generative Adversarial Network (CGAN) and Conditional Wasserstein Generative Adversarial Network (CWGAN). These two novel deep generative models were leveraged and compared to generate new samples for the lysine modification types with fewer samples. Finally, a random forest algorithm was utilized to predict seven categories. The model exhibited better accuracy (ACC) and

Matthew's correlation coefficient (MCC) for tenfold cross-validation as well as for independent test sets. It is implemented at https://github.com/Lab-Xu/MultiLyGAN.

3.5 Some Other Notable DL-Based Approaches for Various PTM Site Predictors

3.5.1 Deep-KCR: Histone Lysine Crotonylation (Kcr) Site Predictor

Deep-KCR is a DL-based approach for histone lysine crotonylation (Kcr) site prediction [78]. The authors truncated 31 residue long sequence segments with lysine (K) located at the center from the 3734 unique histone proteins in the HeLa cell. After performing CD-HIT with the threshold of sequence identity to 30%, 9964 positive samples and 9964 negative samples after undersampling were obtained. CKSAAP, position weight amino acid composition (PWAA) was used to extract sequence-based features; amino acid index (AAindex), composition transition and distribution (CTD) of physicochemical parameters, and encoding based on group weight (EBGW) were selected to extract physicochemical property-based features and Word2Vec was used to extract numerical space-derived information. These extracted features were combined to form a 3184-dimensional feature set. This feature set was sent into convolutional neural networks (CNN) for classification purposes. The model trained on the fusion feature set achieved better performance (AUC = 0.8671) compared with the models trained on the individual original feature set. Furthermore, the authors used a popular Information Gain feature selection method to the fused feature set and obtained 0.8846 AUC through the same CNN network. The prediction performances of different feature encoding schemes, different classifiers, and while comparing with existing tools Deep-Kcr [78] showed the robustness and generalization ability of the model. The Deep-Kcr can be used to predict novel crotonylation sites. It has a working web server which can be found at http://lin-group.cn/server/Deep-Kcr.

3.5.2 nhKCR: Nonhistone Crotonylation Site Predictor

Chen et al. developed a framework called nhKCR [79] to identify lysine crotonylation (Kcr) sites from a large amount of nonhistone sequence data. The authors intercepted 29-residue-long sequence segments with the lysine (K) located at the center from 4230 unique nonhistone proteins. Those proteins were collected from HeLa cell, lung cell, A549, and HCT116 cell databases. After intercepting 29 long sequence residues, and removing 30% sequence similarity with the CD-HIT they came up to 15,605 positive and 75,111 negative samples which were divided into training and testing sets. nhKCR uses improved CNN termed CNNrgb, where it takes AAindex in red, Binary encoding in green, Blosum62 in blue channel respectively. All three features were normalized with a z-score and reshaped into a 29 × 29 dimensional matrix. The three types of encoding schemes were deemed as the RGB channels of a color image and were processed by a two-dimensional convolution neural network. nhKCR [79] consists of an input layer that takes in 29 × 29 matrices corresponding to the RGB channels of a color image. After that nhKCR includes three sequentially connected two-dimensional convolutional blocks, a fully connected layer, and an output layer for classification. While doing a comparative analysis of the model with different encoding schemes as well as with different machine learning algorithms, the model provided superior performance matrices. When the benchmarking analysis on the independent test set was performed nhKCr outperformed the state-of-the-art predictor Dee-Kcr for Kcr site prediction on nonhistone proteins. The web server for nhKCr is at http://nhKcr.erc.monash.edu/.

3.5.3 DeepRMethylSite: Arginine Methylation Site Predictor

Chaudhari et al. proposed a deep learning–based approach for predicting arginine methylation sites in a protein called DeepRMethylSite [80]. The authors collected 10,429 positive samples from 5725 proteins. Those proteins were obtained from DBPTMv3, PhosphositePlusv6.5.8, and UniProt databases. A sliding window size of 51 with the methylated Arg site in the center, flanked by 25 amino acids upstream and downstream of the methylation site was used for subsequent analysis. Those 51 window polypeptides were integer encoded and were embedded through an embedding layer. The embedded polypeptide chain was fed to the Ensemble model which consists of LSTM and CNN. Two layers of stacked LSTM were created for the LSTM model similarly, two layers of stacked CNN were created for the CNN model. For independent testing purposes or, to compare with other widely available Methylation predictors, the training, validation dataset of PrmePred was used to retrain the model and the trained model was tested with PrmePred's independent test set to access the model performance. DeepRMethylSite exhibited efficiency scores that were on par with or better than those of the existing models. DeepRMethylSite is available at https://github.com/dukkakc/DeepRMethylSite.

3.5.4 DeepNitro: Nitrosylation Sites Predictor

Xie et al. developed a computational tool called DeepNitro [81] based on deep neural networks to identify protein nitration and nitrosylation sites. Based on literature review and searches on the different databases and subsequent refining, DeepNitro uses a training set of 1210 positive and 8043 negative sites for tyrosine (Y) nitration, 66 positives and 155 negative sites for tryptophan (W) nitration, lastly 3409 positive and 17,453 negative sites for S-nitrosylation which occurs at Cystine (C). Similarly, the independent dataset consists of 189 positives, and 1182 negative sites for tyrosine nitration and 485 positives, and 4947 negative sites for S-nitrosylation. DeepNitro uses a peptide length of 41 amino acids for subsequent training and prediction. Four different kinds of conventional feature encoding schemes like, (1) One-hot encoding of the nitration or nitrosylation peptide, (2) physical and chemical properties for the nitration or nitrosylation peptide, (3) k-Space spectrum encoding similar to CKSAAP encoding (4) encoding with the position-specific scoring matrix were used. Accordingly, a given nitration or nitrosylation site peptide was encoded with a 3311 - dimensional feature vector, when all the four feature encoding schemes were used. Feature vectors of the samples of the three different types of nitration PTM were fed to a deep neural network, which consists of one input layer, six hidden layers, and one output layer with two neurons with a softmax activation function. Feature combination excavated that combination of PSSM and k-space feature group was the best feature for Tyrosine, Tryptophan, Cysteine nitration or nitrosylation PTM prediction. When compared with other widely available predictors strictly based on an independent dataset, DeepNitro [81] did better so this new method can discover high-level features and increase interpretability of protein nitration and nitrosylation sites. The web server for DeepNitro is available at http://deepnitro.renlab.org (Table 1).

3.5.5 GPS-Palm: S-palmitoylation Site Predictor

Ning et al. developed a deep learning–based graphic representation system to predict S-palmitoylation sites in proteins called GPS-Palm [82]. They collected 5978 unique S-palmitoylation sites from 2840 proteins which were derived from public databases including dbPTM, SwissPalm, HPRD, and PTMD. After homology elimination, the final benchmark data set contained 3098 known sites in 1618 substrates. GPS-Palms uses a dataset of 3098 palmitoylation site peptide PSP (10,10) as a cysteine (C) residue flanked by 10 residues upstream and 10 residues downstream. In addition, GPS-Pal uses a method for data quality discrimination (DQD) to measure data quality weights (DQWs). The DQD was incorporated into the recently developed group-based prediction system (GPS) 5.0 algorithm, which implemented two additional methods of position weight determination (PWD) and scoring matrix optimization (SMO) for performance improvement. Additionally, GPS-Palm also implements a new method named number-to-image transformation (NIT) which transforms numerical sequence similarity values into images. These images were then fed into a deep learning framework of 11-layer convolutional neural networks (CNNs) for model training. Additionally, the researcher used NIT to encode six sequence-derived features including pseudo amino acid composition (PseAAC), CKSAAP, orthogonal binary coding (OBC), physicochemical properties in the Amino Acid index database (AAindex), autocorrelation function (ACF) and PSSM, and three types of structural features, namely, ASA, SS, and backbone torsion angles (BTA). Ten parallel CNNs (pCNNs) were implemented for training and for integrating up to 2835 individual features, and then a GPS-palm [82] toolbox was developed. When compared with other widely available predictors it

Table 1
Summary of DL-based approaches described in this chapter. The approaches are summarized based on the features it uses, the DL-architecture (MLP: Multilayer perceptron, CNN: Convolutional neural network, LSTM: long-short term memory), the type of PTM site prediction, date it was published, and availability (G represents that the method is available in github and W represents that the method is available as a web server)

Method name	Features	DL-architecture	PTM type	Year	Availability	References
DeepACEt	One-hot, BLOSUM,62 matrix, CKSAAP, Information gain, AAIndex, PSSM	MLP	Acetylation	2019	Yes (W)	[32]
DNNAce	Binary encoding (BE), pseudo–amino acid composition (PseAAC), AAindex, normalized Moreau–Broto autocorrelation (NMBroto), encoding based on grouped weight (EBGW), multivariate mutual information (MMI), BLOSUM62, k nearest neighbor (KNN)	MLP	Acetylation	May 2020	Yes (G)	[33]
Deep-PLA	AAC, CKSAAP, PSSM, BE	CNN	Acetylation	2020 Sep	Yes (G)	[35]
SPRINT-Gly	ASA, Disorder region, HSE, SS, AAS, EI, Physicochemical properties	DNN	N- and O-linked glycosylation sites	2019	Yes (W)	[36]
A Hybrid Deep Learning Model for Predicting Protein Hydroxylation Sites	(PseAAC) and position-specific scoring matrix (PSSM	CNN + LSTM	Hydroxylation	2018 Sep	No	[41]
DeepMal	EAAC, EGAAC, DDE KNN BLOSUM62	CNN	Malonylation	December 2020	Yes (G)	[42]
RF-MaloSite and DL-Malosite	Amino acid properties (AAP), binary encoding (BE) and enhanced encoding features (EAAC)	CNN, RF	Malonylation	2020	Yes(G)	[44]

Kmalo	AAC, PAAC, PSSM, One-hot Encoding, and AAindex	CNN	Malonylation	2020	Yes (W)	[47]
MusiteDeep	One-hot encoding	CNN + Transfer Learning	Phosphorylation	2017	Yes (G)	[51]
DeepPhos	One-hot encoding	CNN	Phosphorylation	2019	Yes (G)	[45]
Chlamy-EnPhosSite	Character-level encoding	LSTM + CNN	Phosphorylation	2021	Yes (G)	[55]
DeepPSP	One-hot encoding	CNN + SENet + BiLSTM	Phosphorylation	2021	Yes (G)	[56]
PROSPECT	One-hot encoding, EGAAC, CKSAAGP	CNN + RF	Histidine (H) Phosphorylation	2020	Yes (W)	[57]
DTL-DephosSite	Character-level encoding	BiLSTM + Transfer Learning	Dephosphorylation	2021	Yes (G)	[58]
CNN-SuccSite	CKSAAP PSSM	CNN	Succinylation	2019 Nov	Yes (Not working)	[62]
DeepSuccinylSite	Character-level encoding, one-hot encoding	CNN	Succinylation	April 2020	Yes (G)	[63]
MDCAN-Lys	Physiochemical properties, structural properties	CNN + Attention	Succinylation	June 2021	No	[66]
DeepUbi	One-hot encoding, IPCP, CKSAAP, PseAAC	CNN	Ubiquitination	2019	Yes (G)	[69]
DeepUbiquitylation	One-hot encoding, physicochemical properties, PSSM	1D CNN + MLP	Ubiquitination	2018	Yes (G)	[70]

(continued)

Table 1 (continued)

Method name	Features	DL-architecture	PTM type	Year	Availability	References
UbiComb	Character-level encoding and physicochemical properties	LSTM + 1D CNN	Ubiquitination	2021	Yes (W)	[71]
DL-plant-ubsites-prediction	Word-level embedding	1D CNN	Ubiquitination	2020	Yes (G)	[72]
DeepTL-Ubi	One-hot encoding	CNN + Transfer Learning	Ubiquitination	2020	Yes (G)	[73]
MusiteDeep	One-hot encoding	CNN + CapsNet + Transfer Learning	Multiple PTM	2020	Yes (G)	[74]
pCysMod	Binary encoding profiles, AAC, PSSM, CKSAAP	MLP	Multiple cysteine PTMs	2021	Yes (W)	[75]
MUscADEL	Character-level encoding	Bi-LSTM + RNN	Multiple lysine PTMs	2018	Yes (W)	[76]
MultiLyGAN	AAindex, CKSAAP, PWM, Reduced Alphabet, FoldAmyloid, Binary Encoding, PC-PseAAC, SC-PseAAC, ASA, SS, BTA, HSE, CN	CGAN, CWGAN + RF	Multiple lysine PTMs	2021	Yes (G)	[77]
Deep-KCR	PWAA, AAindex, CTD, EBGW, Word2Vec	CNN	Histone lysine crotonylation	2021	Yes (W)	[78]
nhKCR	AAindex in red, Binary encoding in green, Blosum62 in blue channel	CNNrgb	Lysine crotonylation	2021	Yes (W)	[79]

DeepRMethylSite	Character-level encoding	CNN + LSTM	Methylation	2020	Yes (G)	[80]
DeepNitro	One-Hot encoding, CKSAAP, PSSM, Physical and Chemical Properties	MLP	Nitration and Nitrosylation sites	2018	Yes (W)	[81]
GPS-Palm	PseAAC, CKSAAP, OBC, AAindex, ACF, PSSM, ASA, SS, BTA	CNN	S-palmitoylation	2021	Yes (W)	[82]
Procleave	Protrusion and depth index, Solvent accessibility, Packing, Molecular surface accessibility, Secondary structure features, Solvent exposure properties, and B-factor	CRF, SVM, RF	Protein cleavage	February 2020	Yes (W)	[85]
DeepCalpain	AAC, BE, PSSM, CKSAAP	MLP	Protein cleavage	August 2019	Yes (W)	[86]
DeepCleave	One-hot encoding	CNN	Protein cleavage	February 2020	Yes (W)	[90]
DeepDigest	Word-level encoding	CNN LSTM	Protein cleavage	March 2020	No	[91]

exhibited a 31.34% improvement of AUC values for general prediction of S-palmitoylation sites. In addition, GPS-palm also generated two species-specific models for predicting human and mouse-specific sites. GPS-palm is available as a web server at http://gpspalm.biocuckoo.cn/.

4 Recent Advances in Proteolytic Cleavage Prediction

4.1 Recent ML-Based Approaches for Proteolytic Cleavage Prediction

Proteolytic cleavage is a form of posttranslational modification that has a direct impact on various biological functions including protein regulation. The prediction of protease-specific substrates and their cleavage sites is a fundamental step to understand their effects more deeply. In the last decades, several machine learning and deep learning–based tools have been developed to identify the proteolytic cleavage sites in a given sequence. Li et al. [28] reviewed 19 different computational tools (scoring functions and machine learning–based) developed in the last 20 years. An extensive study was performed on these tools and they were compared using a common independent test set. Among eight different scoring function–based tools and 11 different machine learning–based tools, PROSPERous [24] achieved the best result as compared to other tools. Most of the machine learning–based (ML-based) tools have used one of the traditional classification algorithms including random forests, support vector machines, naive Bayes classifier, K-nearest neighbor, and logistic regression. PROSPERv [83], Cascleave [22, 23], and iProt-Sub [84] are some of the popular ML-based approaches for protease-specific substrates and their cleavage sites prediction. In recent years, some deep learning–based (DL-based) tools have become popular because of their remarkable improvements over ML-based tools in predicting cleavage sites in the protein sequences. Here, we review Procleave, one of the recent ML-based approaches, for proteolytic cleavage prediction.

4.1.1 Procleave

Procleave [85] is one of the latest ML-based protease-specific substrates and cleavage site prediction tools that uses sequence information in addition to 3D structural information. It uses the LOWESS data-smoothing optimization algorithm to represent the structural features of cleavage sites. Then, a conditional random field (CRF) is used to encode optimal approximations of all structural features, sequence information, and chemical group-based features. Procleave uses the MEROPS database to extract cleavage sites and map those sequences to PDB structures by using the BLAST search to generate a 3D structural substrate dataset. Procleave incorporates 3D structural features of substrate cleavage segments in addition to the sequence information that previous tools have used. Protrusion and depth index, Solvent accessibility,

Packing, Molecular surface accessibility, Secondary structure features, Solvent exposure properties, and B-factor are the structural attributes extracted as features. A binary-encoding scheme is used to extract sequence features that convert every amino acid into a 20-dimensional vector. In addition to sequence features and structural features, 20 amino acids were clustered into different eight chemical groups to extract the chemical features. To remove the similar sequences from the positive and negative training set, the CD-HIT method with 70% threshold accuracy is applied. In addition, CD-HIT with 30% of threshold similarity is applied in the independent test set that removes even small redundant information between two sequences. After preprocessing the dataset, CRF model, SVM, and Random Forest were trained with different combinations of features. The CRF model with all the combinations of three types of features achieved the best performance as compared to other models and previously existing tools. A publicly accessible Procleave web server that can predict 27 different proteases is available at http://procleave.erc.monash.edu/.

4.2 Deep Learning–Based Approaches for Proteolytic Cleavage Prediction

Here we review a few DL-based approaches for proteolytic cleavage prediction

4.2.1 DeepCalpain

DeepCalpain [86] is a calpain cleavage prediction tool based on deep neural network (DNN) algorithms. Since calpain-mediated cleavage is associated with the functioning of a broad range of cellular processes, identifying calpain cleavage in a given protein has become an important task. In the dataset preparation process, a total of 442 experimentally identified cleavage sites as positive data and the other 160,698 negative data were extracted from 169 sequences found in UniProt. The transfer learning approach was adopted to improve the training performance to avoid the problems due to the limited size of training data. The window size of each peptide is taken as 31 where the calpain cleavage site in the center was surrounded by 15 amino acids upstream and downstream. The four various types of features, namely, amino acid composition (AAC), binary encoding profiles (BE), PSSM, and composition of k-spaced amino acid pairs (CKSAAP) were prepared from the collected data. The neural network architecture consists of an input layer, a hidden layer, an output layer, and some other fully connected layers. Then, the model was trained and evaluated using four-, six-, eight-, and tenfold cross-validations. While comparing the cross-validation results with other existing tools: GPS-CCD [87], LabCaS [88], and PoPS [89], DeepCalpain outperformed other methods in terms of AUC. One of the things to note is that DeepCalpain does not use an independent test set while comparing its performance with other existing methods. The DeepCalpain tool can be accessed at http://deepcalpain.cancerbio.info.

4.2.2 DeepCleave

DeepCleave [90] is a protease-specific substrate and cleavage site predictor model based on CNN with transfer learning. It is the first deep learning–based model for predicting caspase and matrix metalloproteinase substrate cleavage sites that do not require manual feature extraction. The transfer learning approach is used to extend generic protease family models for the prediction of 12 specific proteases. It also helps to address the problem of having a small dataset of protease-specific cleavage where a model is trained on the larger dataset and weights are reused while training on the other smaller dataset. The data used for training the model is extracted from the MEROPS database and a CD-HIT algorithm is applied to remove the similar sequences. The total available data is split in the ratio of 7:3 as training and testing set and sequence similarity algorithm is applied to remove the similar sequences between these two sets. Since the size of cleavage site data is smaller as compared to the size of noncleavage sites, the training set is balanced by using a bootstrapping method. After preparing the training and test set, the one-hot encoding technique is used to convert an amino acid to a series of 21 binary vectors. Then, the three-layer CNN model is trained to predict whether the given residue is a cleavage site or a noncleavage site. The attention layer is also incorporated inside CNN architecture to make the model focus on only relevant features from a large pool of features generated from convolution operation. The model is trained up to 20 epochs using a training set and tested with an independent test set. The results are better than all previously developed models while evaluated using multiple evaluation metrics. DeepCleave can be accessed at http://deepcleave.erc.monash.edu/.

4.2.3 DeepDigest

DeepDigest [91] is a digestibility prediction tool for predicting eight different proteases (trypsin, ArgC, chymotrypsin, GluC, LysC, AspN, LysN, and LysargiNase) developed by integrating the layers of two deep learning architectures: convolutional neural networks (CNNs) and long-short term memory (LSTM) networks. The convolutional layers are used to extract local sequence features called low-level features and an LSTM layer incorporates the inter–amino acid dependencies. A large dataset covering the eight different proteases was collected for four organisms including *E. coli*, yeast, mouse, and human. A window size of 31 was considered while preparing peptides to train the model. Unlike handcrafted features used in other similar ML-based tools, DeepDigest uses a word-embedding technique where each amino acid in the peptide is mapped into a 21-dimensional feature. While training, the constructed feature vectors from the embedding layer are fed to two convolution layers followed by an LSTM layer. The proposed model was evaluated using tenfold cross-validation and independent testing dataset. Compared with the results of traditional machine learning algorithms (SVM, RF, LR), DeepDigest has shown superior performance for all the eight proteases on a variety of datasets in terms of AUCs, F1 scores, and Matthews Correlation Coefficients (MCCs). The results are also compared with an existing tool called MCpred [92], where DeepDigest has achieved better results comparing all the calculated performance evaluation metrics. Additionally, some other interesting characteristics of proteases are found during this study.

4.2.4 MPSC: Predicting Proteolysis in Complex Proteomes Using Deep Learning

MPSC (Manchester proteome susceptibility calculator) [93] is an integrated web app for predicting cleavage sites for 14 tissue proteases and ultraviolet radiation (UVR)/ROS susceptibility in a single app. Firstly, this study has examined the ability of PROSPER [83] and DeepCleave [90] to identify MMP9-determined cleavage sites in specific proteins: decorin (DCN), vitronectin (VTN), and complex ECM (extracellular matrix) proteome. The MEROPS database [17] consisting of 301 cleavage sites from 53 protein substrates is used to build this model. To address the problem of having a small amount of training data, the transfer learning approach is used where the model is initially trained in some other similar data before training on the actual data. Since the number of positive sequences is very small as compared to the negative dataset, the weight balancing method is used to avoid the class imbalance problem. The model with four bi-LSTM layers, one dense layer, and one fully connected layer was trained using 85% of the total data. Finally, the model is tested on 15% of the total data (independent test set). While comparing the results of the proposed model with the results of previously existing tools including PROSPER and DeepCleave, it has achieved better results in terms of AUC, F1 Score, and MCC. The final product of this work is implemented as a web server which is publicly available at https://www.manchesterproteome.manchester.ac.uk/#/MPSC.

5 Conclusion and Discussion

Just like other areas of science, the field of posttranslational modification has witnessed a plethora of methods based on deep learning–based approaches. One of the major challenges in the field is the lack of a gold-standard benchmark dataset for these various types of PTM modifications where each method is trained and tested on the same dataset. Having that gold-standard benchmark dataset for various types of PTM would be a very important task in the field and some research work in this area is immediately required. This will also help develop and benchmark various deep learning–based approaches. In addition, we are likely to see advances in various areas of the field of posttranslational modification prediction due to deep learning, however, we believe that the most advances will be in the following area.

5.1 Development of Multimodal Approaches for PTM Prediction

There has been a limited number of approaches that exploit information other than protein sequence (protein structural features generated from sequence and others) for the prediction of post-translational modification sites mainly due to the lack of available protein structures. Due to remarkably accurate models obtained by AlphaFold2 [94] and subsequent availability of these models (350K) from EMBL, we are likely to see a lot of multimodal approaches that make use of this suddenly available structural information. In addition, with various advances in the field of word-embedding, natural language processing, and transformers, we are also likely to see approaches that take these natural language processing types of encoding (character-level encoding, word-level encoding, word-embedding (word2vec), etc.) as input.

5.2 Positive--Unlabeled Learning Approaches

In the seminal review paper, Blom et al. [2] very meticulously summarized that "The most significant problem when compiling such a database is that there is almost never any negative data that can be included. In other words, confirming that a specific serine or threonine is *not* phosphorylated can be extremely useful when making prediction methods. Unfortunately, such information is very rarely published. To conclusively prove that a site is negative under all conditions is impossible, but to know that it is negative even in some contexts would be useful." Hence, one of the significant challenges remaining in the field is how to create the negative dataset as negative results are not reported in the literature. In that regard, perhaps the problem is better solved as a positive-unlabeled learning problem rather than a positive-negative learning problem and we might see some approaches solving the PTM prediction problem as a PU-learning problem.

5.3 Deep Transfer Learning–Based Approaches

Despite numerous technological advances in the field of MS and other related technologies, the amount of available data for training deep learning models are still rare. In that regard, we are likely to see a lot of advances in approaches based on transfer learning (or domain adaptation) in the area. There are already some existing approaches that use the concepts of deep transfer learning [51, 58, 74] and we are likely to see more approaches that rely on transfer learning.

5.4 Multitask Multilabel Deep Learning–Based Approaches

Multitask deep learning–based approaches are quite popular in various domains as they try to exploit knowledge transfer. In addition, in PTM prediction the same residue can undergo a multitude of modifications (e.g., Lysine (K) can undergo many types of modifications). In that regard, we are likely to see some advances in the development of multitask multilabel deep learning–based approaches in the field of protein posttranslational modification site prediction.

5.5 Interpretable and Explainable DL-Based Approaches

Biologists are not only interested in a method that works but are more interested in why that particular method works. In that regard, the black-box nature of DL-based approaches has been a roadblock in the field. Consequently, we are likely to see some advances in interpretable and explainable DL-based approaches for protein posttranslational modification site prediction.

References

1. Macek B, Forchhammer K, Hardouin J, Weber-Ban E, Grangeasse C, Mijakovic I (2019) Protein post-translational modifications in bacteria. Nat Rev Microbiol 17(11): 651–664. https://doi.org/10.1038/s41579-019-0243-0
2. Blom N, Sicheritz-Ponten T, Gupta R, Gammeltoft S, Brunak S (2004) Prediction of post-translational glycosylation and phosphorylation of proteins from the amino acid sequence. Proteomics 4(6):1633–1649. https://doi.org/10.1002/pmic.200300771
3. Witze ES, Old WM, Resing KA, Ahn NG (2007) Mapping protein post-translational modifications with mass spectrometry. Nat Methods 4(10):798–806. https://doi.org/10.1038/nmeth1100
4. Nakai K, Kanehisa M (1988) Prediction of in-vivo modification sites of proteins from their primary structures. J Biochem 104(5): 693–699. https://doi.org/10.1093/oxfordjournals.jbchem.a122535
5. Blom N, Gammeltoft S, Brunak S (1999) Sequence and structure-based prediction of eukaryotic protein phosphorylation sites. J Mol Biol 294(5):1351–1362. https://doi.org/10.1006/jmbi.1999.3310
6. Plewczynski D, Tkacz A, Godzik A, Rychlewski L (2005) A support vector machine approach to the identification of phosphorylation sites. Cell Mol Biol Lett 10(1):73–89
7. Eisenhaber B, Eisenhaber F (2010) Prediction of posttranslational modification of proteins from their amino acid sequence. Methods Mol Biol 609:365–384. https://doi.org/10.1007/978-1-60327-241-4_21
8. Trost B, Kusalik A (2011) Computational prediction of eukaryotic phosphorylation sites. Bioinformatics 27(21):2927–2935. https://doi.org/10.1093/bioinformatics/btr525
9. Audagnotto M, Dal Peraro M (2017) Protein post-translational modifications: in silico prediction tools and molecular modeling. Comput Struct Biotechnol J 15:307–319. https://doi.org/10.1016/j.csbj.2017.03.004
10. Ramazi S, Allahverdi A, Zahiri J (2020) Evaluation of post-translational modifications in histone proteins: a review on histone modification defects in developmental and neurological disorders. J Biosci 45:135
11. Wen B, Zeng WF, Liao Y, Shi Z, Savage SR, Jiang W, Zhang B (2020) Deep learning in proteomics. Proteomics 20(21–22): e1900335. https://doi.org/10.1002/pmic.201900335
12. Barrett AJ, Rawlings ND, Woessner JF (2004) Handbook of proteolytic enzymes, 2nd edn. Academic, Amsterdam
13. Klein T, Eckhard U, Dufour A, Solis N, Overall CM (2018) Proteolytic cleavage-mechanisms, function, and "omic" approaches for a near-ubiquitous posttranslational modification. Chem Rev 118(3):1137–1168. https://doi.org/10.1021/acs.chemrev.7b00120
14. Rogers LD, Overall CM (2013) Proteolytic post-translational modification of proteins: proteomic tools and methodology. Mol Cell

Proteomics 12(12):3532–3542. https://doi.org/10.1074/mcp.M113.031310

15. Denadai-Souza A, Bonnart C, Tapias NS, Marcellin M, Gilmore B, Alric L, Bonnet D, Burlet-Schiltz O, Hollenberg MD, Vergnolle N, Deraison C (2018) Functional proteomic profiling of secreted serine proteases in health and inflammatory bowel disease. Sci Rep 8(1):7834. https://doi.org/10.1038/s41598-018-26282-y
16. Uliana F, Vizovisek M, Acquasaliente L, Ciuffa R, Fossati A, Frommelt F, Goetze S, Wollscheid B, Gstaiger M, De Filippis V, Auf dem Keller U, Aebersold R (2021) Mapping specificity, cleavage entropy, allosteric changes and substrates of blood proteases in a high-throughput screen. Nat Commun 12(1):1693. https://doi.org/10.1038/s41467-021-21754-8
17. Rawlings ND, Barrett AJ, Thomas PD, Huang X, Bateman A, Finn RD (2018) The MEROPS database of proteolytic enzymes, their substrates and inhibitors in 2017 and a comparison with peptidases in the PANTHER database. Nucleic Acids Res 46(D1):D624–D632. https://doi.org/10.1093/nar/gkx1134
18. Igarashi Y, Eroshkin A, Gramatikova S, Gramatikoff K, Zhang Y, Smith JW, Osterman AL, Godzik A (2007) CutDB: a proteolytic event database. Nucleic Acids Res 35(Database issue):D546–D549. https://doi.org/10.1093/nar/gkl813
19. Garay-Malpartida HM, Occhiucci JM, Alves J, Belizario JE (2005) CaSPredictor: a new computer-based tool for caspase substrate prediction. Bioinformatics 21(Suppl 1):i169–i176. https://doi.org/10.1093/bioinformatics/bti1034
20. Verspurten J, Gevaert K, Declercq W, Vandenabeele P (2009) SitePredicting the cleavage of proteinase substrates. Trends Biochem Sci 34(7):319–323. https://doi.org/10.1016/j.tibs.2009.04.001
21. Wee LJ, Tan TW, Ranganathan S (2007) CASVM: web server for SVM-based prediction of caspase substrates cleavage sites. Bioinformatics 23(23):3241–3243. https://doi.org/10.1093/bioinformatics/btm334
22. Song J, Tan H, Shen H, Mahmood K, Boyd SE, Webb GI, Akutsu T, Whisstock JC (2010) Cascleave: towards more accurate prediction of caspase substrate cleavage sites. Bioinformatics 26(6):752–760. https://doi.org/10.1093/bioinformatics/btq043
23. Wang M, Zhao XM, Tan H, Akutsu T, Whisstock JC, Song J (2014) Cascleave 2.0, a new approach for predicting caspase and granzyme cleavage targets. Bioinformatics 30(1):71–80. https://doi.org/10.1093/bioinformatics/btt603
24. Song J, Li F, Leier A, Marquez-Lago TT, Akutsu T, Haffari G, Chou KC, Webb GI, Pike RN, Hancock J (2018) PROSPERous: high-throughput prediction of substrate cleavage sites for 90 proteases with improved accuracy. Bioinformatics 34(4):684–687. https://doi.org/10.1093/bioinformatics/btx670
25. Song J, Tan H, Boyd SE, Shen H, Mahmood K, Webb GI, Akutsu T, Whisstock JC, Pike RN (2011) Bioinformatic approaches for predicting substrates of proteases. J Bioinforma Comput Biol 9(1):149–178. https://doi.org/10.1142/s0219720011005288
26. duVerle DA, Mamitsuka H (2012) A review of statistical methods for prediction of proteolytic cleavage. Brief Bioinform 13(3):337–349. https://doi.org/10.1093/bib/bbr059
27. Bao Y, Marini S, Tamura T, Kamada M, Maegawa S, Hosokawa H, Song J, Akutsu T (2019) Toward more accurate prediction of caspase cleavage sites: a comprehensive review of current methods, tools and features. Brief Bioinform 20(5):1669–1684. https://doi.org/10.1093/bib/bby041
28. Li F, Wang Y, Li C, Marquez-Lago TT, Leier A, Rawlings ND, Haffari G, Revote J, Akutsu T, Chou KC, Purcell AW, Pike RN, Webb GI, Ian Smith A, Lithgow T, Daly RJ, Whisstock JC, Song J (2019) Twenty years of bioinformatics research for protease-specific substrate and cleavage site prediction: a comprehensive revisit and benchmarking of existing methods. Brief Bioinform 20(6):2150–2166. https://doi.org/10.1093/bib/bby077
29. Suo SB, Qiu JD, Shi SP, Sun XY, Huang SY, Chen X, Liang RP (2012) Position-specific analysis and prediction for protein lysine acetylation based on multiple features. PLoS One 7(11):e49108. https://doi.org/10.1371/journal.pone.0049108
30. Hou T, Zheng G, Zhang P, Jia J, Li J, Xie L, Wei C, Li Y (2014) LAceP: lysine acetylation site prediction using logistic regression classifiers. PLoS One 9(2):e89575. https://doi.org/10.1371/journal.pone.0089575
31. Lee TY, Hsu JB, Lin FM, Chang WC, Hsu PC, Huang HD (2010) N-Ace: using solvent accessibility and physicochemical properties to identify protein N-acetylation sites. J Comput Chem 31(15):2759–2771. https://doi.org/10.1002/jcc.21569
32. Wu M, Yang Y, Wang H, Xu Y (2019) A deep learning method to more accurately recall known lysine acetylation sites. BMC

Bioinformatics 20(1):49. https://doi.org/10.1186/s12859-019-2632-9
33. Yu B, Yu ZM, Chen C, Ma AJ, Liu BQ, Tian BG, Ma Q (2020) DNNAce: prediction of prokaryote lysine acetylation sites through deep neural networks with multi-information fusion. Chemometr Intell Lab 200. https://doi.org/10.1016/j.chemolab.2020.103999
34. Chen G, Cao M, Luo K, Wang L, Wen P, Shi S (2018) ProAcePred: prokaryote lysine acetylation sites prediction based on elastic net feature optimization. Bioinformatics 34(23):3999–4006. https://doi.org/10.1093/bioinformatics/bty444
35. Yu K, Zhang Q, Liu Z, Du Y, Gao X, Zhao Q, Cheng H, Li X, Liu ZX (2020) Deep learning based prediction of reversible HAT/HDAC-specific lysine acetylation. Brief Bioinform 21(5):1798–1805. https://doi.org/10.1093/bib/bbz107
36. Taherzadeh G, Dehzangi A, Golchin M, Zhou Y, Campbell MP (2019) SPRINT-Gly: predicting N- and O-linked glycosylation sites of human and mouse proteins by using sequence and predicted structural properties. Bioinformatics 35(20):4140–4146. https://doi.org/10.1093/bioinformatics/btz215
37. Ismail HD, Newman RH, Kc DB (2016) RF-Hydroxysite: a random forest based predictor for hydroxylation sites. Mol BioSyst 12(8):2427–2435. https://doi.org/10.1039/c6mb00179c
38. Xu Y, Wen X, Shao XJ, Deng NY, Chou KC (2014) iHyd-PseAAC: predicting hydroxyproline and hydroxylysine in proteins by incorporating dipeptide position-specific propensity into pseudo amino acid composition. Int J Mol Sci 15(5):7594–7610. https://doi.org/10.3390/ijms15057594
39. Ehsan A, Mahmood MK, Khan YD, Barukab OM, Khan SA, Chou KC (2019) iHyd-PseAAC (EPSV): identifying hydroxylation sites in proteins by extracting enhanced position and sequence variant feature via Chou's 5-step rule and general pseudo amino acid composition. Curr Genomics 20(2):124–133. https://doi.org/10.2174/1389202920666190325162307
40. Qiu WR, Sun BQ, Xiao X, Xu ZC, Chou KC (2016) iHyd-PseCp: identify hydroxyproline and hydroxylysine in proteins by incorporating sequence-coupled effects into general PseAAC. Oncotarget 7(28):44310–44321. https://doi.org/10.18632/oncotarget.10027
41. Long H, Liao B, Xu X, Yang J (2018) A hybrid deep learning model for predicting protein hydroxylation sites. Int J Mol Sci 19(9). https://doi.org/10.3390/ijms19092817
42. Wang MH, Cui XW, Li S, Yang XH, Ma AJ, Zhang YS, Yu B (2020) DeepMal: accurate prediction of protein malonylation sites by deep neural networks. Chemometr Intell Lab 207. https://doi.org/10.1016/j.chemolab.2020.104175
43. Zhang Y, Xie R, Wang J, Leier A, Marquez-Lago TT, Akutsu T, Webb GI, Chou KC, Song J (2019) Computational analysis and prediction of lysine malonylation sites by exploiting informative features in an integrative machine-learning framework. Brief Bioinform 20(6):2185–2199. https://doi.org/10.1093/bib/bby079
44. Al-Barakati H, Thapa N, Hiroto S, Roy K, Newman RH, Kc D (2020) RF-MaloSite and DL-Malosite: methods based on random forest and deep learning to identify malonylation sites. Comput Struct Biotechnol J 18:852–860. https://doi.org/10.1016/j.csbj.2020.02.012
45. Luo F, Wang M, Liu Y, Zhao XM, Li A (2019) DeepPhos: prediction of protein phosphorylation sites with deep learning. Bioinformatics 35(16):2766–2773. https://doi.org/10.1093/bioinformatics/bty1051
46. Taherzadeh G, Yang Y, Xu H, Xue Y, Liew AW, Zhou Y (2018) Predicting lysine-malonylation sites of proteins using sequence and predicted structural features. J Comput Chem 39(22):1757–1763. https://doi.org/10.1002/jcc.25353
47. Chung CR, Chang YP, Hsu YL, Chen S, Wu LC, Horng JT, Lee TY (2020) Incorporating hybrid models into lysine malonylation sites prediction on mammalian and plant proteins. Sci Rep 10(1):10541. https://doi.org/10.1038/s41598-020-67384-w
48. Dou Y, Yao B, Zhang C (2014) PhosphoSVM: prediction of phosphorylation sites by integrating various protein sequence attributes with a support vector machine. Amino Acids 46(6):1459–1469. https://doi.org/10.1007/s00726-014-1711-5
49. Ismail HD, Jones A, Kim JH, Newman RH, Kc DB (2016) RF-Phos: a novel general phosphorylation site prediction tool based on random forest. Biomed Res Int 2016:3281590. https://doi.org/10.1155/2016/3281590
50. Song J, Wang H, Wang J, Leier A, Marquez-Lago T, Yang B, Zhang Z, Akutsu T, Webb GI, Daly RJ (2017) PhosphoPredict: a bioinformatics tool for prediction of human kinase-specific phosphorylation substrates and sites by integrating heterogeneous feature selection. Sci Rep 7(1):6862. https://doi.org/10.1038/s41598-017-07199-4

51. Wang D, Zeng S, Xu C, Qiu W, Liang Y, Joshi T, Xu D (2017) MusiteDeep: a deep-learning framework for general and kinase-specific phosphorylation site prediction. Bioinformatics 33(24):3909–3916. https://doi.org/10.1093/bioinformatics/btx496
52. Xue Y, Li A, Wang L, Feng H, Yao X (2006) PPSP: prediction of PK-specific phosphorylation site with Bayesian decision theory. BMC Bioinformatics 7:163. https://doi.org/10.1186/1471-2105-7-163
53. Gao J, Thelen JJ, Dunker AK, Xu D (2010) Musite, a tool for global prediction of general and kinase-specific phosphorylation sites. Mol Cell Proteomics 9(12):2586–2600. https://doi.org/10.1074/mcp.M110.001388
54. Xue Y, Ren J, Gao X, Jin C, Wen L, Yao X (2008) GPS 2.0, a tool to predict kinase-specific phosphorylation sites in hierarchy. Mol Cell Proteomics 7(9):1598–1608. https://doi.org/10.1074/mcp.M700574-MCP200
55. Thapa N, Chaudhari M, Iannetta AA, White C, Roy K, Newman RH, Hicks LM, Kc DB (2021) A deep learning based approach for prediction of Chlamydomonas reinhardtii phosphorylation sites. Sci Rep 11(1):12550. https://doi.org/10.1038/s41598-021-91840-w
56. Guo L, Wang Y, Xu X, Cheng KK, Long Y, Xu J, Li S, Dong J (2021) DeepPSP: a global-local information-based deep neural network for the prediction of protein phosphorylation sites. J Proteome Res 20(1):346–356. https://doi.org/10.1021/acs.jproteome.0c00431
57. Chen Z, Zhao P, Li F, Leier A, Marquez-Lago TT, Webb GI, Baggag A, Bensmail H, Song J (2020) PROSPECT: a web server for predicting protein histidine phosphorylation sites. J Bioinforma Comput Biol 18(4):2050018. https://doi.org/10.1142/S0219720020500183
58. Chaudhari M, Thapa N, Ismail H, Chopade S, Caragea D, Kohn M, Newman RH, Kc DB (2021) DTL-DephosSite: deep transfer learning based approach to predict dephosphorylation sites. Front Cell Dev Biol 9:662983. https://doi.org/10.3389/fcell.2021.662983
59. Xu Y, Ding YX, Ding J, Lei YH, Wu LY, Deng NY (2015) iSuc-PseAAC: predicting lysine succinylation in proteins by incorporating peptide position-specific propensity. Sci Rep 5:10184. https://doi.org/10.1038/srep10184
60. Jia J, Liu Z, Xiao X, Liu B, Chou KC (2016) iSuc-PseOpt: identifying lysine succinylation sites in proteins by incorporating sequence-coupling effects into pseudo components and optimizing imbalanced training dataset. Anal Biochem 497:48–56. https://doi.org/10.1016/j.ab.2015.12.009
61. Jia J, Liu Z, Xiao X, Liu B, Chou KC (2016) pSuc-Lys: predict lysine succinylation sites in proteins with PseAAC and ensemble random forest approach. J Theor Biol 394:223–230. https://doi.org/10.1016/j.jtbi.2016.01.020
62. Huang KY, Hsu JB, Lee TY (2019) Characterization and identification of lysine succinylation sites based on deep learning method. Sci Rep 9(1):16175. https://doi.org/10.1038/s41598-019-52552-4
63. Thapa N, Chaudhari M, McManus S, Roy K, Newman RH, Saigo H, Kc DB (2020) DeepSuccinylSite: a deep learning based approach for protein succinylation site prediction. BMC Bioinformatics 21(Suppl 3):63. https://doi.org/10.1186/s12859-020-3342-z
64. Hasan MM, Kurata H (2018) GPSuc: global prediction of generic and species-specific succinylation sites by aggregating multiple sequence features. PLoS One 13(10):e0200283. https://doi.org/10.1371/journal.pone.0200283
65. Ning Q, Zhao X, Bao L, Ma Z, Zhao X (2018) Detecting succinylation sites from protein sequences using ensemble support vector machine. BMC Bioinformatics 19(1):237. https://doi.org/10.1186/s12859-018-2249-4
66. Wang H, Zhao H, Yan Z, Zhao J, Han J (2021) MDCAN-Lys: a model for predicting succinylation sites based on multilane dense convolutional attention network. Biomol Ther 11(6). https://doi.org/10.3390/biom11060872
67. Hasan MM, Yang S, Zhou Y, Mollah MN (2016) SuccinSite: a computational tool for the prediction of protein succinylation sites by exploiting the amino acid patterns and properties. Mol BioSyst 12(3):786–795. https://doi.org/10.1039/c5mb00853k
68. Ning W, Xu H, Jiang P, Cheng H, Deng W, Guo Y, Xue Y (2020) HybridSucc: a hybrid-learning architecture for general and species-specific succinylation site prediction. Genomics Proteomics Bioinformatics 18(2):194–207. https://doi.org/10.1016/j.gpb.2019.11.010
69. Fu H, Yang Y, Wang X, Wang H, Xu Y (2019) DeepUbi: a deep learning framework for prediction of ubiquitination sites in proteins. BMC Bioinformatics 20(1):86. https://doi.org/10.1186/s12859-019-2677-9
70. He F, Wang R, Li J, Bao L, Xu D, Zhao X (2018) Large-scale prediction of protein ubiquitination sites using a multimodal deep architecture. BMC Syst Biol 12(Suppl 6):109.

https://doi.org/10.1186/s12918-018-0628-0
71. Siraj A, Lim DY, Tayara H, Chong KT (2021) UbiComb: a hybrid deep learning model for predicting plant-specific protein ubiquitylation sites. Genes (Basel) 12(5). https://doi.org/10.3390/genes12050717
72. Wang H, Wang Z, Li Z, Lee TY (2020) Incorporating deep learning with word embedding to identify plant ubiquitylation sites. Front Cell Dev Biol 8:572195. https://doi.org/10.3389/fcell.2020.572195
73. Liu Y, Li A, Zhao XM, Wang M (2021) DeepTL-Ubi: a novel deep transfer learning method for effectively predicting ubiquitination sites of multiple species. Methods 192: 103–111. https://doi.org/10.1016/j.ymeth.2020.08.003
74. Wang D, Liu D, Yuchi J, He F, Jiang Y, Cai S, Li J, Xu D (2020) MusiteDeep: a deep-learning based webserver for protein post-translational modification site prediction and visualization. Nucleic Acids Res 48(W1):W140–W146. https://doi.org/10.1093/nar/gkaa275
75. Li S, Yu K, Wu G, Zhang Q, Wang P, Zheng J, Liu ZX, Wang J, Gao X, Cheng H (2021) pCysMod: prediction of multiple cysteine modifications based on deep learning framework. Front Cell Dev Biol 9:617366. https://doi.org/10.3389/fcell.2021.617366
76. Chen Z, Liu X, Li F, Li C, Marquez-Lago T, Leier A, Akutsu T, Webb GI, Xu D, Smith AI, Li L, Chou KC, Song J (2019) Large-scale comparative assessment of computational predictors for lysine post-translational modification sites. Brief Bioinform 20(6):2267–2290. https://doi.org/10.1093/bib/bby089
77. Yang Y, Wang H, Li W, Wang X, Wei S, Liu Y, Xu Y (2021) Prediction and analysis of multiple protein lysine modified sites based on conditional wasserstein generative adversarial networks. BMC Bioinformatics 22(1):171. https://doi.org/10.1186/s12859-021-04101-y
78. Lv H, Dao FY, Guan ZX, Yang H, Li YW, Lin H (2021) Deep-Kcr: accurate detection of lysine crotonylation sites using deep learning method. Brief Bioinform 22(4). https://doi.org/10.1093/bib/bbaa255
79. Chen YZ, Wang ZZ, Wang Y, Ying G, Chen Z, Song J (2021) nhKcr: a new bioinformatics tool for predicting crotonylation sites on human nonhistone proteins based on deep learning. Brief Bioinform 22:bbab146. https://doi.org/10.1093/bib/bbab146
80. Chaudhari M, Thapa N, Roy K, Newman RH, Saigo H, KC BD (2020) DeepRMethylSite: a deep learning based approach for prediction of arginine methylation sites in proteins. Mol Omics 16(5):448–454. https://doi.org/10.1039/d0mo00025f
81. Xie Y, Luo X, Li Y, Chen L, Ma W, Huang J, Cui J, Zhao Y, Xue Y, Zuo Z, Ren J (2018) DeepNitro: prediction of protein nitration and nitrosylation sites by deep learning. Genomics Proteomics Bioinformatics 16(4):294–306. https://doi.org/10.1016/j.gpb.2018.04.007
82. Ning W, Jiang P, Guo Y, Wang C, Tan X, Zhang W, Peng D, Xue Y (2021) GPS-Palm: a deep learning–based graphic presentation system for the prediction of S-palmitoylation sites in proteins. Brief Bioinform 22(2):1836–1847. https://doi.org/10.1093/bib/bbaa038
83. Song J, Tan H, Perry AJ, Akutsu T, Webb GI, Whisstock JC, Pike RN (2012) PROSPER: an integrated feature-based tool for predicting protease substrate cleavage sites. PLoS One 7(11):e50300. https://doi.org/10.1371/journal.pone.0050300
84. Song J, Wang Y, Li F, Akutsu T, Rawlings ND, Webb GI, Chou KC (2019) iProt-Sub: a comprehensive package for accurately mapping and predicting protease-specific substrates and cleavage sites. Brief Bioinform 20(2): 638–658. https://doi.org/10.1093/bib/bby028
85. Li F, Leier A, Liu Q, Wang Y, Xiang D, Akutsu T, Webb GI, Smith AI, Marquez-Lago T, Li J, Song J (2020) Procleave: predicting protease-specific substrate cleavage sites by combining sequence and structural information. Genomics Proteomics Bioinformatics 18(1):52–64. https://doi.org/10.1016/j.gpb.2019.08.002
86. Liu ZX, Yu K, Dong J, Zhao L, Liu Z, Zhang Q, Li S, Du Y, Cheng H (2019) Precise prediction of calpain cleavage sites and their aberrance caused by mutations in cancer. Front Genet 10:715. https://doi.org/10.3389/fgene.2019.00715
87. Liu Z, Cao J, Gao X, Ma Q, Ren J, Xue Y (2011) GPS-CCD: a novel computational program for the prediction of calpain cleavage sites. PLoS One 6(4):e19001. https://doi.org/10.1371/journal.pone.0019001
88. Fan YX, Zhang Y, Shen HB (2013) LabCaS: labeling calpain substrate cleavage sites from amino acid sequence using conditional random fields. Proteins 81(4):622–634. https://doi.org/10.1002/prot.24217
89. Boyd SE, Garcia de la Banda M, Pike RN, Whisstock JC, Rudy GB (2004) PoPS: a computational tool for modeling and predicting protease specificity. Proc IEEE Comput

Syst Bioinform Conf:372–381. https://doi.org/10.1109/csb.2004.1332450

90. Li F, Chen J, Leier A, Marquez-Lago T, Liu Q, Wang Y, Revote J, Smith AI, Akutsu T, Webb GI, Kurgan L, Song J (2020) DeepCleave: a deep learning predictor for caspase and matrix metalloprotease substrates and cleavage sites. Bioinformatics 36(4):1057–1065. https://doi.org/10.1093/bioinformatics/btz721

91. Yang J, Gao Z, Ren X, Sheng J, Xu P, Chang C, Fu Y (2021) DeepDigest: prediction of protein proteolytic digestion with deep learning. Anal Chem 93(15):6094–6103. https://doi.org/10.1021/acs.analchem.0c04704

92. Lawless C, Hubbard SJ (2012) Prediction of missed proteolytic cleavages for the selection of surrogate peptides for quantitative proteomics. OMICS 16(9):449–456. https://doi.org/10.1089/omi.2011.0156

93. Ozols M, Eckersley A, Platt CI, Stewart-McGuinness C, Hibbert SA, Revote J, Li F, Griffiths CEM, Watson REB, Song J, Bell M, Sherratt MJ (2021) Predicting proteolysis in complex proteomes using deep learning. Int J Mol Sci 22(6). https://doi.org/10.3390/ijms22063071

94. Jumper J, Evans R, Pritzel A, Green T, Figurnov M, Ronneberger O, Tunyasuvunakool K, Bates R, Zidek A, Potapenko A, Bridgland A, Meyer C, Kohl SAA, Ballard AJ, Cowie A, Romera-Paredes B, Nikolov S, Jain R, Adler J, Back T, Petersen S, Reiman D, Clancy E, Zielinski M, Steinegger M, Pacholska M, Berghammer T, Bodenstein S, Silver D, Vinyals O, Senior AW, Kavukcuoglu K, Kohli P, Hassabis D (2021) Highly accurate protein structure prediction with AlphaFold. Nature 596(7873):583–589. https://doi.org/10.1038/s41586-021-03819-2

INDEX

A

Acetylation 5, 11, 21, 23, 146, 147, 187, 190, 198, 205–210, 213, 214, 224, 249, 254, 272, 276, 289, 290, 303, 304, 308
AlphaFold 276
AlphaFold2 316
Amadori ketoamine 126
Amino acid composition (AAC) 67, 68, 71, 72, 76, 79, 81–85, 90, 92–94, 99, 129, 164–167, 173, 290–293, 297, 299, 303, 307–311, 313
AMS 4.0 106
Andromeda 2, 18
Arabidopsis thaliana 17, 23, 301
Area under curve (AUC) 112, 114, 117, 118, 129, 167, 168, 180, 181, 213, 214, 290, 293, 299, 305, 312, 313, 315
Artificial neural network (ANN) 126, 207, 208, 226, 228
AScore 18, 19
Average cumulative hydrophobicity (ACH) 67, 78, 79

B

BioVec 106
Bottom-up proteomics 2–4, 12

C

Capillary electrophoresis (CE) 139, 178
Cascleave 288, 310
CaSPredictor 288
CAZy 139, 143
CD-Hit 127, 210, 299, 305, 313, 314
CDK 107, 108, 112–115, 118
Chlamy-EnPhosSite 295, 309
Chromatography 1, 2, 5, 7–9, 11, 12, 47, 58, 126, 137, 178
CK2 ......107, 108, 112–114, 118
C-mannosylation 142
CNN-SuccSite 297, 309
Co-modification 150, 152
Composition of K-spaced amino acid pairs (CKSAAP) 79, 98, 126, 289, 290, 297, 299, 304, 305, 307–311, 313
Composition, transition, and distribution (CTD) 67, 68, 77, 90, 98, 99, 305, 310
Conjoint triad 74, 75, 98, 99, 164, 165
Convolutional neural network (CNN) 107, 110, 290–298, 300–302, 305, 306, 308–311, 314
CrosstalkMapper 261–272
CutDB 288

D

dbPTM 21, 179, 224, 225, 249, 307
Deacon Active Site Profiler (DASP) 159, 160
Deep learning ......viii, 19, 107, 121, 206, 209, 210, 289–294, 296, 297, 299, 302–304, 307, 308, 314–316
Deep neural network (DNN) 178, 180, 290–292, 300, 303, 307, 308, 313
DeepACEt 289, 290, 308
DeepCalpain 311, 313, 314
DeepCleave 311, 314, 315
DeepDigest 311, 314, 315
Deep-KCR 305, 310
DeepMal 292, 308
DeepNitro 306, 307, 311
DeepPhos 294, 295, 309
Deep-PLA 290, 291, 308
DeepPSP 295, 309
DeepRMethylSite 306, 311
DeepSuccinylSite 297, 298, 309
DeepTL-Ubi 301, 302, 310
DeepUbi 299, 309
DeepUbiquitylation 299–300, 309
Dephosphorylation ......vii, 45–47, 51–53, 56, 59, 60, 62, 63, 296, 297
Dephosphorylation assay 51, 56, 58, 59
Dipeptide composition (DPC) 72, 99
DL-MaloSite 292, 293, 308
DNNAce 290, 308
DTL-DephosSite 296, 297, 309

Dukka B. KC (ed.), *Computational Methods for Predicting Post-Translational Modification Sites*, Methods in Molecular Biology, vol. 2499, https://doi.org/10.1007/978-1-0716-2317-6,

E

Efficiently Learning an Encoder that Classifies Token Replacements Accurately (ELECTRA) 105–122
Electron transfer dissociation (ETD) 14–16
Embedding layer 110, 211, 293, 296, 299, 301, 304, 306, 315
EnsembleGly 178
Evolutionary features 300
Experimental validation 206, 244
Explainable 317

F

False discovery 2, 17, 54
Feature selection 69, 96, 128, 164, 165, 180, 208, 289, 293, 305
Features Extraction from Protein Sequence (FEPS) .... viii, 90–99, 164, 293
Fine-tuning108–110, 112–115, 117, 118, 296, 301, 302
Fragmentation pattern 3
Fully-connected layer 110, 211, 212, 292, 303, 304, 306, 315
Functional Analysis Tools for Post-translational modification (FAT-PTM) .... viii, 145–153
Functional post-translational modification (PTM) .... vii, ix, 146, 221–257

G

GalNAc 137, 142
Geary autocorrelation 67, 81, 90, 99
GenBank 159, 161
General phosphorylation site prediction 106–110, 112, 114, 117
Generative Adversarial Network 304
GitHub 203, 263, 279, 292, 293, 308
Glutarylation 205, 206, 210, 213, 217, 303, 304
Glycan .... viii, 8, 9, 14, 135–142, 178
Glycan Pathway Predictor (GPP) 140, 178
Glycation .... viii, 125–132, 205–207, 210, 213, 214, 303, 304
Glycobiology 135–138, 140
GlycoEnzDB 139, 140
Glycoenzymes 136, 139–142
GlycoEP 178
GlycoGene DataBase (GGDB) 138, 139
Glycogene expression profile 136
Glycogenes 139, 140
GlycoMine 178
GlyConnect 21, 137, 139
GlycoPP 178
GlycoProtDB 137, 179
GlycoSim 140, 142
Glycosylation .... viii, 5, 6, 8, 9, 12, 14, 21, 23, 128, 135–143, 146, 177–179, 181, 182, 187, 286, 291
Glycosyltransferases 8, 136, 139
GlyGen 137, 139
GlyNN 126, 129
Gly-PseAAC 126, 128, 129
GlyTouCan 137, 139
GPS 2.0 .... 106, 120
GPS-Palm 21, 307, 311, 312
GPS-SUMO 106
GPS-YNO2 106

H

Half-sphere exposure (HSE) 179, 180, 291, 304, 308, 310
Heatmap 59, 265–267
High Quality Indices (HQI) 67, 78, 79
High-performance liquid chromoatography (HPLC) 12, 178
Histones 11, 243–245, 251, 263–272, 276, 290, 305, 310

I

iFeature 89, 90, 92
iHyd-PseAAC 291, 292
iHyd-PSeCp 291, 292
INcremental Feature groUp SEection (INFUSE) 128, 129
Information gain 89, 99, 165, 208, 289, 305, 308
Interpretable 122, 317
iProtGly-SS 126–132
iPTMNet .... viii, 146, 188–191, 193–197, 202

K

KEGG GLYCAN 139
Kinases .... viii, 44, 105, 106, 232
Kinase-specific phosphorylation 107–110, 112, 114, 122, 294
Kinase-specific phosphorylation site prediction 105–122, 294, 295
Kmalo 293, 309
K-nearest neighbor (KNN) 166, 169, 172, 207, 208, 290, 292, 308, 310

L

Lecting frontier DataBase (LfDB) 138
Linear Code for Reaction Rules (LiCoRR) 140, 141
Linux-apache-mysql-php (LAMP) 213

Liquid chromatography-tandem mass spectrometry (LC-MS/MS) 1–4, 8, 12, 22, 25, 26, 51, 52
Localization rate 2, 20, 57
Logistic regression 208, 242, 289, 310
Long short-term memory (LSTM) 122, 206, 212, 291, 294, 295, 298, 300, 306, 308–311, 314, 315
Long short-term memory RNN (LSTM-RNN) 211, 303
LuciPHOr2 18–20
Lysine 4, 9, 11, 48, 54, 61, 62, 70, 126–130, 146, 147, 205–217, 249, 251, 254, 265, 266, 268, 276, 289, 290, 292, 297–299, 301, 303–305, 310, 317
Lysine crotonylation 305, 310

M

Malonylation 128, 205–208, 210, 213, 217, 292, 293, 303, 304, 308, 309
Manchester proteome susceptibility calculator (MPSC) 315
Mascot 2, 18, 20, 47, 54, 55, 61
Mascot Delta Score 19
Masked language modeling (MLM) 106, 107, 110, 113, 115–117, 121
Mass spectrometry .... vii, 1, 2, 24, 26, 44, 47, 61, 126, 135, 137, 146, 178, 188, 222, 249, 251, 261, 262, 275, 286
Maximum relevance and minimum redundancy (mRMR) 208
MDCAN-Lys 298, 309
MEROPS 288, 310, 314, 315
Metabolic regulation 158
Methylation 5, 11, 21, 23, 68, 187, 190, 205–207, 209, 210, 213, 217, 224, 254, 266–269, 272, 303, 304, 306, 311
Middle-down 22, 261, 262
Moran autocorrelation 67, 81, 90, 99
Moreau–Broto autocorrelation 290
MS/MS spectra 2, 13, 14, 16, 17, 19, 26
Multilabel 317
MultiLyGAN 304, 310
Multiple sequence alignment (MSA) 86, 122, 226, 228–230, 252, 277
Multitask 317
MUscADEL .... viii, 205–217, 303, 304, 310
MusiteDeep 107, 110, 112, 114, 118, 120, 121, 294, 296, 302, 303, 309, 310

N

Naïve Bayes (NB) 166, 168, 169, 172, 276, 278
Natural language processing (NLP) 106, 107, 114, 115, 121, 122, 211, 316
NetCGlyc 142
NetNGlyc 142, 178
NetOGlyc 142, 178
N-GlycoGo 178
N-GlyDE 178
N-linked glycosylation 178, 180, 181, 224, 291, 302

O

O-linked glycosylation .... viii, 177–184, 291, 302, 308
One-hot encoding 73, 121, 293–295, 297, 299, 306, 309–311, 314
Overlapping properties (OP) 67, 75
Oxidation 4, 6–8, 21, 23, 55, 146, 158

P

PCTPred 276
PCysMod 303, 310
Pearson correlation coefficient (PCC) 81, 208, 304
Peptide fragment spectra 2, 19
Peptide libraries 25, 26, 44, 48, 50, 57, 58
Peroxiredoxins .... viii, 155–173
PhosPhAt 4.0 146
Phosphatases .... vii, 5, 44–46, 48, 49, 51, 52, 56, 58, 59, 62
Phosphoglycerylation 128
PhosphoGRID 146
Phosphopeptide library .... vii, 45, 48–51, 58
Phosphoproteome 7, 12
Phosphorylation .... vii, 4–7, 12, 14, 19–21, 23, 44, 52, 54–57, 60, 67–69, 73, 96, 105–110, 112, 113, 121, 122, 145–149, 153, 157, 187, 190, 194, 196–198, 201, 224, 227, 232, 243, 244, 249, 251, 252, 254, 276, 279, 280, 286, 287, 294–296, 302, 309
Phosphorylation site database (PHOSIDA) 21
PhosphoSitePlus 21, 206, 210, 224, 225, 227, 236, 239, 249, 252, 253, 257, 301, 304
PKA 107, 108, 112–115, 118
PKC 107, 108, 112–115, 118
Position-specific scoring matrix (PSSM) 68, 86–88, 90, 131, 132, 178, 179, 183, 208, 289–293, 295, 297, 300, 303, 307–311, 313
Position weight amino acid composition (PWAAC) 67, 72, 73, 305
Positive-unlabeled learning 316
Post-translational modification (PTM) vii–ix, 1–27, 44, 45, 57, 68–70, 105, 122, 125, 146–152, 187–203, 206–217, 221–257, 262–272, 275–282, 285–287, 289, 292, 297, 302–310, 315–317
Precision/recall analysis 236, 239, 256

PreGly 126, 127, 129
Pretraining 106–112, 114–118, 120–122, 301
PREX 159
PROFEAT 89, 91
Propy 89, 91
PROSPECT 296, 309
PROSPERous 288, 310
Protein Lysine Modifications Database (PLMD) 21, 206, 298–301
Protein-protein interaction 145
Proteolysis 3, 4, 287, 315
Proteolytic cleavage .... vii, viii, 285–289, 310–315
Protr/ProtrWeb 89
PseAAC-Builder 89, 91, 92
Pseudo-amino acid compostion (PseAAC) 68, 81, 82, 84, 89–91, 99, 290, 292, 299, 307–309, 311
PtmRS 18–20
PTM-Score 18–20
PTM-X .... viii, 276–279, 281, 282
Pupylation 128

Q

Quasi Sequence-Order Numbers 67

R

Random forest 2, 142, 161, 165, 208, 276, 281, 292, 293, 296, 304, 310, 313
Rcpi 89, 92
Relative accessible surface area (rASA) 180
Relative entropy (RE) 88, 89, 98, 99
RELIEFF 208
Restful API .... viii, 187–203
RF-Hydroxysite 291
RNN 294, 310

S

SAPH-ire .... viii, 224–229, 231–234, 236–239, 243–247, 250, 252–254, 256, 257
Seq2Feature 89, 90, 92
Sequence-order-coupling numbers (SOCN) 67, 76
SEQUEST 2, 18, 20
Serine/threonine-specific kinase 107
Shannon entropy 68, 88, 93
SitePrediction 288
SLoMo 18, 19
S-nitrosylation 7, 23, 147, 190, 303, 306
Solid-phase peptide synthesis (SPPS) 46, 50
SPIDER 3.0 .... 128, 130, 131, 179, 183
SPOT-disorder 179, 180, 183
SPRINT-Gly .... viii, 178, 179, 181, 183, 184, 291, 308
Stoichiometry 222, 251
Structural features 157, 173, 179, 208, 304, 307, 310, 313, 316
Substitution matrices 68, 90, 106, 209
Succinylation 128, 205, 206, 210, 213, 297, 298, 303, 304, 309
Sumoylation 147, 190, 205, 207, 209, 210, 213, 217, 224, 302–304
Support vector machine (SVM) 2, 68, 126, 128, 129, 142, 166, 168, 169, 172, 178, 180, 207, 208, 228, 289, 293, 311, 313, 315

T

Targeted MS 24–26
Theoretical spectra 2
Time-of-flight (TOF) 13, 25
Transfer learning 290, 293–296, 302, 309, 310, 313–316
Transformer 109, 110, 121, 316
Tripeptide composition (TPC) 72
Trypsin proteolysis 4

U

UbiComb 300, 301, 310
Ubiquitination 5, 9, 21, 23, 147, 190, 205–207, 210, 213, 217, 224, 249, 251, 254, 299–304, 309, 310
UniCarbKB 179
UniPep 21, 179
Uniprot 17, 21, 107, 108, 127, 163, 179, 191, 192, 196–198, 210, 224–227, 279, 280, 289, 306, 313
UniprotKB 139, 169, 173, 190, 297

W

Word2Vec 106, 301, 305, 310, 316

X

Xgboost 164–166, 168, 173

MIX
Papier aus verantwortungsvollen Quellen
Paper from responsible sources
FSC® C105338

If you have any concerns about our products,
you can contact us on
ProductSafety@springernature.com

In case Publisher is established outside the EU,
the EU authorized representative is:
Springer Nature Customer Service Center GmbH
Europaplatz 3, 69115 Heidelberg, Germany

Printed by Libri Plureos GmbH
in Hamburg, Germany